中国履行《生物多样性公约》第五次国家报告

CHINA'S FIFTH NATIONAL REPORT ON THE IMPLEMENTATION OF THE CONVENTION ON BIOLOGICAL DIVERSITY

中华人民共和国环境保护部
Ministry of Environmental Protection of the People's Republic of China

中国环境出版社·北京

图书在版编目（CIP）数据

中国履行《生物多样性公约》第五次国家报告=China's fifth national report on the implementation of the convention on biological diversity：汉、英 / 中华人民共和国环境保护部.—北京：中国环境出版社，2014.5
ISBN 978-7-5111-1829-5

Ⅰ.①中… Ⅱ.①中… Ⅲ.①生物多样性－国际公约－研究报告－中国－汉、英②生物多样性－环境保护－研究报告－中国－汉、英 Ⅳ.①Q16②X176

中国版本图书馆CIP数据核字（2014）第079176号

出版人	王新程
策划编辑	王素娟
责任编辑	赵楠婕
责任校对	扣志红
封面设计	彭　杉
内文制作	杨曙荣

出版发行　中国环境出版社
　　　　　（100062　北京市东城区广渠门内大街16号）
　　　　　网　　址：http://www.cesp.com.cn
　　　　　电子邮箱：bjgl@cesp.com.cn
　　　　　联系电话：010-67112765　编辑管理部
　　　　　　　　　　010-67162011　生态（水利水电）图书出版中心
　　　　　发行热线：010-67125803　010-67113405（传真）

印　刷	北京中科印刷有限公司
经　销	各地新华书店
版　次	2014年9月第1版
印　次	2014年9月第1次印刷
开　本	787×1092　1/16
印　张	23.25
字　数	556千字
定　价	152.00元

【版权所有。未经许可，请勿翻印、转载，违者必究。】
如有缺页、破损、倒装等印装质量，请寄回本社更换。

中国履行《生物多样性公约》第五次国家报告

主持部门	环境保护部
参与部门	国家发展和改革委员会　教育部　科学技术部 财政部　国土资源部　住房和城乡建设部 水利部　农业部　商务部　海关总署 国家工商行政管理总局 国家质量监督检验检疫总局 国家新闻出版广电总局 国家林业局　国家知识产权局　国家旅游局 中国科学院　国家海洋局　国家中医药管理局 国务院扶贫办
项目承担单位	环境保护部南京环境科学研究所

目 录 CONTENTS

执行概要 ... 1
一、中国的生物多样性及其战略意义 ... 1
二、国家生物多样性保护目标 ... 1
三、主要保护行动 ... 2
四、生物多样性面临的威胁、生物多样性保护存在的主要问题及优先重点 ... 7

第1章 中国生物多样性的现状和面临的威胁 ... 11
1.1 生物多样性对经济社会发展的重要意义 ... 11
1.2 中国生物多样性的现状 ... 12
1.3 生物多样性面临的主要威胁 ... 15
1.4 生物多样性丧失对经济社会发展带来的影响 ... 16

第2章 国家生物多样性战略与行动计划及其执行情况 ... 18
2.1 《中国生物多样性保护战略与行动计划》的制订 ... 18
2.2 国家生物多样性保护目标 ... 18
2.3 履行《生物多样性公约》的主要行动 ... 23
2.4 国家生物多样性战略和行动计划的总体进展评估 ... 41

第3章 生物多样性保护融入部门和跨部门规划的情况 ... 43
3.1 国家发展和改革委员会 ... 43
3.2 教育部门 ... 44
3.3 科技部门 ... 45
3.4 国土资源部门 ... 46
3.5 住房和城乡建设部门 ... 48
3.6 水利部门 ... 49

3.7 农业部门 50
3.8 商务部门 55
3.9 海关 56
3.10 工商行政管理部门 57
3.11 质量监督检验检疫部门 58
3.12 林业部门 59
3.13 知识产权部门 62
3.14 旅游部门 63
3.15 海洋部门 64
3.16 中医药管理部门 67
3.17 扶贫开发 68
3.18 履行其他相关公约 69

第4章 2020年生物多样性目标的实施进展以及对千年发展目标的贡献 75

4.1 2020年生物多样性目标的评估指标 75
4.2 2020年生物多样性目标评估指标的数据分析 77
4.3 中国实现全球《生物多样性战略计划》和2020年生物多样性目标的总体进展评估 95
4.4 对实现千年发展目标的贡献 95
4.5 在执行《生物多样性公约》方面取得的经验 96

第5章 生物多样性保护面临的主要问题与优先行动 98

5.1 面临的主要问题 98
5.2 优先行动 99

附 录 101

附录1 有关缔约方和国家报告编写的情况 101
附录2 中国履行《生物多样性公约》第五次国家报告参与编写人员名单 103
附表1 国家生物多样性战略与行动计划的进展评估 109

附表2　中国实现全球《生物多样性战略计划》(2011—2020)
　　　　和2020年生物多样性目标的进展评估...................116
附表3　干旱和半干旱地区生物多样性工作方案执行情况..........130
附表4　保护区工作方案执行情况................................131
附表5　全球生物分类倡议能力建设战略和全球植物保护
　　　　战略的执行情况..142

参考文献..147

Contents

Executive Summary 153

    I   China's Biodiversity and Its Strategic Importance 153
    II   National Targets for Biodiversity Conservation 154
    III Main Actions 154
    IV Threats to Biodiversity and Main Issues and Priorities for
         Biodiversity Conservation 161

Part 1   Current Status of and Threats to China's Biodiversity 167

    1.1 Importance of Biodiversity for Social and Economic Development
         167
    1.2 Current Status of China's Biodiversity 168
    1.3 Main Threats to Biodiversity in China 172
    1.4 Economic and Social Implications of Biodiversity Loss 174

**Part 2   National Biodiversity Strategy and Action Plan and Its
Implementation 176**

    2.1 Development of China's Updated NBSAP 176
    2.2 National Targets for Biodiversity Conservation 177
    2.3 Main Actions to Implement the Convention on Biological Diversity
         185
    2.4 Overall Assessment of Progress in Implementing NBSAP 209

Part 3   Sectoral and Cross-sectoral Integration of Biodiversity ... 212

    3.1 Development and Reform Commission 212
    3.2 Education 214
    3.3 Science and Technology 216

3.4	Land and Resources	217
3.5	Housing, Urban and Rural Development	219
3.6	Water Resources Management	221
3.7	Agriculture	223
3.8	Commerce	229
3.9	Customs	231
3.10	Industry and Commerce Administration	232
3.11	Quality Supervision, Inspection and Quarantine	234
3.12	Forestry	236
3.13	Intellectual Property Office	241
3.14	Tourism	242
3.15	Oceanic Administration	243
3.16	Chinese Medicine Management	247
3.17	Poverty Reduction and Development	248
3.18	Implementation of Other Related Conventions	250

Part 4 | Progress in Implementation of 2020 Biodiversity Targets and Contributions to Millennium Development Goals 258

4.1	Indicators for Assessing Progress Towards 2020 Biodiversity Targets	258
4.2	Data Analysis of Indicators for Assessing the 2020 Biodiversity Targets	261
4.3	Overall Assessment of China's Progress in Implementing the Strategic Plan for Biodiversity and Achieving the 2020 Biodiversity Targets	282
4.4	Contributions to Achievement of Millennium Development Goals	283
4.5	China's Experiences in the Implementation of the Convention	284

Part 5 | Main Issues and Priority Actions for Biodiversity Conservation in China 286

5.1	Main Issues	286
5.2	Priority Actions	287

Appendix 291

Appendix I Information Concerning Party and Process of Preparing National Report............291

Appendix II List of Personnel Involved in the Preparation of China's Fifth National Report on the Implementation of the CBD 294

Annex I Assessment of Progress in Implementing the Updated NBSAP............302

Annex II Assessment of China's Progress in Implementing the Strategic Plan for Biodiversity 2011-2020 and the 2020 Biodiversity Targets 313

Annex III Implementation of the Programme of Work on Biodiversity of Arid and Semi-arid Lands 332

Annex IV Implementation of the Programme of Work on Protected Areas............334

Annex V Implementation of the Capacity-building Strategy for the Global Taxonomy Initiative and the Global Strategy for Plant Conservation 350

References 357

执行概要

根据《生物多样性公约》第 26 条和缔约方大会 X/10 号决议,环境保护部会同中国履行《生物多样性公约》工作协调组成员单位和其他相关机构,编制了《中国履行〈生物多样性公约〉第五次国家报告》。在编制过程中,召开了 5 次相关领域专家参加的研讨会,对第五次国家报告相关问题及其初稿进行了讨论;向履约协调组成员单位等部门征求了意见。第五次国家报告经修改完善后由环境保护部批准发布。

一、中国的生物多样性及其战略意义

生物多样性是指所有来源的活的生物体中的变异性,这些变异性的来源包括陆地、海洋和其他水生生态系统及其所构成的生态综合体等,包含了物种内部、物种之间和生态系统的多样性。生物多样性是人类赖以生存的条件,是社会经济可持续发展的战略资源,是生态安全和粮食安全的重要保障。生物多样性不仅提供给人类食品、洁净水、药物、木材、能源和工业原料等多种生产生活必需品,而且提供固碳释氧、涵养水源、土壤保持、净化环境、养分循环、休闲旅游等多方面的生态服务。

中国是世界上生物多样性最丰富的 12 个国家之一。中国地域辽阔,拥有复杂多样的生态系统类型。动植物资源极为丰富,其中高等植物种数居世界第三位,脊椎动物种数占世界总种数的 13.7%。中国生物遗传资源丰富,是水稻、大豆等重要农作物的起源地,也是野生和栽培果树的主要起源和分布中心。同时,中国也是生物多样性受到严重威胁的国家之一。生物多样性的丧失会导致严重的后果,引起不断恶化的健康问题、更高的食品风险、日益增加的生态脆弱性、更少的发展机会等。生物多样性保护关系到中国社会经济发展全局,关系到当代及子孙后代的福祉,对于建设生态文明和美丽中国具有重要的意义。

二、国家生物多样性保护目标

中国政府对建设生态文明和美丽中国作出了部署,把生态文明建设放在突出地位,融入经济建设、政治建设、文化建设、社会建设各方面和全过程,着力推进绿色发展、循环发展、低碳发展,形成节约资源和保护环境的空间格局、产业结构、生产方式、生活方式。2010 年年底,国务院发布了《全国主体功能区规划》,将国土空间划分为优化开发、重点开发、限制开发及禁止开发四类主体功能区;将 25 个重点生态功能区列入国家层面的限制开发区域,区域内限制进行大规模高强度的工业化城镇化开发,以保护和修复生态环境、提供生态产品为首要任务;将国家级自然保护区、世界文化与自然遗产、国家级风景名胜区、国家森林公园和国家地质公园列入国家层面的禁止开发区域,禁止进行工业化城镇化开发,保护中国自然文化资源和珍稀动植物基因资源。

针对生物多样性丧失的严峻形势，中国政府于 2010 年 9 月 17 日发布并实施了《中国生物多样性保护战略与行动计划》（2011—2030 年）（简称《战略与行动计划》）。中国从建设生态文明高度制定的相关国家规划与《战略与行动计划》一起，构建了比较全面的国家生物多样性保护目标体系（见表1）。

表1 中国国家生物多样性保护目标体系

1. 近期目标：到 2015 年，力争使重点区域生物多样性下降的趋势得到有效遏制
- 完成 8~10 个生物多样性保护优先区域的本底调查与评估，并实施有效监控；
- 加强就地保护，陆地自然保护区面积占陆地国土面积的比例维持在 15% 左右，使 90% 的国家重点保护物种和典型生态系统类型得到保护；
- 合理开展迁地保护，使 80% 以上的就地保护能力不足和野外现存种群数量极小的受威胁物种得到有效保护；
- 森林覆盖率提高到 21.66%，森林蓄积量比 2010 年增加 6 亿 m^3；
- 初步建立生物多样性监测、评估与预警体系、生物物种资源出入境管理制度以及生物遗传资源获取与惠益共享制度；
- 主要污染物排放总量显著减少，与 2010 年相比化学需氧量、二氧化硫排放均减少 8%，氨氮、氮氧化物排放均减少 10%；
- 资源节约型、环境友好型社会建设取得重大进展

2. 中期目标：到 2020 年，努力使生物多样性的丧失与流失得到基本控制
- 对生物多样性保护优先区域的本底调查与评估全面完成，并实施有效监控；
- 全国森林保有量达到 223 万 km^2 以上，比 2010 年增加约 22.3 万 km^2，全国森林蓄积量增加到 150 亿 m^3 以上，比 2010 年增加约 12 亿 m^3；
- 累计治理"三化"草原 165 万 km^2 以上，全国草原退化趋势得到基本遏制，草原生态环境明显改善，天然草原基本实现草畜平衡；
- 近海生态环境恶化趋势得到根本扭转，海洋生物多样性下降趋势得到基本遏制；
- 水域生态环境逐步得到修复，渔业资源衰退和濒危物种数目增加的趋势得到基本遏制；
- 基本建成布局合理、功能完善的自然保护区体系，国家级自然保护区功能稳定，主要保护对象得到有效保护；
- 生物多样性监测、评估与预警体系、生物物种资源出入境管理制度以及生物遗传资源获取与惠益共享制度得到完善，进一步健全国内相关传统知识的文献化编目和产权保护制度；
- 全社会研究开发投入占国内生产总值的比重提高到 2.5% 以上，力争科技进步贡献率达到 60% 以上；
- 单位国内生产总值能源消耗和二氧化碳排放大幅下降，主要污染物排放总量显著减少

3. 远景目标：到 2030 年，使生物多样性得到切实保护

三、主要保护行动

近年来，中国政府积极履行《生物多样性公约》，主要采取了以下保护行动：

1. 完善法律法规体系和体制机制

初步建立了生物多样性保护法律体系，制定并颁布了一系列有关生物多样性保护的国家、行业和地方标准。2011 年，成立了"中国生物多样性保护国家委员会"，统筹全国生物多样性保护工作。生物物种资源保护部际联席会议、中国履行《生物多样性公约》工作协调组运行良好。大部分省级人民政府加强了环保、农业、林业、海洋等涉及生物多样性保护的机构建设，并成立了跨部门协调机制。

2. 发布实施一系列生物多样性保护规划

2010 年,中国政府发布并实施了《全国主体功能区规划》和《中国生物多样性保护战略与行动计划》(2011—2030 年)。国务院还批准实施了《全国生物物种资源保护与利用规划纲要》、《中国水生生物资源养护行动纲要》、《全国重要江河湖泊水功能区划》(2011—2030)、《全国海洋功能区划》(2011—2020 年)、《全国湿地保护工程"十二五"实施规划》(2011—2015 年)、《全国海岛保护规划》(2011—2020)、《全国畜禽遗传资源保护与利用"十二五"规划》等一系列规划,推动了生物多样性保护工作。开展了生态省、市、县创建活动,已有 15 个省(区、市)开展生态省建设,13 个省颁布生态省建设规划纲要,1 000 多个县(市、区)开展生态县建设,建成 1 559 个国家生态乡镇和 238 个国家级生态村;全国水生态文明城市建设试点工作已经启动,首批确定 46 个全国水生态文明城市建设试点,使生物多样性纳入当地经济社会发展规划中。

3. 加强保护体系建设

建立了以自然保护区为主体,风景名胜区、森林公园、自然保护小区、农业野生植物保护点、湿地公园、沙漠公园、地质公园、海洋特别保护区、种质资源保护区为补充的就地保护体系。截至 2013 年年底,全国建立自然保护区 2 697 个,面积约 146.3 万 km^2,自然保护区面积约占全国陆域面积的 14.8%;建立森林公园 2 855 处,规划面积 17.4 万 km^2;建立国家级风景名胜区 225 处、省级风景名胜区 737 处,面积约 19.4 万 km^2,占中国陆域面积的 2.0%;建立自然保护小区 5 万多处,面积超过 1.5 万 km^2;建成国家级农业野生植物保护点 179 个;已建湿地公园 468 处;建立国家级海洋特别保护区(海洋公园)45 个,总面积 6.68 万 km^2;建立国家级水产种质资源保护区 368 个,面积超过 15.2 万 km^2。

加强濒危物种的拯救和繁育。对濒危野生动植物实施抢救性保护,通过开发濒危物种繁育技术、扩大濒危物种种群、加强野外巡护、栖息地恢复、实施放归自然等一系列措施,使一批极度濒危的陆生野生动植物种逐步摆脱灭绝的风险。同时还采取多种有效措施,不断加强对其他野生动植物的普遍保护。

科学开展迁地保护。建有各级各类植物园 200 个,收集保存了占中国植物区系三分之二的 2 万个物种。建立了 240 多个动物园、250 处野生动物拯救繁育基地。建立了以保种场为主,保护区和基因库为辅的畜禽遗传资源保种体系,对 138 个珍贵、稀有、濒危的畜禽品种实施重点保护。加强了农作物遗传资源的收集和保存设施建设,农作物收集品总量达 42.3 万份,比 2007 年增加了约 3 万份。建立了 400 多处野生植物种质资源保育基地。建立了中国西南野生生物种质资源库,搜集和保存中国野生生物种质资源。

4. 推动生物资源的可持续利用

实施重点保护野生动植物利用管理制度,包括国家重点保护野生动物特许猎捕证制度和驯养繁殖许可证制度、国家重点保护植物采集证制度。实施了森林采伐限额制度、基本草原保护制度、草畜平衡制度、禁牧休牧制度、渔业捕捞许可管理制度、禁渔期和禁渔区制度。加大水生生物资源增殖放流，加强海洋牧场建设。加强对野生动植物繁育利用的规范管理和执法监管，制定了科学严格的技术标准，建立了专用标识制度。对种群恢复较困难的濒危物种，进行人工繁育，开发替代品，减少对濒危物种的压力。不断加大执法力度，严厉查处非法销售、收购国家重点保护野生动植物及其产品的违法违规行为，查获了一批濒危物种重特大走私案件。

5. 大力开展生境保护与恢复

继续实施了天然林资源保护，退耕还林，退牧还草，三北及长江、沿海等防护林建设，京津风沙源治理，岩溶地区石漠化综合治理，湿地保护与恢复，水土流失综合治理等重点生态工程。自2001年以来，重点工程区生态状况明显改善，全国森林资源持续增长，森林面积较10年前增长了23.0%；森林覆盖率比10年前上升了3.8个百分点；森林蓄积量比10年前增长了21.8%。一批国际和国家重要湿地得到了抢救性保护，自然湿地保护率平均每年增加1个多百分点，约一半的自然湿地得到有效保护。在近岸海域修复整治红树林、滩涂等退化湿地面积达2 800多 km^2，投入经费达44.3亿元。治理小流域1.2万条，完成水土流失综合防治面积27.0万 km^2，实施封育保护面积72万 km^2，45万 km^2 的生态环境得到了初步修复。2008年以来，中央财政共安排农村环保专项资金195亿元，支持4.6万个村庄开展环境综合整治，8 700多万农村人口受益。重点生态工程的实施，促进了退化生态系统和野生物种生境的恢复，有效保护了生物多样性。

6. 制定和落实有利于生物多样性保护的鼓励措施

为避免对生物多样性和环境造成消极影响，中国政府于2007年取消了553项高耗能、高污染、资源性产品的出口退税，包括濒危动植物及其制品、皮革、部分木板和一次性木制品等。

中国政府对实施重点生态工程的农户提供补助。按照核定的退耕还林实际面积向退耕户提供补助。截至2012年年底，中央已累计投入3 247亿元，1.2亿农民直接受益，户均已累计获得7 000元补助。

对于天然林资源保护工程，国家给予森林管护补助和造林育林补助，对森工企业职工养老保险社会统筹、森工企业社会性支出及森工企业下岗职工基本生活保障费用实行补助。天然林一期工程累计投入1 186亿元。2010年年底，国务院决定，2011—2020年实施天然林资源保护二期工程，总资金约2 440亿元。

设立森林生态效益补偿基金，对重点公益林营造、抚育、保护和管理支出给予一定补助，2013年中央财政下拨149亿元。各地对地方公益林也进行了补偿。

对于退牧还草工程，国家给予草原围栏建设资金补助和饲料粮补助，2003—2012年中央累计投入资金175.7亿元，工程惠及450多万名农牧民。2011年起建立草原生态保护补助奖励机制，至今累计安排286亿元，草原禁牧补助实施面积达82万 km^2，享受草畜平衡奖励的草原面积达173.7万 km^2。

设立国家重点生态功能区转移支付资金，2013年转移支付资金达到423亿元。

7. 推动生物安全管理体系建设

完善外来入侵物种防控的体制机制，初步形成了林业有害生物监测预警网络体系和农业外来入侵物种监测预警网络。开展了外来入侵物种集中灭除行动。建立了农业转基因生物安全评价制度、生产许可制度、经营许可制度、产品标识制度和进口审批制度，开展林木转基因工程活动审批，实现了转基因技术研发与应用的全过程管理。

8. 严格控制环境污染

中国政府将主要污染物排放总量显著减少作为经济社会发展的约束性指标，着力解决突出环境问题。近10年来，中国主要污染物年均浓度总体呈下降趋势，单位国内生产总值污染物排放强度下降了55%以上。2004年以来，单位GDP二氧化碳排放强度下降了15.2%。中国政府严格执行环境影响评价制度，2008年以来国家层面拒批332个、总投资1.1万亿元涉及高污染、高能耗、资源消耗型、低水平重复建设和产能过剩项目。

9. 推动公众参与

中国将生物多样性相关知识纳入中小学教育课程，并在全国普通高校开展生物多样性相关学位教育。截至2012年，中国共培养相关专业人才55.6万余人。各有关部门、各地加大对生物多样性保护宣传力度，特别是开展了"2010国际生物多样性年中国行动"宣传活动，各类宣传活动影响受众9亿多人次，此后每年都开展媒体培训宣传和促进企业参与生物多样性保护的大型宣传活动，带动社会公众参与热情，提高公众生物多样性保护意识。

以上保护行动取得明显成效，主要包括：

（1）森林资源持续增长，森林面积较10年前增长了23.0%；森林蓄积量较10年前增长了21.8%。

（2）治理小流域1.2万条，完成水土流失综合防治面积27.0万km^2；实施封育保护面积72万km^2，其中45万km^2的生态环境得到了初步修复；2006年以来增加湿地保护面积1.8万km^2，恢复湿地1 000km^2。

（3）一些国家重点保护野生动植物种群数量稳中有升，分布范围越来越大，生境质量不断改善。大熊猫数量从20世纪80年代的1 000多只增加到现在的1 590只，朱鹮数量从20世纪80年代的7只增加到目前的1 800多只，红豆杉、兰科植物、苏铁等保护植物种群不断扩大。

（4）截至2013年年底，建立自然保护区2 697个，总面积约146.3万km^2，自然保护区面积约占全国陆域面积的14.8%。另外还建立了大量风景名胜区、森林公园、自然保护小区、农业野生植物保护点、湿地公园、地质公园、海洋特别保护区、种质资源保护区。自然保护区有效保护了中国90%的陆地生态系统类型、85%的野生动物种群和65%的高等植物群落，涵盖了25%的原始天然林、50%以上的自然湿地、30%的典型荒漠地区和近3%的主张管辖海域。

（5）主要污染物年排放量总体呈下降趋势。2000年以来，单位GDP污染物排放强度下降了55%以上。2004年以来，单位GDP二氧化碳排放强度下降了15.2%。

总之，中国政府加大了生物多样性保护力度，通过完善保护政策、加强保护体系建设、恢复退化生态系统、控制环境污染、强化科学技术研究、推动公众参与、增加资金投入等措施，使生态破坏加剧的趋势有所减缓，部分区域生态系统功能得到恢复，一些重点保护物种种群有所增长。《战略与行动计划》的实施开局良好并取得积极进展；其中取得很大进展的行动有1项，取得较大进展的行动有15项，取得一定进展的行动有14项（见表2）。

表2 国家生物多样性战略与行动计划的实施进展评估

行动	进展评估	行动	进展评估
1. 制定促进生物多样性保护和可持续利用政策	◐	16. 加强畜禽遗传资源保种场和保护区建设	◐
2. 完善生物多样性保护与可持续利用的法律体系	◐	17. 科学合理地开展物种迁地保护体系建设	◐
3. 建立健全生物多样性保护和管理机构，完善跨部门协调机制	◐	18. 建立和完善生物遗传资源保存体系	◐
4. 将生物多样性保护纳入部门和区域规划、计划	◐	19. 加强人工种群野化与野生种群恢复	◐
5. 保障生物多样性的可持续利用	◐	20. 加强生物遗传资源的开发利用与创新研究	◐
6. 减少环境污染对生物多样性的影响	◐	21. 建立生物遗传资源及相关传统知识保护、获取和惠益共享的制度和机制	◐
7. 开展生物物种资源和生态系统本底调查	◐	22. 建立生物遗传资源出入境查验和检验体系	◐
8. 开展生物遗传资源和相关传统知识的调查编目	◐	23. 提高对外来入侵物种的早期预警、应急与监测能力	◐
9. 开展生物多样性监测和预警	◐	24. 建立和完善转基因生物安全评价、检测和监测技术体系与平台	◐
10. 促进和协调生物遗传资源信息化建设	◐	25. 制订生物多样性保护应对气候变化的行动计划	◐
11. 开展生物多样性综合评估	◐	26. 评估生物燃料生产对生物多样性的影响	◐
12. 统筹实施和完善全国自然保护区规划	◐	27. 加强生物多样性保护领域的科学研究	◐
13. 加强生物多样性保护优先区域的保护	◐	28. 加强生物多样性保护领域的人才培养	◐
14. 开展自然保护区规范化建设，提高自然保护区管理质量	◐	29. 建立公众广泛参与机制	◐
15. 加强自然保护区外生物多样性的保护	◐	30. 推动建立生物多样性保护伙伴关系	◐

注：● 表示全部实现；◕ 表示有很大进展；◐ 表示有较大进展；◑ 表示有一定进展；○ 表示没有进展。

在实现全球 2020 年生物多样性目标（共 20 个）方面，除目标 2、目标 16 和目标 18 因缺乏相应指标无法评估外，目标 1、目标 3、目标 4、目标 5、目标 7、目标 8、目标 10、目标 11、目标 14、目标 15、目标 17、目标 19、目标 20 的相关评估指标均有不同程度的改善，表明这些目标的实施正沿着正确的轨道推进，特别是目标 3（鼓励措施）、目标 5（减少生境退化和丧失）、目标 8（控制环境污染）、目标 11（强化保护区系统和有效管理）、目标 14（恢复和保障重要生态系统服务）、目标 15（增强生态系统的复原力和碳储量）进展较大；但目标 5 中的草原生态系统保护和目标 6（可持续渔业）、目标 9（防治外来入侵物种）、目标 12（保护受威胁物种）、目标 13（保护遗传资源）的相关评估指标大多呈现恶化的趋势，表明虽然已开展了大量工作，但尚需采取更加有效的策略和措施才能实现这些目标（见表 3）。

四、生物多样性面临的威胁、生物多样性保护存在的主要问题及优先重点

尽管中国政府采取多种措施保护生物多样性，但生物多样性下降的趋势尚未得到根本遏制。无脊椎动物受威胁（极危、濒危和易危）的比例为 34.7%，脊椎动物受威胁的比例为 35.9%；受威胁植物有 3 767 种，约占评估高等植物总数的 10.9%；需要重点关注和保护的高等植物达 10 102 种，占评估高等植物总数的 29.3%。遗传资源丧失的问题突出，根据第二次全国畜禽遗传资源调查的结果，超过一半以上的地方品种的群体数量呈下降趋势。

1. 威胁

造成生物多样性下降的直接原因是：

（1）野生生物生境的退化或丧失。湿地和草地开垦，海岸线开发，交通、水电等大型工程建设，使野生动植物生境遭到破坏，种群繁衍面临直接威胁。

（2）自然资源的过度利用。草原过度放牧造成草原退化、沙化。高强度捕捞加剧了海洋渔业资源的衰退。尽管采取了一系列执法行动，但野生动植物非法贸易的现象仍然存在，甚至在有些地区还很猖獗。

（3）环境污染。江河湖海的水体污染严重，直接威胁水生生物多样性。化肥、杀虫剂、除草剂的使用，造成日趋严重的面源污染。中国管辖海域水环境状况总体较好，但近岸海域水污染依然严重。海洋环境污染引起赤潮等多种海洋生态灾害，对海洋生物多样性造成严重损害。

（4）单一品种的大规模种植。栽培的农作物集中在少数几个品种，使许多传统品种遭到淘汰，甚至永远消失。

（5）外来物种入侵。中国是世界上遭受外来入侵物种危害最严重的国家之一。外来入侵物种有 500 余种，给环境和经济带来巨大损失。

表3 中国实现全球2020年生物多样性目标的进展评估

目标	指标	变化趋势	目标	指标	变化趋势
1. 提高对生物多样性的认知	通过Google或百度检索到中国生物多样性的条目	✓	10. 减少珊瑚礁和其他脆弱生态系统的压力	污染物削减量	✓
3. 鼓励措施	生态补偿和重点生态工程投资	✓		森林蓄积量	✓
4. 可持续生产和消费	污染物削减量	✓		减少的水土流失面积	✓
	可持续消费的指标	⋯		珊瑚礁的生物多样性	⋯
5. 减少生境退化和丧失	森林生态系统面积和蓄积量	✓		气候变化对生物多样性的影响	⋯
	湿地生态系统面积	✓	11. 强化保护区系统和有效管理	保护区的数量和面积	✓
	草地生态系统面积	✗		保护区的生态代表性和管理有效性	⋯
	天然草原产草量	✓	12. 保护受威胁物种	红色名录指数	✗
	减少的沙漠生态系统面积	✓	13. 保护遗传资源	地方品种资源量	✗
	生态退化	⋯	14. 恢复和保障重要生态系统服务	农村居民家庭人均纯收入和减少的贫困人口数量	✓
6. 可持续渔业	海洋营养指数	✓		森林蓄积量	✓
	鱼类红色名录指数	✗		减少的水土流失面积	✓
	渔业对生物多样性的影响	⋯		减少的沙化土地面积	✓
7. 可持续农业、水产养殖业和林业	森林蓄积量	✓	15. 增强生态系统的复原力和碳储量	森林蓄积量	✓
	天然草原产草量	✓		减少的水土流失面积	✓
	农林业对生物多样性的影响	⋯		减少的沙化土地面积	✓
8. 控制环境污染	污染物削减量	✓	17. 实施《战略与行动计划》	政策和规划的实施	✓
			19. 发展和应用科学技术成果	有关生物多样性的论文	✓
				通过Google或百度检索到中国生物多样性的条目	✓
9. 防治外来入侵物种	每20年新发现的外来入侵物种种数*	✓	20. 大幅度增加资金	重点生态工程投资	✓

注：✓表示增加；✗表示下降；⋯表示没有足够数据。*表示外来入侵物种对生物多样性造成的不利影响在加大。

（6）气候变化。气候变化使生物物候、分布和迁移发生改变，使一些物种在原栖息地消失；使有害生物的分布范围改变，对生物多样性危害加剧。

2. 主要问题

中国生物多样性保护存在的主要问题是：①法制和体制有待进一步完善；②保护意识有待进一步提高；③保护与开发利用的矛盾突出；④经费投入不足；⑤科学研究相对滞后。

3. 优先重点

中国生物多样性保护工作仍然任重道远。今后一段时期是中国生物多样性保护的关键时期，只有下更大的决心，采取更加有效的措施，投入更多的资源才能从根本上扭转生物多样性丧失的趋势。今后应优先开展如下重点工作：

（1）完善生物多样性保护法律法规，加大执法监督力度

修订《中华人民共和国环境保护法》、《中华人民共和国野生动物保护法》《野生植物保护条例》和《自然保护区条例》；制定《湿地保护条例》《外来入侵物种管理条例》《遗传资源管理条例》和《林业转基因生物安全管理条例》等法律法规。健全自然资源资产产权制度和用途管制制度，实行最严格的源头保护制度、损害赔偿制度和生态环境损害责任终身追究制度；尽快建立生态补偿机制，把生物多样性保护纳入生态补偿政策，特别是对生物多样性保护优先区域加快建立相应的补偿机制，在资金和制度上给予支持。加强执法能力建设，加大对破坏生物多样性违法活动的打击力度，加大对物种资源出入境的执法检查力度。

（2）促进公众参与，提高公众保护意识

开展多种形式的生物多样性保护宣传教育活动，充分发挥各社团组织和企业的作用，不断提高全民保护意识。探索建立社会监督生物多样性保护的机制和政策。提高公民科学素质，推动公众参与生物多样性保护活动，形成全社会共同推进生物多样性保护和可持续利用的氛围。

（3）落实《全国主体功能区规划》和《中国生物多样性保护战略与行动计划》

建立国土空间开发保护制度，优化国土空间开发格局，提出针对各主体功能区的生物多样性保护政策措施；划定生态保护红线，确保国土生态安全。切实执行《中国生物多样性保护战略与行动计划》，强化对生物多样性优先区域的监管，将生物多样性保护纳入国家、部门和地方相关规划。加强对开发建设活动的环境管理，落实生态恢复责任。建立评估监督机制，促进各项规划的有效实施。

（4）进一步完善就地保护网络，加大保护力度

优化自然保护区和风景名胜区空间结构，科学构建生物多样性保护网络体系，建立国家公园体制。继续实施天然林资源保护、退耕还林、退牧还草、"三北"防护林及长江流域等防护林建设、京津风沙源治理、岩溶地区石漠化综合治理、湿地保护与恢复、自然保护区建设、水土流失综合治理等重点生态工程，启动生物多样性保护重大工程。

（5）加强机制与机构能力建设，提高管理水平

加强中国生物多样性保护国家委员会的统筹协调能力，继续发挥中国履行《生物多样性公约》工作协调组和生物物种资源保护部际联席会议的作用。进一步加强各有关部门中生物多样性保护相关机构的能力建设，尤其要加强对地方生物多样性保护工作的支持力度，促进管理能力不断提高。

（6）建立生物多样性调查、监测和发布制度

定期开展生物多样性调查，建立生物多样性监测预警体系，及时掌握生物多样性的动态变化，发布生物多样性红色名录，有针对性地实施对重要物种和生态系统的有效监控。

第 1 章
中国生物多样性的现状和面临的威胁

1.1 生物多样性对经济社会发展的重要意义

生物多样性是指所有来源的活的生物体中的变异性，这些来源包括陆地、海洋和其他水生生态系统及其所构成的生态综合体等，这包含物种内部、物种之间和生态系统的多样性（《生物多样性公约》中文版）。通俗地讲，生物多样性就是指地球上所有的植物、动物和微生物等物种及其所拥有的基因，以及与环境所构成的生态系统。

生物多样性是人类赖以生存的条件，是经济社会可持续发展的物质基础，是生态安全和粮食安全的保障，也是文学艺术创造和科学技术发明的重要源泉之一。生物多样性具有多方面的价值和功能。第一产业的农、林、牧、渔各业直接以生物资源作为经营的主要对象，它为人们提供了必要的生活物质基础。第二产业的许多行业也直接以生物资源及其产品为原料，特别是制药业，世界上 50% 以上的药物成分来源于天然动植物中。复杂多样的生态系统不仅是人类生存栖息的环境，同时还提供多方面的生态服务。据估计，2000 年中国森林生态系统在产品提供、固碳释氧、涵养水源、土壤保持、净化环境、养分循环、休闲旅游、维持生物多样性等方面的服务价值约为 1.4 万亿元 /a，相当于同期中国国内生产总值的 14.2%（赵同谦等，2004）。草原是地球的碳库，中国草原生态系统总的碳储量约为 440.9 亿 t；草原是天然蓄水库和能量库，80% 的黄河水量、30% 的长江水量、50% 以上的东北河流的水量直接来源于草原地区。中国草原生态系统的总价值达到 12 403 亿元人民币（相当于 1 497.9 亿美元）（谢高地，2001），约每公顷草地 3 100 元，远超过其生产所直接创造的价值。我国湿地贮存着约 2.7 亿 t 淡水，占全国可利用淡水资源总量的 96%，在涵养水源、调节水文、净化水质、补充地下水及抗旱防涝中发挥着重要作用。湿地为全球 20% 的已知物种提供了生存环境，维护着丰富的生物多样性，是宝贵的种质和基因资源库。湿地是巨大的"储碳库"，占陆地生态系统碳总储量的 35%。昆虫授粉对中国水果和蔬菜生产发挥了十分巨大的作用，2008 年昆虫授粉对中国水果和蔬菜产生的经济价值为 521.7 亿美元，占全国 44 种水果和蔬菜总产值的 25.5%（安建东等，2011）。对于生物多样性特别丰富的地区，如海南岛生态系统的生态调节功能价值是其产品价值的 8 倍多（欧阳志云等，2004）；西双版纳州的生态系统服务价值是其区域生产总值的 11 倍以上（景兆鹏等，2012）。

由于生物多样性对一个国家或地区经济社会发展起着重要作用，目前越来越受到国际社会的广泛关注，成为继气候变化之后又一国际环境热点问题。

1.2 中国生物多样性的现状

中国是世界上生物多样性最丰富的 12 个国家之一，也是北半球生物多样性最丰富的国家，拥有森林、灌丛、草甸、草原、荒漠、苔原、湿地、海域等众多自然生态系统（环境保护部，2011）。根据全国生态环境十年变化（2000—2010 年）遥感调查与评估项目统计，面积排列前四位的生态系统分别是草地、森林、农田和荒漠，四类生态系统面积之和占全国生态系统总面积的 82.9%（表 1-1，图 1-1）。

表 1-1　2010 年中国陆域生态系统分布和面积百分比

生态系统类型	面积 / 万 km^2	占全国生态系统总面积的百分比 /%
森林	193.3	20.3
灌丛	69.2	7.3
草地	283.7	29.9
湿地	35.7	3.8
荒漠	127.7	13.4
农田	182.4	19.2
城镇	25.6	2.7
其他	32.0	3.4

注：上述数据中不包括中国台湾地区。

中国有高等植物 3 万余种，居世界第三位，仅次于巴西和哥伦比亚；有脊椎动物 6 000 余种，占世界总种数的 13.7%（环境保护部，2011）。中国维管束植物和哺乳动物丰富度总体上呈现南方高、北方低，山区高、平原低的分布特征，热点地区主要位于岷山、邛崃山、横断山、喜马拉雅山东南段、秦岭、大巴山、武陵山、武夷山、西双版纳、桂西南边境地区、海南中南部山区 [图 1-2（a）、（b）]。中国大部分鸟类是迁徙性的，春季北迁至繁殖地，而秋季南迁至越冬地，因而鸟类分布具有明显的迁徙特征，热点地区主要分布在环渤海地区、中国台湾地区、两广沿海地区、鄱阳湖区、藏东南、横断山脉及滇西北高黎贡山和西双版纳地区等地 [图 1-2（c）]。中国两栖、爬行动物物种丰富度以秦岭—淮河以南地区比较丰富，热点地区主要位于武夷山、西双版纳、桂西南山区、南岭和海南中南部山区等地 [图 1-2（d）、（e）]。中国内陆水域鱼类物种丰富度以长江流域和珠江流域最为丰富，淮河流域和黑龙江流域次之，热点地区主要位于长江上游干流及其支流嘉陵江、乌江、珠江、闽江、鄱阳湖和洞庭湖等地 [图 1-2（f）]（徐海根等，2013）。

中国海域物种多样性丰富，已记录到海洋生物 28 000 多种，其中原核生物界 9 门 574 种、原生生物界 15 门 4 894 种、真菌界 5 门 371 种、植物界 6 门 1 496 种、动物界 24 门 21 398 种，约占全球海洋物种数的 11%。

中国生物遗传资源丰富，是水稻、大豆等重要农作物的起源地，也是野生和栽培果树的主要起源中心。据不完全统计，中国有栽培作物 1 339 种，其野生近缘种达 1 930 个，果树种类居世界第一。中国是世界上家养动物品种最丰富的国家之一，有家养动物品种 576 个（环境保护部，2011）。

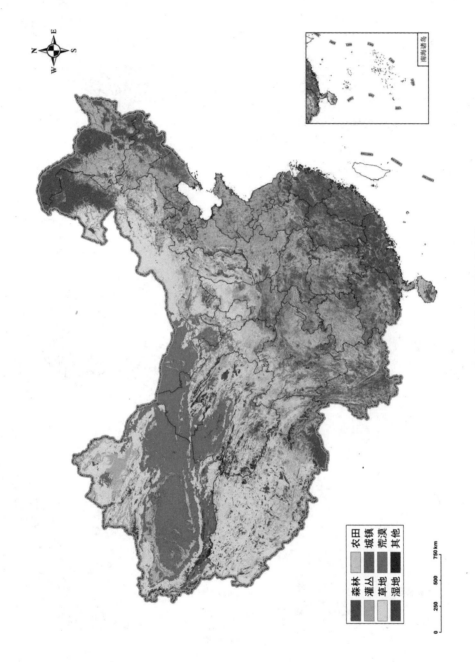

图 1-1 2010 年中国陆域生态系统分布图

数据来源：全国生态环境十年变化（2000—2010 年）遥感调查与评估项目，欧阳志云等提供。

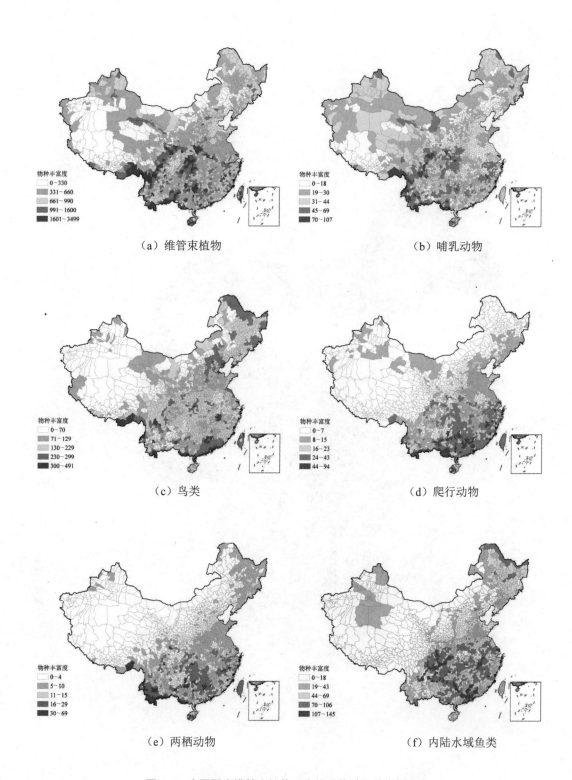

图 1-2 中国野生维管束植物和脊椎动物种数的空间分布

中国无脊椎动物受威胁（极危、濒危和易危）的比例为34.7%，脊椎动物受威胁的比例为35.9%（汪松等，2004）；中国受威胁植物有3 767种，约占评估高等植物总数的10.9%；需要重点关注和保护的高等植物达10 102种，占评估高等植物总数的29.3%（环境保护部和中国科学院，2013）。中国遗传资源丧失的问题突出。根据第二次全国畜禽遗传资源调查的结果，全国有15个地方畜禽品种资源未发现，超过一半以上的地方品种的群体数量呈下降趋势（国家畜禽遗传资源委员会组，2011）。

1.3 生物多样性面临的主要威胁

中国生物多样性面临的威胁是多方面的。主要压力是，随着人口的快速增长和工业化、城市化进程的加快，引起野生生物生境的退化或丧失、自然资源的过度利用、严重的环境污染、单一品种的大规模种植、外来物种入侵和气候变化等问题。

（1）野生生物生境的退化或丧失

导致野生生物濒危的主要因素是生境的丧失或退化（魏辅文，2010；环境保护部和中国科学院，2013）。20世纪50年代至90年代的湿地开垦造成湿地面积大幅度减少。近年来，虽然内陆水域面积有所增长，但滩涂围垦面积仍在扩大。2008—2012年，全国填海造地面积达650.6 km^2。由于滩涂围垦，中国的红树林资源下降了约2/3，直接造成了部分重要保护物种栖息和繁殖场所遭到破坏。自20世纪50年代以来，中国共开垦草地约19.3万 km^2，全国现有耕地的18.2%源于草地开垦（樊江文等，2002）。近些年来，草地开垦的事件仍有发生。铁路和公路建设使野生动植物栖息环境破碎化，种群繁衍面临直接威胁。中国水电装机容量突破2.3亿 kW，居世界首位。兴修水利和建闸筑坝造成湖泊、江河的隔断，彻底改变了河道的自然状态，对鱼类繁殖造成灾难性的后果（许存泽，2006）。

（2）自然资源的过度利用

野生生物资源的过度开发、利用，造成生物种群数量急剧下降，导致生物资源的枯竭和衰退。中国草原过度放牧现象严重，全国重点天然草原的牲畜平均超载率为28%（农业部草原监理中心，2012）。长期过度放牧破坏了草原植被，造成草原退化、沙化。目前，全国90%的草原存在不同程度的退化和沙化。中国海洋捕捞渔业在整个渔业体系中举足轻重，海洋年捕捞量达1 500万 t（农业部渔业局，2011）。高强度捕捞加剧了海洋渔业资源的衰退，小型鱼、低龄鱼、低值鱼比例增加，渔业资源营养级降低。野生动植物由于具有药用、食用、观赏等多方面的经济价值，往往成为非法贸易的对象。虽然中国采取了一系列执法行动，但是非法贸易的现象仍然存在，甚至在有些地区还很猖獗。

（3）环境污染

环境污染物能产生多种毒性，阻碍生物的正常生长发育，使生物丧失生存或繁衍的能力。化肥、杀虫剂、除草剂的使用，也造成日趋严重的面源污染。自20世纪50年代以来，昆明滇池由于水体污染导致富营养化，高等水生植物种类丧失了36%，鱼类种类丧失了25%（吕利军等，2009）。中国管辖海域水环境状况总体较好，但近岸海域海水污染依然严重。海洋环境污染引起赤潮等多种海洋生态灾害，对海洋生物多样性造成严重损害。

（4）单一品种的大规模种植

随着新品种的开发和广泛使用，栽培的作物集中在少数几个品种，单一品种的推广面积大幅度提高，而许多拥有重要基因资源的传统品种遭到淘汰，甚至永远消失。

（5）外来物种入侵

外来物种入侵是导致生物多样性丧失的主要原因之一。中国幅员辽阔，跨越近50个纬度、5个气候带，多样化的生态系统使中国更易遭受外来入侵物种的侵害，来自世界各地的大多数外来物种都可能在中国找到合适的生境。中国是世界上遭受外来入侵物种危害最严重的国家之一，目前有外来入侵物种500余种（徐海根等，2011），松材线虫、湿地松粉蚧、松突圆蚧、美国白蛾、松干蚧、稻水象甲、美洲斑潜蝇、非洲大蜗牛等外来入侵物种对农林业生产、环境和生物多样性造成严重的不利影响。据估计，外来入侵物种每年对中国环境和经济造成的损失约1 199亿元（徐海根等，2004）。

（6）气候变化

气候变化使生物物候、分布和迁移发生改变，使一些物种在原栖息地消失。青海湖地区气候呈现暖干化趋势，与20世纪中期相比，豆雁等26种鸟从湖区消失（马瑞俊等，2006）。气候变化使有害生物的分布范围改变，危害加剧。例如，气候变暖使加拿大一枝黄花的分布范围扩大（吴春霞等，2008）。气候变化使海洋生物的群落结构发生改变。我国黄海主要冷水动物种数和种群密度随水温的升高正在下降，黄海冷水底栖生物区系多样性较半世纪前显著降低（刘瑞玉，2011）。

1.4 生物多样性丧失对经济社会发展带来的影响

生物多样性既为人们提供了必要的生活物质、工业原料和天然药物，又在保护环境、维护生态安全，特别是净化空气、保证水质、改良土壤等方面发挥着关键作用，是人类赖以生存的条件和经济社会可持续发展的物质基础，关系到当代及子孙后代的福祉。千年生态系统评估认为，生物多样性的丧失会直接或间接地引起不断恶化的健康问题、更高的食品风险、日益增加的脆弱性和更少的发展机会（Millennium Ecosystem Assessment，2005）。

（1）对人身财产安全的直接影响

生物多样性的丧失，将会增加生态系统的脆弱性。生物多样性多个组分的丧失，特别是在景观水平的功能多样性和生态系统多样性的下降，将导致生态系统稳定性的下降。红树林和珊瑚礁都是生物多样性的丰富源泉，也是非常好的抵御洪水和暴风雨的自然缓冲物。红树林和珊瑚礁遭到破坏，会增加沿海地区洪水泛滥的概率，给沿海地区居民的海产养殖和居所带来严重危害。人类对森林的过度采伐，往往造成水土流失，是造成泥石流的重要原因之一。

（2）对粮食安全带来的影响

生物多样性的丧失将使食物的多样性减少，使人类仅仅依赖少数几种主要的食物，从而打破人们均衡的饮食结构，影响人类健康。例如，传粉昆虫的丧失，会导致依赖昆虫授粉的作物产量的下降，导致农业生产系统稳定性的下降（Garibaldi et al.，2011）。农作物的野生近缘种在农业生产中发挥了重要作用。20 世纪 70 年代，中国著名水稻育种专家袁隆平院士利用海南发现的野生稻不育株与栽培种杂交，成功地实现了水稻的杂交制种，为中国和世界粮食安全作出了卓越贡献。而如果当时野生稻消失，就无法创造出如此巨大的科技成果。遗憾的是，中国目前野生稻的自然群落正在迅速减少，有些分布点甚至濒临灭绝。野生稻的灭绝不仅仅是一个物种的消失，而与人类的粮食安全休戚相关。

（3）对医药产业带来的影响

中国有中药资源 1.2 万多种，在世界上位居前列。珍稀药用资源往往具有自然分布范围小、再生能力差、生长周期长等特点。长期以来人们对野生药用资源的过度利用，导致不少野生药用资源的蕴藏量急剧减少，直至濒临灭绝。医药产业的发展需要建立在丰富的药用资源的基础上，如果没有了这些资源，医药产业就失去了发展的根基。尽管现在实验室研制的药品越来越多，但目前全球仍有大量人群使用这些药用资源治疗疾病。这些药用资源一旦灭绝将产生严重的后果。

（4）对未来发展带来的影响

生物多样性的丧失，会减少当地居民的发展机会。生物多样性的丧失和生态系统的破坏，将可能导致原本旅游资源丰富的地区丧失大量游客，当地居民会失去发展旅游业的机会。在一些情况下，生物多样性的丧失是不可逆转的，这种情况将会使我们的子孙后代也失去发展的机会。

保护生物多样性，是保障生态安全的必然要求，也是维护自然生产力、实现可持续发展的重要方面，具有双重意义。保护生物多样性，对于当前中国加快经济发展方式转变、建设生态文明、实现可持续发展，进而实现中华民族伟大复兴的中国梦，具有重要意义。

第2章
国家生物多样性战略与行动计划及其执行情况

2.1 《中国生物多样性保护战略与行动计划》的制订

《中国生物多样性保护战略与行动计划》（2011—2030年）于2010年9月15日由国务院第126次常务会议审议通过，2010年9月17日由环境保护部发布实施。该《战略与行动计划》明确了中国今后20年生物多样性保护的指导思想、战略目标和战略任务，并在全国范围内划分了35个生物多样性保护优先区域（图2-1），提出了需要重点开展的10个优先领域、30个优先行动和39个优先项目。

《战略与行动计划》的制订历经了三年多时间，开展了多项专题研究，召开了多次工作会议、咨询会议和国际研讨会，多次向国务院20多个部门和31个省级人民政府征求意见。因此，《战略与行动计划》体现了广泛的代表性和参与性，是中国履行《生物多样性公约》（以下简称《公约》）工作协调组和生物物种资源保护部际联席会议各成员单位共同努力的结果，也是国内外众多机构参与其中精诚合作的典范。

中国在1994年就发布了《中国生物多样性保护行动计划》（以下简称《行动计划》）。由于当时制定时间较早，《公约》的一些重点内容，特别是公约的第三大目标（公平和公正地分享由使用遗传资源产生的惠益）未能在原《行动计划》中体现。《公约》生效后的许多优先领域，如外来入侵物种、遗传资源及相关传统知识的获取与惠益分享、转基因生物安全等内容也未能在原《行动计划》中适当体现。同时，原《行动计划》缺少国家战略的内容。根据中国经济社会发展和国际生物多样性保护要求，中国政府提出了新形势下生物多样性保护的战略与行动计划。更新后的战略与行动计划明确了三个不同时期的目标，划定了35个保护优先区域，新增了战略的内容，还考虑了外来入侵物种、气候变化、遗传资源获取与惠益分享、传统知识、转基因生物安全等问题。《战略与行动计划》的颁布和实施，已经并将对中国乃至国际生物多样性保护产生积极而深远的影响。

2.2 国家生物多样性保护目标

2010年10月，在日本召开的《公约》缔约方大会第十次会议通过了全球《生物多样性战略计划（2011—2020）》。该战略计划确定了2020年全球生物多样性目标（也称爱知目标，以下简称2020年目标），为全球生物多样性保护确定了路线图和时间表，为制定国家目标提供了灵活的框架。2020年目标由5个战略目标和20个具体目标组成（表2-1）。

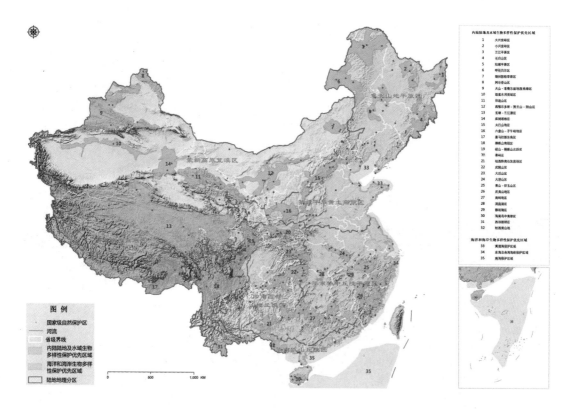

图 2-1 中国 35 个生物多样性保护优先区域

2010 年年底，国务院发布了《全国主体功能区规划》，将中国国土空间划分为四大主体功能区，即优化开发区域、重点开发区域、限制开发区域和禁止开发区域；将 25 个重点生态功能区列入限制开发区域，区域内限制进行大规模高强度的工业化城镇化开发，以保护和修复生态环境、提供生态产品为首要任务；同时，将国家级自然保护区、世界文化与自然遗产、国家级风景名胜区、国家森林公园和国家地质公园列入禁止开发区域，禁止进行工业化城镇化开发，保护中国自然文化资源和珍稀动植物基因资源。

2012 年 11 月召开的中国共产党第十八次全国代表大会对建设生态文明作出了部署，提出了"建设美丽中国"的宏大愿景，把生态文明建设放在突出地位，融入经济建设、政治建设、文化建设、社会建设各方面和全过程。要求"坚持节约资源和保护环境的基本国策，坚持节约优先、保护优先、自然恢复为主的方针，着力推进绿色发展、循环发展、低碳发展，形成节约资源和保护环境的空间格局、产业结构、生产方式、生活方式，从源头上扭转生态环境恶化趋势"。

《战略与行动计划》提出了"保护优先、持续利用、公众参与、惠益共享"的基本原则，确立了近期（2015 年）、中期（2020 年）和远期（2030 年）目标（表 2-1）。《战略与行动计划》虽然是在 2010 年《公约》缔约方大会第十次会议达成全球《生物多样性战略计划（2011—2020）》和 2020 年目标前发布的，但在其编制过程中充分考虑了《公约》2010 年后战略计划的草案。因此，《战略与行动计划》基本体现了《公约》2010 年后战略计划的目标和任务。

中国政府从建设生态文明和美丽中国高度提出的战略思想和战略目标,与《战略与行动计划》一起,勾画了比较全面的中国国家生物多样性保护目标体系(表 2-1)。但针对 2020 年目标之目标 7、目标 9、目标 10、目标 13、目标 16 和目标 19,相应的国家目标不够具体、明确,缺乏有效的措施和手段。今后,中国应高度重视农林业可持续管理、外来入侵物种防治、遗传资源的保护管理及其惠益分享、气候变化对珊瑚礁和其他脆弱生态系统的影响等问题,同时应进一步加大资金投入,加强生物多样性保护领域的科学技术研发和推广应用。

表 2-1　2020 年全球生物多样性目标与中国国家生物多样性目标的对应关系

全球目标	国家目标	所依据的国家战略与规划
全球《生物多样性战略计划(2011—2020)》		
远景目标:到 2050 年,生物多样性受到重视、得到保护、恢复及合理利用,维持生态系统服务,实现一个可持续的健康的地球,所有人都能共享重要惠益。 2020 年任务:采取有效和紧急的行动,以阻止生物多样性的丧失,并确保到 2020 年生态系统有复原能力并继续提供主要服务,从而保障地球生命的多样性,为人类福祉和消除贫困作出贡献	远景目标:到 2030 年,使生物多样性得到切实保护。 2020 年目标:到 2020 年,努力使生物多样性的丧失与流失得到基本控制。 2015 年目标:到 2015 年,力争使重点区域生物多样性下降的趋势得到有效遏制	《战略与行动计划》[c]
2020 年全球生物多样性目标(爱知目标)		
目标 1:最迟到 2020 年,人们认识到生物多样性的价值,并知道采取何种措施来保护和可持续利用生物多样性	● 扎实开展环境宣传活动,普及环境保护知识,增强全民环境意识。 ● 到 2030 年,保护生物多样性成为公众的自觉行动	《全国环境宣传教育行动纲要》(2011—2015)[d] 《战略与行动计划》[c]
目标 2:最迟到 2020 年,生物多样性的价值已被纳入国家与地方发展和扶贫战略及规划进程,并正在被酌情纳入国民经济核算体系和报告系统	● 把资源消耗、环境损害、生态效益纳入经济社会发展评价体系,建立体现生态文明要求的目标体系、考核办法和奖惩机制	《中国共产党第十八次全国代表大会报告》[a]
目标 3:最迟到 2020 年,消除、淘汰或改革危害生物多样性的鼓励措施(包括补贴),以尽量减少或避免消极影响,制定和执行有助于保护和可持续利用生物多样性的积极鼓励措施,遵照《公约》和其他相关国际义务,并顾及国家社会经济条件	● 加快建立生态补偿机制,加大对重点生态功能区的均衡性转移支付力度,研究设立国家生态补偿专项资金,推行资源型企业可持续发展准备金制度	《中华人民共和国国民经济和社会发展第十二个五年规划纲要》[b]
目标 4:最迟到 2020 年,所有级别的政府、商业和利益相关方都已采取措施,实现或执行了可持续的生产和消费计划,并将利用自然资源造成的影响控制在安全的生态限值范围内	● 到 2015 年,资源节约型、环境友好型社会建设取得重大进展。 ● 着力推进绿色发展、循环发展、低碳发展,形成节约资源和保护环境的空间格局、产业结构、生产方式、生活方式	《中华人民共和国国民经济和社会发展第十二个五年规划纲要》[b] 《中国共产党第十八次全国代表大会报告》[a]

全球目标	国家目标	所依据的国家战略与规划
目标 5：到 2020 年，使所有自然生境，包括森林的丧失速度至少减少一半，并在可行情况下降低到接近零，同时大幅度减少退化和破碎化程度	● 到 2015 年森林覆盖率提高到 21.66%，森林蓄积量比 2010 年增加 6 亿 m^3。 ● 到 2020 年，全国草原退化趋势得到基本遏制，草原生态环境明显改善。 ● 到 2020 年，近海生态环境恶化趋势得到根本扭转、海洋生物多样性下降趋势得到基本遏制。 ● 到 2020 年，水域生态环境逐步得到修复，渔业资源衰退和濒危物种数目增加的趋势得到基本遏制	《中华人民共和国国民经济和社会发展第十二个五年规划纲要》[b] 《全国草原保护建设利用总体规划》[c] 《国家海洋事业发展"十二五"规划》[c] 《中国水生生物资源养护行动纲要》[c]
目标 6：到 2020 年，以可持续和合法的方式管理和捕捞所有鱼群、无脊椎动物种群及水生植物，并采用基于保证生态系统安全的方式，避免过度捕捞，同时对所有枯竭物种制订了恢复的计划和措施，使渔业不对受威胁鱼群和脆弱生态系统产生有害影响，确保渔业对种群、物种和生态系统的影响在安全的生态限值范围内	● 到 2020 年，水域生态环境逐步得到修复，渔业资源衰退和濒危物种数目增加的趋势得到基本遏制。 ● 到 2020 年，近海生态环境恶化趋势得到根本扭转、海洋生物多样性下降趋势得到基本遏制	《中国水生生物资源养护行动纲要》[c] 《国家海洋事业发展"十二五"规划》[c]
目标 7：到 2020 年，农业、水产养殖业及林业用地实现可持续管理，确保生物多样性得到保护	● 到 2020 年，全国森林保有量达到 223 万 km^2 以上，比 2010 年增加约 22.3 万 km^2；全国森林蓄积量增加到 150 亿 m^3 以上，比 2010 年增加约 12 亿 m^3 ● 到 2020 年，畜牧业生产方式不断转变，草原可持续发展能力有效增强。 ● 到 2020 年，捕捞能力和捕捞产量与渔业资源可承受能力大体相适应	《全国林地保护利用规划纲要（2010—2020）》[c] 《全国草原保护建设利用总体规划》[c] 《中国水生生物资源养护行动纲要》[c]
目标 8：到 2020 年，污染（包括营养物过剩造成的污染）被控制在不对生态系统功能和生物多样性构成危害的范围内	● 到 2015 年，主要污染物排放总量显著减少，与 2010 年相比化学需氧量、二氧化硫排放均减少 8%，氨氮、氮氧化物排放均减少 10%。 ● 到 2020 年，单位国内生产总值能源消耗和二氧化碳排放大幅下降，主要污染物排放总量显著减少	《中华人民共和国国民经济和社会发展第十二个五年规划纲要》[b] 《中国共产党第十八次全国代表大会报告》[a]
目标 9：到 2020 年，查明外来入侵物种及其入侵路径并确定其优先次序，优先物种得到控制或根除，并制定措施对入侵路径加以管理，以防止外来入侵物种的引进和种群建立	● 到 2020 年，全国林业有害生物成灾率控制在 4%	《全国林业有害生物防治建设规划（2011—2020）》[c]
目标 10：到 2015 年，尽可能减少由气候变化或海洋酸化对珊瑚礁和其他脆弱生态系统的多重人为压力，维护它们的完整性和功能	● 到 2020 年，单位国内生产总值能源消耗和二氧化碳排放大幅下降。 ● 到 2020 年，基本建成布局合理、功能完善的自然保护区体系，国家级自然保护区功能稳定，主要保护对象得到有效保护	《中国共产党第十八次全国代表大会报告》[a] 《战略与行动计划》[c]

全球目标	国家目标	所依据的国家战略与规划
目标 11：到 2020 年，至少有 17% 的陆地和内陆水域以及 10% 的沿海和海洋区域，尤其是对于生物多样性和生态系统服务具有特殊重要性的区域，通过有效而公平管理的、生态上有代表性和连通性好的保护区系统和其他基于区域的有效保护措施得到保护，并被纳入更广泛的陆地景观和海洋景观	• 到 2015 年，陆地自然保护区总面积占陆地国土面积的比例维持在 15% 左右，使 90% 的国家重点保护物种和典型生态系统类型得到保护。 • 海洋保护区占管辖海域面积的比例由 2010 年的 1.1% 提升到 2015 年的 3%。 • 到 2020 年，海洋保护区总面积达到中国管辖海域面积的 5% 以上，近岸海域海洋保护区面积占到 11% 以上。 • 到 2020 年，基本建成布局合理、功能完善的自然保护区体系，国家级自然保护区功能稳定，主要保护对象得到有效保护	《战略与行动计划》c 《国家海洋事业发展"十二五"规划》c 《全国海洋功能区划（2011—2020）》c 《战略与行动计划》c
目标 12：到 2020 年，防止了已知受威胁物种的灭绝，且其保护状况，尤其是其中减少最严重的物种的保护状况得到改善和维持	• 到 2015 年，使 80% 以上的就地保护能力不足和野外现存种群数量极小的受威胁物种得到有效保护。 • 到 2020 年，国家级自然保护区功能稳定，主要保护对象得到有效保护。 • 到 2020 年，使绝大多数的珍稀濒危物种种群得到恢复和增殖，生物物种受威胁的状况进一步缓解	《战略与行动计划》c 《全国生物物种资源保护与利用规划纲要》c
目标 13：到 2020 年，保持了栽培植物、养殖和驯养动物及野生近缘物种，包括其他社会经济以及文化上宝贵的物种的遗传多样性，同时制定并执行了减少遗传侵蚀和保护其遗传多样性的战略	• 到 2020 年，努力使生物多样性的丧失与流失得到基本控制。基本建成布局合理、功能完善的自然保护区体系，主要保护对象得到有效保护。 • 修订《国家畜禽遗传资源保护名录》，对列入保护名录的珍贵、稀有、濒危的畜禽遗传资源实施重点保护，确保受保护的品种不丢失、主要经济性状不降低	《战略与行动计划》c 《全国畜禽遗传资源保护和利用"十二五"规划》
目标 14：到 2020 年,提供重要服务（包括与水相关的服务）以及有助于健康、生计和福祉的生态系统得到恢复和保障，同时顾及了妇女、土著和地方社区以及贫穷和弱势群体的需要	• 到 2020 年，生态系统稳定性增强，人居环境明显改善。 • 到 2020 年，天然草原基本实现草畜平衡，草原植被明显恢复，草原生产能力显著提高。 • 到 2020 年，近海生态环境恶化趋势得到根本扭转，海洋生物多样性下降趋势得到基本遏制	《中国共产党第十八次全国代表大会报告》a 《全国草原保护建设利用总体规划》c 《国家海洋事业发展"十二五"规划》c
目标 15：到 2020 年，通过养护和恢复行动，生态系统的复原力以及生物多样性对碳储存的贡献得到加强，包括恢复了至少 15% 退化的生态系统，从而有助于减缓和适应气候变化及防止荒漠化	• 到 2020 年，与 2010 年相比新增森林面积 5.2 万 km^2，森林蓄积量净增加 11 亿 m^3，森林碳汇增加 4.16 亿 t。 • 到 2020 年，累计治理"三化"草原 165 万 km^2 以上，草原植被明显恢复，草原生产能力显著提高。 • 到 2020 年，水域生态环境逐步得到修复	实施天然林资源保护二期工程c 《全国草原保护建设利用总体规划》c 《中国水生生物资源养护行动纲要》c

全球目标	国家目标	所依据的国家战略与规划
目标16：到2015年，《生物多样性公约》关于获取遗传资源和公正、公平分享其利用所产生惠益的《名古屋议定书》已经根据国家立法生效并实施	● 到2020年，生物遗传资源获取与惠益共享制度得到完善	《战略与行动计划》c
目标17：到2015年，各缔约方已经制定、作为政策工具通过和开始执行了一项有效、参与性的最新国家生物多样性战略与行动计划	已发布《战略与行动计划》	
目标18：到2020年，与生物多样性保护和可持续利用有关的土著和地方社区的传统知识、创新和做法以及他们对生物资源的习惯性利用得到尊重，并纳入和反映到公约的执行中，这些应与国家立法和国际义务相一致并由土著和地方社区在各级层次的充分和有效参与	● 到2020年，进一步健全国内相关传统知识的文献化编目和产权保护制度	《全国生物物种资源保护与利用规划纲要》c
目标19：到2020年，已经提高、广泛分享和转让并应用了与生物多样性及其价值、功能、状况和变化趋势以及有关其丧失可能带来的后果的知识、科学基础和技术	● 到2020年，全社会研究开发投入占国内生产总值的比重提高到2.5%以上，力争科技进步贡献率达到60%以上，本国人发明专利年度授权量和国际科学论文被引用数均进入世界前5位。 ● 扎实开展环境宣传活动，普及环境保护知识，增强全民环境意识	《国家中长期科学和技术发展规划纲要（2006—2020）》c 《全国环境宣传教育行动纲要（2011—2015）》d
目标20：最迟到2020年，依照"资源调集战略"商定的进程，用于有效执行《战略计划》而从各种渠道筹集的财务资源将较目前水平有大幅提高	● 拓宽投入渠道，加大国家和地方资金投入，引导社会、信贷、国际资金参与生物多样性保护，形成多元化投入机制	《战略与行动计划》c

注：a表示中国共产党全国代表大会审议通过；b表示全国人民代表大会审议通过；c表示国务院颁布或批准发布；d表示环境保护部等六部门联合颁布。

2.3 履行《生物多样性公约》的主要行动

2.3.1 法律法规

中国政府在现有50余部有关生物多样性法律法规的基础上，近年来颁布了《中华人民共和国海岛保护法》、《中华人民共和国植物新品种保护条例》等多部法律法规，制定并颁布了一系列有关生物多样性保护的国家、行业和地方标准，使保护和利用生物多样性的法律法规体系日臻完善。

2.3.2 跨部门工作机制

为履行《生物多样性公约》，国务院批准成立了由环境保护部牵头、24个部门组成的中国履行《生物多样性公约》工作协调组，在环境保护部成立了履约办公室。为加强生物物种资源保护与管理，国务院批准成立了以环境保护部牵头、17个部委组成的生物物种资源保护部际联席会议制度，在环保部设立了联席会议办公室。为组织实施好"2010国际生物多样性年"的相关活动，中国政府成立了由国务院分管总理亲任主席、25个部门共同组成的"2010国际生物多样性年中国国家委员会"。2011年，国务院批准将"2010国际生物多样性年中国国家委员会"更名为"中国生物多样性保护国家委员会"，并在环境保护部设立秘书处。环保部牵头、各部门参与的履约协调机制各有特色，互相促进，在推进中国生物多样性保护工作中发挥着重要作用。

大部分省（自治区、直辖市）人民政府加强了环保、农业、林业、海洋等涉及生物多样性保护的机构建设，并成立了跨部门的协调机制。

案例2.1 云南省生物多样性保护：从滇西北走向全省

云南省是中国生物多样性最丰富的省份。滇西北地处青藏高原与云贵高原的过渡地带，为生物多样性保护优先区域，包括云南省18个县（市、区）。为加强滇西北生物多样性保护，2008年2月，云南省人民政府在丽江召开了滇西北生物多样性保护工作会议，颁布了《关于加强滇西北生物多样性保护的若干意见》，发布了《滇西北生物多样性保护丽江宣言》（以下简称《丽江宣言》）。随后，制定了《滇西北生物多样性保护规划纲要》、《滇西北生物多样性保护行动计划》，建立了云南省生物多样性保护联席会议制度。2010年5月，云南省政府在腾冲召开了生物多样性保护联席会议和"国际生物多样性年云南行动座谈会"，成立了云南省生物多样性保护基金会，发布了《2010国际生物多样性年云南行动腾冲纲领》（以下简称《腾冲纲领》），并成立了云南生物多样性研究院。2012年4月，云南省政府在西双版纳召开云南省生物多样性保护联席会议，发布《云南省生物多样性保护西双版纳约定》（以下简称《版纳约定》），提出了全面推进生物多样性保护的10项措施。2013年，《云南省生物多样性保护战略与行动计划》（2012—2030）已由省政府批准发布，该《战略与行动计划》是云南省未来20年生物多样性保护工作的纲领性文件。从《丽江宣言》《腾冲纲领》《版纳约定》到《云南省生物多样性保护战略与行动计划》，云南省生物多样性保护力度越来越大。

"十一五"期间，云南省各级部门累计投入生物多样性保护与可持续利用的资金近70亿元。"十二五"以来，云南省生物多样性保护的投入进一步加大，云南省生物多样性保护基金会已接受捐款3 230万元。2013年3月，云南省政府决定由省级财政设立5 000万元的生物多样性保护专项资金。

> **案例 2.2　四川省生物多样性保护战略与行动计划（2011—2020）**
>
> 　　四川省位于长江上游，是世界 25 个生物多样性热点地区之一。四川省编制并实施战略与行动计划，对中国履行《生物多样性公约》意义重大。在环境保护部、联合国开发计划署和美国大自然保护协会的支持下，2007 年四川省环境保护厅和林业厅会同多个省级部门，启动了《四川省生物多样性保护战略与行动计划》的编制工作。在各有关部门、单位和专家的共同努力下，经过多次调研、协商和征求意见，历时两年多时间，完成了《四川省生物多样性保护战略与行动计划》的编制工作。2011 年 12 月，这一战略与行动计划经四川省人民政府第 89 次常务会审议批准发布。
>
> 　　该行动计划首次在四川全省划定了 13 个生物多样性保护优先区域，确定了政策法规、基础信息、保护野生生物及其栖息地、监测及保护研究等 9 个优先领域以及 46 个优先行动。
>
> 　　目前，四川省各相关部门在生物多样性保护基础信息系统建设、汶川地震灾后四川生物多样性保护能力恢复、野生生物及其栖息地保护、濒危珍稀野生生物抢救性保护等优先领域开展了大量工作，有效推进了行动计划的实施。

2.3.3　调查和监测

（1）国家森林资源清查

　　国家森林资源清查，以省为总体，每 5 年为一个周期，采用抽样技术系统布设 41.5 万个地面固定样地和 284 万个遥感判读样地，在统一时间内，按统一的要求查清各省（自治区、直辖市）和全国森林资源现状，掌握其消长变化。清查成果内容丰富、信息广泛、数据可靠，是反映全国和省级森林资源状况最权威的数据。中国已完成七次森林资源清查，2009—2013 年开展了第八次清查，即将发布最新清查结果。

（2）生物多样性调查

　　2006—2008 年，实施了近海近岸海洋生物调查。通过调查基本摸清了中国海洋生物的"家底"，出版了《中国海洋物种和图集》。该专著收录了中国海域海洋生物 28 000 余种，编绘了 18 000 余种物种形态图。

　　在重点地区开展了植物、动物和微生物多样性的专项调查，主要包括泛喜马拉雅地区、青藏高原及新疆地区、罗霄山脉地区、南方丘陵山区、西南民族地区、东南沿海主要丘陵平原生态敏感区、热带岛屿和海岸带、长江流域、西北干旱区、黄土高原、东北大小兴安岭地区、东北草原区等。2011—2012 年，中国对云南、广西、贵州三省的 26 县（市、区）的物种资源进行了系统的野外调查，发现动植物新类群或疑似新类群 19 个，中国分布新记录 3 个，省级分布新记录 49 个。

　　"十一五"期间，完成了第二次全国畜禽遗传资源调查，出版了《中国畜禽遗传资源志》。调查中有 15 个地方畜禽品种资源未发现，超过一半以上的地方品种的群体数量呈下降趋势。

2009年起组织开展了第二次全国湿地资源调查，利用"3S"技术，参照《湿地公约》标准，对面积在 0.08 km² 以上的湿地进行了全面调查。

中国还进行了全国生态环境十年（2000—2010）变化遥感调查与评估项目和农业生物资源调查，正在开展第二次全国重点陆生野生动物资源调查、第二次重点保护野生植物资源调查、第四次大熊猫种群栖息地资源调查和中药资源普查试点工作。

（3）生物多样性监测

制定了基于分层抽样的全国生物多样性监测网络构建方案，起草了植物、哺乳动物、鸟类、两栖爬行动物、鱼类、土壤动物、蝴蝶、大型真菌等多个生物类群的监测技术指南，并开展了监测技术培训工作。

中国森林生物多样性监测网络（CForBio）（http://www.cfbiodiv.org）于2004年建立，涵盖了不同纬度带的森林植被类型，包括针阔混交林、落叶阔叶林、常绿落叶阔叶混交林、常绿阔叶林以及热带雨林。截至 2012 年，中国森林生物多样性监测网络包括 12 个大型监测样地，每个样地面积在 9～25 hm²。

自 2011 年起，开展了鸟类和两栖动物监测示范，在全国不同地区和生态系统设置监测样地 200 多个，设置样线 450 余条、样点 430 多个。

自 2004 年起，在近岸海域部分生态脆弱区和敏感区建立了 18 个海洋生态监控区，开展系统的生物多样性监测、评价和保护，监测面积达 5.2 万 km²，涵盖海湾、河口、滨海湿地、珊瑚礁、红树林和海草床等典型海洋生态系统。

自 2005 年起，在青海省"三江源"地区建立了生态监测体系，该体系包括 5 个生态监测系统，14 个生态监测综合站点、496 个基础监测点、3 个水土保持监测小区、2 个水文水资源巡测站、4 个水文资源巡测队和 2 个自动气象站。

对 15 个农业野生植物保护点开展监测，已获取连续 5 年的监测数据。

（4）生物多样性评价

2007—2012 年，完成了全国生物多样性评价工作。该项工作以县域为单元，首次系统采集了全国 34 039 种野生维管束植物和 3 865 种野生脊椎动物的县域分布数据，建立了国家生物多样性信息系统，基本掌握了全国陆域生物多样性现状、空间分布特征及主要威胁因素，识别了全国生物多样性热点地区，发现全国存在较大的保护空缺，初步解决了长期以来中国生物多样性本底不清这一难题。为科学、准确评价湿地生态系统的状况，中国又制定了湿地生态系统健康、功能、价值评价指标体系。

2.3.4　就地保护

中国建立了以自然保护区为主体，风景名胜区、森林公园、自然保护小区、农业野生植物保护点、湿地公园、地质公园、海洋特别保护区、种质资源保护区为补充的保护体系。截至 2013 年年底，全国共建立各种类型、不同级别的自然保护区 2 697 个，面积约 146.3 万 km²，自然保护区面积约占全国陆域面积的 14.8%。国家级自然保护区 407 个，面积约 94.0 万 km^2，占全国自然保护区总面积的 64.3%，占陆域面积的 9.8%。2008 年以来，海洋保护区数量尤其是国家级海洋保护区数量有较大幅度的增长，截至 2012 年年底，共建有各级、各类海洋保护区 240 多处，总面积达到 8.7 万 km^2，占到中国主张管辖海域的近 3%。

案例 2.3 武夷山强制保护模式

武夷山分布着世界同纬度带现存最完整、最典型、面积最大的中亚热带原生性森林生态系统，1999 年 12 月被列入《世界遗产名录》。通过创新模式，武夷山探索出"用占总面积 10% 的生态产业发展，换取占总面积 90% 的生物多样性保护"的可持续发展新经验。

（1）加强制度建设，推进旅游规范化管理。先后出台了《福建省武夷山景区保护管理办法》和《武夷山九曲溪保护管理暂行规定》等法规，极大地推动了武夷山生态环境保护和旅游业可持续发展。

（2）首创联合保护制度，构建完整的保护体系。与省地市政府相关部门、周边各级政府成立联合保护委员会，形成保护区周边长 200 多 km 的联合保护防线；聘用 272 名生态公益林护林员，吸收近 10% 的区内群众直接参与资源管护工作；建立专业森林消防队，成立武警执勤点。在多方努力下，已连续 25 年无森林火灾、无重大林政案和无重大森林病虫害发生，成为全国生态文明建设的重要窗口和典范。

（3）推进"原生态"强制保护措施，实现保护与经济协调发展。近年来，将约占保护区总面积 10% 的集体林划为固定生产区域，供区内群众发展毛竹、茶叶、养蜂等非资源消耗型生态产业。同时，确保占总面积 90% 的森林资源和生物多样性得到有效保护。这个模式被联合国教科文组织誉为"中国自然保护区较好解决保护与发展矛盾问题的一个成功典范"。保护区的森林覆盖率已从建区初的 92.1% 上升到 96.3%，真正做到了保护与发展的双赢。

（4）创建科技平台，展示生物多样性成果。2010 年，对武夷山"十年科考"及其研究成果进行数字化，成功建成了国内较为完整的生物多样性研究信息平台，集成应用了数据库、GIS、虚拟动画和音视频多媒体等技术，完整、直观地展示了武夷山珍稀动植物的分布及生态，对武夷山生物多样性研究与科普教育具有重要意义。

案例2.4 黄山封闭轮休保护模式

黄山，1990年12月被联合国教科文组织列入《世界文化与自然遗产名录》，2004年2月入选世界地质公园，是国家首批"5A"级旅游区。黄山风景名胜区管委会以景区封闭轮休为主要切入点，探索出一套对世界遗产地进行完善保护与适度开发的可持续发展新模式。

（1）强化景区制度化管理。安徽省人民代表大会和政府相关部门先后通过了《黄山风景名胜区管理条例》、《黄山风景名胜区总体规划》（2007—2025）、《黄山风景区生态环境保护规划》，从制度上规范了旅游区的开发与建设。

（2）创造性引入景点轮休制度。先后对天都峰、莲花峰、始信峰等主要景点，分别实施为期2～4年的封闭轮休保护措施。

（3）创新景区接待服务模式。2007年以来，外迁黄山风景名胜区管委会机关和部分职工宿舍，提出"山上游山下住"的构想。核心景区已经实现用电为主、液化气为辅的能源消费格局。

（4）建立保护投入长效机制。设立地质遗迹保护专项资金，从每张门票收入中提取10%作为遗产地保护专项经费，"十一五"期间，累计投入遗产保护资金达6亿多元。

（5）加强国际合作与交流。2008年，世界旅游组织在黄山建立世界遗产地旅游可持续发展观测站。2009年以来，黄山先后加入世界自然保护联盟（IUCN）、全球可持续旅游委员会（GSTC）、世界旅游业理事会（WTTC）和亚太旅游协会（PATA）等国际保护和旅游组织。2010年黄山荣获世界旅游业理事会（WTTC）颁发的"全球旅游目的地管理奖"。2011年年底，黄山作为亚洲的唯一代表，跻身首批全球目的地可持续旅游标准试验区，并与国际组织专家共同制定了《全球目的地可持续旅游标准》。

截至2012年年底，建立森林公园2 855处，规划总面积17.4万 km²，其中，国家级森林公园764处、省级森林公园1 315处；建立国家级风景名胜区225处，面积约10.4万 km²，省级风景名胜区737处，面积约9.0万 km²，两者面积占中国陆域面积的2.0%；建立自然保护小区5万多处，面积1.5多万 km²；建成国家级农业野生植物保护点179个；已建湿地公园468处。2007—2012年，建立国家级水产种质资源保护区368个，面积15.2万多 km²。

> **案例 2.5　四川省都江堰市自然保护与经济协调发展模式**
>
> 　　都江堰市地处四川盆地西缘山地，位于北纬30°45′～31°22′、东经107°25′～107°47′，面积1 208 km²。都江堰市地貌类型多样，海拔高差大，云雾多、湿度大、日照少、霜期短，物种丰富。森林覆盖率2003年为50.1%，2012年增加到58.9%。
>
> 　　都江堰市政府高度重视生物多样性保护工作，1992年建立了面积达310 km²的龙溪—虹口自然保护区，并将周边117 km²的乡镇划为外围保护带，参照保护区管理。1993年该保护区升级为省级自然保护区，1997年升级为国家级自然保护区。在联合国开发计划署、联合国基金、野生动植物保护国际（FFI）、中国环境与发展国际合作委员会生物多样性工作组的支持下，2003年编制完成了《都江堰市生物多样性保护策略与行动计划》。该行动计划指导着都江堰市生物多样性保护工作，推进当地社会经济协调发展。2006年青城山、赵公山等约195 km²的土地又被列入四川大熊猫栖息地世界自然遗产。目前，都江堰市有622 km²的土地属于严格保护区域，基本实现了行动计划确定的生物多样性保护目标。
>
> 　　都江堰市林业局、龙溪—虹口国家级自然保护区管理局大力支持山区农村经济发展，成立了"三木药材"合作社、林下种养殖合作社、高山野菜合作社等农村经济合作组织，实现林地多样化经营，减少了对森林资源的破坏。"三木药材"合作社现有社员2 300余户，带动周边林农18 000余户，基地面积100多km²，产值达6亿元以上。
>
> 　　都江堰市依托良好的自然资源发展旅游业，旅游收入已达77.4亿元。龙溪—虹口国家级自然保护区外围保护带虹口乡，依托良好的自然环境和保护区品牌，2000年在外围保护带范围建立了虹口景区，2011年创建为4A级景区。2013年景区接待游客72.3万人次，旅游收入8 604万元。
>
> 　　龙溪—虹口国家级自然保护区周边农民依托保护区的品牌和外围保护带良好的自然环境，开展农家乐经营，现已发展到192家。这些农家乐经营业主及相关从业人员都从传统的耕作、采集、伐木等生产生活方式转变为乡村旅游接待，经济效益明显提升，2012年人均收入已达10 542元。同时，因较高的经济收入带来意识的转变，这些农民都自觉地参与到自然保护活动中来。

　　自然保护区已成为中国主体功能区中的关键区域，是"禁止开发区"的主体内涵，有效保护了中国90%的陆地生态系统类型、85%的野生动物种群和65%的高等植物群落，涵盖了25%的原始天然林、50%以上的自然湿地和30%的典型荒漠地区，对维护中国生态安全以及促进经济社会可持续发展发挥了重要作用。

2.3.5 迁地保护

（1）植物园和动物园

植物园是实施植物物种资源迁地保护最主要的基地。据不完全统计，目前已建有各级各类植物园 200 个，收集保存了占中国植物区系 2/3 的 2 万个物种。建立了野生植物种质资源保育基地 400 多处，成立了苏铁种质资源保护中心和兰科植物种质资源保护中心，分别收集保存苏铁类、兰科类植物 240 余种和 500 余种。据不完全统计，中国建立了 240 多个动物园（含动物展区）、250 处野生动物拯救繁育基地。

（2）作物遗传资源

截至 2012 年 12 月，中国农作物收集品总量已达 42.3 万份，比 2007 年增加了约 3 万份。为了妥善保存收集的遗传资源，中国政府加强了保存设施建设。一方面对原有的 1 座国家长期库、1 座国家复份库、10 座国家中期库、32 个国家种质圃（含 2 个试管苗库）进行扩建和设施改造，另一方面新建了 7 个国家级种质圃。对保存的重要作物遗传资源进行了核心种质库的构建工作。目前已构建了水稻、小麦、玉米、大豆、棉花、大麦、谷子等农作物的核心种质，建立了水稻、小麦、玉米、大豆等农作物的微型核心种质，发掘了大量重要功能基因。

（3）牧草种质资源

初步建立了牧草种质资源保护利用工作体系，初步查清了中国牧草种质资源的数量、种类、分布及利用状况。建立了牧草中期库 2 座，短期库（工作库）8～10 个，种质资源圃 5 处，保存草种质材料 24 万多份，完成鉴定材料 18 783 份，为异地保存牧草种质资源及其遗传多样性提供了基本条件。

（4）畜禽遗传资源

初步建立了以保种场为主、保护区和基因库为辅的畜禽遗传资源保种体系，对 138 个珍贵、稀有、濒危的畜禽品种实施重点保护。通过实施良种工程项目，新建、改建、扩建了 120 多个重点资源保种场、保护区和基因库。通过实施畜禽种质资源保护项目，每年使 100 多个地方品种得到有效保护，同时还开展了分子水平的遗传资源评价。截至 2012 年 8 月，共建成国家级畜禽遗传资源基因库、保护区和保种场 150 个。

（5）林木种质资源

对杉木、松类、杨树、侧柏、云杉、桦树、蒙古栎、鹅掌楸、水青冈、桤木、桉树、梅花、蜡梅、丁香、牡丹、竹类等近百个重点树种（属）的遗传多样性及变异状况进行了分析评价，获得了这些树种遗传变异和多样性分布的重要数据，制定了遗传改良和林木种质资源保存策略。在全国 31 个省(区、市)设立了省级林木种苗管理站，295 个地(市)、1 569 个县（市）设立了林木种苗管理机构，承担着林木种质资源管理职能，形成了较为完备的林木种质资源管理体系。建立了一批林木种质资源异地保存专项库和综合库，保存树种2 000多种，其中重点树种120多种。正在制定《全国林木种质资源调查收集与保存利用规划》，指导林木种质资源保护工作。

（6）野生生物种质资源

截至2012年年底，中国西南野生生物种质资源库已经搜集保存10 096种植物的种子材料76 864份、844种植物非种子离体繁殖材料9 123份和437种活体植物材料45 980株；搜集保存354种动物种质材料13 805份，主要是珍稀特有野生脊椎动物种质资源；搜集保存319种大型真菌330份和815种微生物种质资源8 235份；保存1 311种生物的12 155份DNA材料。

（7）海洋生物遗传资源

构建了海洋生物种质资源库。位于中国海洋大学的大型海藻种质资源库收集保藏了61个物种近500个株系的大型海藻种质资源，位于国家海洋局第三海洋研究所的中国海洋微生物菌种保藏中心保存了14 000多株菌种。开展了多种海洋生物的基因组或转录组测序，2010年7月，中国科学家宣布绘制完成了牡蛎全基因组序列图谱，这是世界上首张贝类全基因组序列图谱。2010年7月，中国科学家宣布半滑舌鳎全基因组序列测定完成，这是世界上首个鲽形目鱼类基因组序列图谱。

2.3.6 重点生态工程

中国政府陆续实施了天然林资源保护、退耕还林、"三北"及长江流域等防护林建设、京津风沙源治理工程、岩溶地区石漠化综合治理工程等重点生态工程。自2001年以来，重点工程区生态状况明显改善，全国森林资源持续增长，共完成造林面积48.2万 km^2，森林面积较10年前增长了23.0%。目前，森林覆盖率达20.4%，比10年前上升了3.8个百分点；森林蓄积量达137.2亿 m^3，比10年前增长了21.8%，重点生态工程促进了野生物种生境的恢复和物种数量、种类的增加。

中国政府还实施了天然草原退牧还草工程。自2003年启动实施该项工程以来，截至2012年，共建设草原围栏60.6万 km^2，其中禁牧围栏26.2万 km^2、休牧围栏31.7万 km^2、划区轮牧围栏2.7万 km^2、补播严重退化草地15.3万 km^2。工程区平均植被覆盖度为64%，比非工程区提高12个百分点；亩均鲜草产量达212 kg，比非工程区提高70%左右；植被结构趋于稳定，生物多样性得到改善，优良牧草比例明显提高。

在一些重点地区大力开展水土流失综合治理工程。2009—2012年，全国共治理小流域1.2万条，完成水土流失综合防治面积27.0万 km^2。继续推进水土流失封育保护。全国累计实施封育保护面积72万 km^2，其中有45万 km^2生态环境得到了初步修复。特别是青海"三江源"、新疆内陆河流域、西藏等重点修复区，封育保护效果日益明显，生态功能得到有效保护。

自2006年以来实施各类湿地保护项目205个，一批国际和国家重要湿地得到了抢救性保护，自然湿地保护率平均每年增加1个多百分点，约一半的自然湿地得到有效保护；湿地保护管理能力明显增强，湿地工程区的民生得到进一步改善。

积极实施滨海湿地生态修复与重建。正在推进对滨海芦苇湿地、红树林、珊瑚礁、海草床、碱蓬湿地的修复与系统重建工作，2010年以来利用中央分成海域使用金支出项目，投入近38.75亿元，修复红树林、滩涂等重要湿地面积达2 800多 km^2。

组织开展了生态省、市、县创建活动。目前，已有15个省（区、市）开展生态省建设，13个省颁布生态省建设规划纲要，1 000多个县（市、区）开展生态县建设。命名了38个国家生态县（市、区），建成1 559个国家生态乡镇和238个国家级生态村。2008年以来，中央财政共安排农村环保专项资金195亿元，实施"以奖促治、以奖代补"政策措施，支持4.6万个村庄开展环境综合整治，8 700多万农村人口受益。

2.3.7 环境污染控制

（1）扎实推进环境污染减排

中国政府将主要污染物排放总量显著减少作为经济社会发展的约束性指标，着力解决突出环境问题。2000—2010年，主要污染物年均浓度总体呈下降趋势。尤其是2006年以来，将二氧化硫和化学需氧量两项主要污染物指标的排放量削减10%列为国民经济和社会发展的约束性指标，大力推进主要污染物总量控制。2006年以来，工业废水中化学需氧量，废气中二氧化硫、烟尘、工业粉尘，工业固体废物的排放量持续下降。近10年，单位GDP污染物排放强度大幅下降了55%以上。2004年以来，单位GDP能耗下降19.6%，单位GDP二氧化碳排放强度下降15.2%。但中国单位GDP污染物排放强度和能耗仍然很高，废水排放量仍在增加。

（2）严格实施规划环评和项目环评

严格建设项目环评，采取"区域限批"、"行业限批"等措施，2008年以来，国家层面拒批332个、总投资1.1万亿元，涉及高污染、高能耗、消耗资源型、低水平重复建设和产能过剩项目。这对调整产业结构、优先经济增长发挥了重要作用。

（3）深化污染防治

颁布并实施《重点流域水污染防治规划》（2011—2015）、《重点区域大气污染防治"十二五"规划》，发布新修订的《环境空气质量标准》；加强地下水污染防治，积极落实《全国地下水污染防治规划》，编制《华北平原地下水污染防治工作方案》；深入推进江河湖泊污染治理，全国七大水系好于Ⅲ类水质比例由2005年的41%提高到2012年的64%，劣Ⅴ类水质比例由27%下降到12.3%；强力推进历史遗留铬渣治理，全国堆存长达数10年甚至半个世纪的670万t铬渣基本处置完毕。

案例2.6 辽河保护区的创新管理模式

长期以来，由于过度开发，辽河水污染、生物多样性丧失等问题突出，被列入国家重点治理的河流之一。2010年辽宁省政府划定辽河保护区，成立辽河保护区管理局。辽宁省人大颁布实施了《辽宁省辽河保护区条例》，依法授权辽河保护区管理局统一负责辽河保护区内的污染防治、资源保护和生态建设等管理工作，承担水利、环保、国土、交通、林业、农业、渔业等部门的相关监督管理和行政执法职责。

辽河保护区依辽河干流而设，全长538 km，总面积1 869.2 km²，涉及辽宁省14个县（区）。辽河保护区管理局以工程、生态和管理三措并举，开展了如下工作：

（1）实施污染源头治理。在辽河流域集中新建、改建城市污水处理厂134座，建设乡镇污水处理厂121个、垃圾处理场34个。

（2）建设生态控制工程。采用河道清淤、生态护岸、恢复水生植物等措施综合整治河道167 km；新建生态蓄水工程16座，新增湿地53.3 km²。

（3）封育主行洪保障区。辽河主行洪保障区全部退耕还河，收租土地386.7 km²，并实施补植护堤林22.7 km²。

（4）实行严格的工作责任制。坚持依法行政，严格管理。关闭河道内采砂点123处，清理并无害化处理垃圾2 000多m³。

经过三年多的工作，辽河治理和保护工作取得明显成效：①水质呈逐年持续好转态势。2009年年底干流提前一年全部消灭劣V类水体，2012年80%时段达到了Ⅳ类水体，部分时段、区段达到了Ⅲ类水体标准。②生态系统已呈正向演替。辽河保护区植被覆盖率由13.7%提高到63%，生物多样性明显增加，辽河入海口的斑海豹种群在逐步扩大，河刀鱼已开始洄游，沙塘鳢、银鱼繁殖数量显著增加。

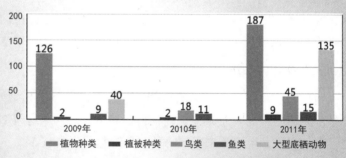

辽河保护区生物多样性恢复状况对比图

开原西孤家子湿地治理前（2010年）

开原西孤家子湿地治理后（2011年）

2.3.8 外来入侵物种防控

（1）完善管理体制机制

制定了国家外来入侵物种应急预案。全国18个省（自治区、直辖市）成立了外来入侵物种管理办公室或建立了联席会议制度，27个省（自治区、直辖市）发布了有关外来入侵物种管理的应急预案。全国已组织启动了12次二级以上应急响应。湖南省2011年发布了《湖南省外来物种管理条例》。发布了57项外来入侵物种监测、评估、防控技术规范，以及《第二批外来入侵物种名单》和《国家重点管理外来入侵物种名录（第一批）》；制订并发布了植物检疫行业标准352项、国家标准104项、国际标准2项。

（2）强化监测预警能力建设

初步形成了林业有害生物监测预警网络体系和农业外来入侵物种监测预警网络。开展了外来入侵物种调查，掌握了重点区域外来入侵物种的分布和危害；初步构建了外来物种风险评估技术体系，并完成了1 500多种外来物种风险评估。在全国开展了邮检工作。

（3）开展外来入侵物种清除活动

在全国22个省（直辖市、自治区）600多个县市开展了"全国十省百县"灭毒除害行动，分别开展了以豚草、水花生等20种外来入侵物种为重点的集中灭除，动员人员4 272多万人次，累计铲除（防治）外来入侵物种面积达5.73多万 $km^2 \cdot$次，重点区域铲除（防治）率达75%以上，有效控制了外来入侵物种的扩散和蔓延。2010年2月，印发了《松材线虫病防治技术方案（修订版）》，进一步加强了松材线虫的防治。松材线虫病发生面积从最高峰的846.7 km^2下降到2011年的453.3 km^2。

（4）开展宣传和培训

利用广播、电视、报刊、网络等多种媒体，开展了外来入侵物种防治技术与管理宣传工作，先后出版《农业外来入侵物种知识100问》等材料20多万册。举办了全国生物安全管理培训班、全国外来入侵物种应急管理培训会议。

2.3.9 转基因生物安全管理

中国政府十分重视转基因生物安全管理工作，主要开展了如下工作：

（1）建立了严格规范的管理制度

国务院颁布了《农业转基因生物安全管理条例》，规定对农业转基因生物安全管理实行安全评价制度、生产许可制度、经营许可制度、产品标识制度和进口审批制度。中国还发布了相关配套规章，实现了转基因技术研发与应用的全过程管理。国家林业局颁布了《开展林木转基因工程活动审批管理办法》，对开展林木转基因工程活动进行规范化管理。国务院批准建立了部际联席会议制度，负责研究、协调农业转基因生物安全管理工作中的重大政策和法规问题。

(2) 建立了科学健全的评价体系

安全评价工作由不同领域专家组成的农业转基因生物安全委员会负责。该委员会委员由有关部委推荐，农业部聘任，目前有委员 64 名。评价中遵循科学、个案、熟悉、逐步的原则，对农业转基因生物实行分级、分阶段安全评价。

(3) 加强技术支撑能力建设

中国政府注重转基因生物安全评价和检测技术能力建设，目前已有 39 个转基因生物安全评价和检测机构经过国家计量认证和农业部审查认可，研究制定了 82 项转基因生物安全技术标准，开展了转基因生物长期生态检测，部分成果获得国际科学界的高度评价，为我国转基因生物安全监管提供了有力的技术支撑。

2.3.10 鼓励措施

(1) 取消对生物多样性有消极影响的补贴政策

为避免对生物多样性和环境造成的消极影响，中国于 2007 年取消了 553 项高耗能、高污染、资源型产品的出口退税，包括濒危动植物及其制品、皮革、部分木板和一次性木制品等。

(2) 建立矿山环境治理和生态恢复保证金

2006 年，财政部会同国土资源部、原环保总局出台了建立矿山环境治理和生态恢复责任机制的指导意见，要求按矿产品销售收入的一定比例，提取矿山环境治理和生态恢复保证金。目前已有 30 个省（自治区、直辖市）建立了矿山环境恢复治理保证金制度。截至 2012 年年底，已有 80% 的矿山缴纳了保证金，累计 612 亿元，占应缴总额的 62%。

(3) 为退耕还林户提供补助

自 1999 年起实施退耕还林工程，国家按照核定的退耕还林实际面积向退耕户提供补助。退耕还林者享有退耕土地上的林木所有权、期限最长 70 年的承包经营权，并按国家有关规定享受税收优惠。2007 年，国务院下发《关于完善退耕还林政策的通知》，完善对退耕农户直接补助政策。长江流域和南方地区每年补助现金 1 575 元 /hm^2，黄河流域及北方地区每年补助现金 1 050 元 /hm^2；还生态林补助 8 年，还经济林补助 5 年。2008—2011 年，中央累计安排专项资金 462 亿元。截至 2012 年年底，中央已累计投入 3 247 亿元，2 279 个县 1.24 亿农民直接受益，户均已累计获得 7 000 元政策补助。

(4) 为实施天然林资源保护工程提供相关补助

2000 年在 17 个省启动天然林资源保护工程，国家给予森林管护补助和造林育林补助，对森工企业职工养老保险社会统筹、森工企业社会性支出及森工企业下岗职工基本生活保障费用实行补助。天然林一期工程累计投入 1 186 亿元。2010 年年底，国务院决定，2011—2020 年实施天然林资源保护二期工程，工程范围增加 11 个县（区、市）。人工造林补助标准为 4 500 元 /hm², 封山育林补助标准 1 050 元 /hm², 飞播造林补助标准 1 800 元 /hm², 教育补助标准每人每年 3 万元, 长江上游和黄河上中游、东北内蒙古等重点林区的卫生补助每人每年 1.5 万元和 1 万元。对国有林, 中央财政安排森林管护费每公顷每年 75 元; 对集体林, 属于国家级公益林的, 2011—2012 年, 中央财政安排森林生态效益补偿基金每公顷每年 150 元, 自 2013 年起, 标准提高到每年 225 元; 属于地方公益林的, 主要由地方财政安排补偿基金, 中央财政每公顷每年补助森林管护费 45 元。天然林保护二期工程总投资资金约 2 440 亿元。

（5）为实施退牧还草工程提供补助

2003 年起在内蒙古、四川、青海等 8 省区和新疆生产建设兵团实施退牧还草工程, 国家给予草原围栏建设资金补助和饲料粮补助。2011 年提升了相关中央投资补助比例和标准, 青藏高原地区围栏建设补助为每公顷 300 元, 其他地区为 240 元; 补播草种补助每公顷 300 元。人工饲草地建设补助每公顷 2 400 元, 舍饲棚圈建设补助每户 3 000 元。退牧还草工程 2003—2012 年中央累计投入资金 175.7 亿元。工程惠及 174 个县（旗、团场）、90 多万农牧户、450 多万名农牧民。

（6）建立草原生态保护补助奖励机制

2011 年起在内蒙古、西藏、青海等 8 个主要草原牧区省（自治区）全面建立草原生态保护补助奖励机制, 对禁牧草原按每公顷每年 90 元的标准给予补助, 对落实草畜平衡制度的草场按每公顷每年 22.5 元的标准给予奖励, 给予牧民生产性补贴（包括牧草良种补贴每年每公顷 150 元和牧民生产资料综合补贴每年每户 500 元）, 并对牧民进行培训以促进牧民转移就业。草原生态奖励补助资金从 2011 年的 136 亿元增加到 2012 年的 150 亿元, 累计安排 286 亿元。截至 2012 年年底, 草原禁牧补助实施面积达 82 万 km², 享受草畜平衡奖励的草原面积达 173.7 万 km²。

（7）为湿地保护提供补助

2010 年, 财政部会同国家林业局启动了湿地保护补助工作, 将 27 个国际重要湿地、43 个湿地类型自然保护区、86 个国家湿地公园纳入补助范围。各地加大财政补助力度, 逐步将重要湿地纳入生态补偿范围。

（8）建立森林生态效益补偿基金

2004 年建立中央森林生态效益补偿基金, 对国家级公益林的营造、抚育、保护和管理支出给予一定补助, 由中央财政预算安排。其中, 国有国家级公益林每公顷每年补助 75 元, 集体和个人所有的国家级公益林补偿标准为每公顷每年 225 元, 目前补偿范围已达 92.4 万 km²。2013 年, 中央财政共下拨森林生态效益补偿基金 149 亿元。各地对地方公益林也进行了补偿。

（9）初步建立国家重点生态功能区生态补偿机制

2008 年中央财政设立国家重点生态功能区转移支付资金以来, 转移支付范围不断扩大、资金量不断增加。2013年，转移支付范围包括492个县域和1 367个禁止开发区域，转移支付资金423亿元。2013年，将云南、贵州、四川、新疆四个省（自治区）风景名胜区列入生态补偿试点。

案例2.7 新安江流域生态补偿试点

新安江发源于安徽省黄山市境内，流入浙江省千岛湖国家级风景名胜区，经富春江、钱塘江入东海，流域总面积达11 674 km²。新安江是流入千岛湖水量最大的河流，而千岛湖是浙江重要的饮用水水源地。为了保护新安江水环境，多年来上游地区的安徽省黄山市等地以牺牲自身发展为代价，延缓了工业化、城镇化进程。

2011年3月和9月，财政部、环保部先后印发了《关于启动实施新安江流域水环境补偿试点工作的函》和《关于开展新安江流域水环境补偿试点的实施方案》，2011年安排补偿资金3亿元，专项用于新安江上游水环境保护和水污染治理，其中中央财政安排2亿元，浙江省安排1亿元。安徽、浙江两省在全国率先建立跨省流域水环境补偿机制，加强对新安江和千岛湖流域水资源的保护。

2.3.11 科学研究

中国政府鼓励并支持有关保护和持续利用生物多样性的研究工作，在"国家科技支撑计划"、"国家重点基础研究发展规划"、"国家高技术发展计划"、"国家自然科学基金"、"公益性行业科研专项等科技计划"中设立有关生物多样性保护与可持续利用的项目。例如"'十一五'国家科技支撑计划"设立了"中国重要生物物种资源监测和保育关键技术与应用示范"、"典型脆弱生态系统重建技术与示范"和"中国陆地生态系统综合监测、评估与决策支持系统"等重点项目；国家重点基础研究发展规划设立了"中国—喜马拉雅地区生物多样性演变和保护研究"、"农业生物多样性控制病虫害和保护种质资源的原理与方法"等项目。国家在自然科技资源平台建设方面，安排了涉及动物、植物、微生物、种质资源等方面的资源调查与收集、信息平台建设、实物和信息共享工作。这些研究工作形成了一系列有价值、有影响的科研成果，为中国生物多样性保护提供了科技支撑。

2.3.12 公众参与

为在中小学生中普及生物多样性保护知识，2011年版的《义务教育生物学课程标准》《义务教育初中科学课程标准》和《普通高中生物课程标准（实验）》将生物多样性相关知识纳入到义务教育课程标准中。通过中小学阶段的课堂教育,中小学生的生物多样性保护知识普遍得到提高。全国有1 908所普通高校设有生物专业,有近50所高校设有生态学本科专业,38所高校可授予生态学硕士学位,22所高校可授予生态学博士学位,学位教育为生物多样性研究与保护奠定了人才基础。截至2012年,中国共培养相关专业人才55.6万余人。

各有关部门、各地区通过利用电视、网络、报刊、广播等媒体,通过举办培训班、大讲堂、发放培训材料等方式,主办各类宣传活动,加大对生物多样性保护重要性和知识的宣传力度。特别是在开展2010国际生物多样性年中国行动宣传活动中,在国家层面组织各类大型宣传活动40项,发放各类宣传品37万余件,通过各类媒体宣传影响受众8.04亿人次;在地方层面,举办大型宣传活动191次,印发宣传材料35万余份,制作生物多样性保护专题片25部。各地还积极动员自然保护区、动物园、植物园、公园、环境保护宣教与科研机构以及电视、报刊、网络等媒体共约2万家单位,面向大中小学生、公众开展了系列宣传,覆盖面达1亿人次。通过宣传教育,公众的参与热情高涨,生物多样性保护意识有了明显的提高,生物多样性的重要性获得广泛认同。

2.3.13 国际合作与交流

中国在多边合作、双边合作、南南合作等方面开展了积极探索,取得了可喜成果,推进了履行《生物多样性公约》和相关议定书的进程。

(1) 多边合作稳步推进

中国是《生物多样性公约》《濒危野生动植物物种国际贸易公约》《关于特别是作为水禽栖息地的国际重要湿地公约》和《联合国气候变化框架公约》等公约的缔约方。中国积极参与这些公约的谈判,认真履行相关义务,积极参与国际多边体系建设,为推动《生物多样性公约》在国际层面的发展和履约进程作出力所能及的贡献,提供资金支持。中国与全球环境基金(GEF)、世界银行、联合国开发计划署、联合国环境规划署等国际机构一起实施了一批生物多样性项目,引进了有益的保护管理理念、技术和资金,促进了国内生物多样性保护工作。"中国生物多样性伙伴关系和行动框架"(CBPF)自2007年11月第32届GEF理事会批准以来,先后实施了9个项目。

(2) 双边合作取得突破

中国与德国、美国、俄罗斯、英国、挪威、加拿大、澳大利亚等50多个国家建立了广泛的对外合作与交流渠道,初步形成了以政府间合作为主的多元化合作体系。为期6年的中国—欧盟生物多样性项目(ECBP)于2011年圆满完成,标志着双边合作取得突破,为推动中国生物多样性保护和可持续利用发挥了重要的作用。

(3) 南南合作取得新进展

近年来,中国政府积极开展生物多样性领域的南南合作,与众多发展中国家签署了生物多样性相关领域的合作协议。举办培训班,支持东南亚、南亚等区域发展中国家的能力建设。成立了中国—东盟环境保护合作中心,这是中国政府建立的首个南南环境合作和区域环境合作平台。

> **案例2.8 中国—欧盟生物多样性项目取得丰硕成果**
>
> 第一,服务战略规划,作出重要贡献。项目为编制《中国生物多样性保护战略与行动计划》(2011—2030)提供了重要支持,同时对地方层面的生物多样性主流化进程产生了深远影响。据统计,在实施项目的省份(直辖市、自治区)中,4/5以上将生物多样性保护要求纳入了"十二五"规划纲要。
>
> 第二,巩固协调机制,促进履约进程。在国家层面,巩固了"中国履行《生物多样性公约》工作协调组",促进了国家发展和改革委员会、国土资源部、农业部、水利部、国家质检总局、国家中医药管理局、国家林业局等相关部门协力推进履约进程。在地方层面,生物多样性保护纳入区域经济社会发展规划中。例如,与生物多样性保护要求一致的《海南省土地利用总体规划》已于2010年获得国务院批准实施。
>
> 第三,支撑高端决策,取得积极成效。项目支持了中国环境与发展国际合作委员会相关战略研究,提出的生态系统服务及生物多样性相关政策建议,已呈报中国政府最高决策层。项目开创性地将生物多样性纳入了国家重点产业发展战略环评,这对建设资源节约型、环境友好型社会,推进加快经济发展方式转变发挥了积极作用。
>
> 第四,加强能力建设,收获丰硕成果。项目提供技术和资金支持,协助各级政府制定了生物多样性相关政策、法规、计划、标准、指南等。到目前为止,已有46项经国务院、相关地方政府、人大或部门批准实施,27项已上报待批,广泛地提高了有关层面履约工作的能力。
>
> 第五,开展宣传教育,提升环境意识。项目建立了宣传教育公共平台,通过网站、宣传片、出版物、通信和形式多样的参与式培训活动,面向政府、企业、社区、学校、媒体等,宣传和普及生物多样性理念与知识,很大程度上提高了全民的生物多样性知识和意识。特别是重点支持了2010国际生物多样性年中国行动,一系列相关宣传活动取得了良好的社会效果。
>
> 第六,致力团队建设,培养履约人才。六年来,从中央到地方,从政府部门到科研机构,一大批国内、国际的决策者、管理者、学者以及媒体工作者们,深度参与了项目的执行和管理,更新了理念,开拓了视野,增强了能力,提高了水平,成为生物多样性领域中富有执行能力和影响力的中坚团队和人才。

案例2.9 生物多样性保护融入土地利用规划与土地整理复垦中

在中国—欧盟生物多样性项目支持下，国土资源部于2008年10月正式启动了"中国土地利用规划与土地整理中的生物多样性保护"项目，并把海南省和贵州省作为项目试点省份。该项目将生物多样性保护融入土地利用规划和土地整理复垦中，制定了专项规划编制大纲和指南，提出了适合中国实际的生态型土地整理工程技术，在土地利用与生物多样性保护领域开展了有益尝试。海南省国土环境资源厅2010年下发了《土地利用总体规划修编中融入生物多样性保护内容的技术导则》，要求在土地利用总体规划中强化生态保护、融入生物多样性保护理念。同时，海南省人民政府批复了《基于生物多样性保护的海南省陵水黎族自治县土地利用总体规划》和《基于生物多样性保护的海南省乐东黎族自治县土地利用总体规划》。贵州省强化土地整治过程中生物多样性保护理念，发布了《贵州省国土资源厅关于在土地综合整治工作中切实加强生态和生物多样性保护的通知》，2010年9月省政府批复了《基于生物多样性保护的关岭自治县土地整理开发复垦规划》和《基于生物多样性保护的荔波县土地整理复垦开发规划》。这些文件和规划的出台优化了土地利用结构和布局，推动了生物多样性保护。

案例2.10 呼伦贝尔草原生物多样性保护和可持续管理项目

位于中国内蒙古自治区呼伦贝尔市的呼伦贝尔草原，由于长期的超载过牧，历史上的过度砍伐森林，采掘业的不合理开采，再加上气候干旱，正面临着大面积退化、沙化的威胁。在中国—欧盟生物多样性项目（ECBP）支持下，在有关国际组织和科研机构的指导下，呼伦贝尔市政府开展了草原生物多样性保护和可持续管理，取得了明显成效。

（1）建立了以副市长牵头的生物多样性保护管理委员会，同时，各旗（市、区）都成立了生物多样性管理委员会，基本上构建了呼伦贝尔市生物多样性保护工作机制，使生物多样性保护纳入政府工作内容。

（2）出台了国内第一个地方生物多样性管理办法，明确地将呼伦贝尔市生物多样性保护纳入了当地的"十二五"规划之中，使得当地草原生物多样性保护有法可依。

（3）编制了《呼伦贝尔草原生物多样性保护规划与行动计划》，按照规划和行动计划逐步开展了呼伦贝尔草原生物多样性保护活动。

（4）签署了中蒙合作备忘录，搭建了与蒙古国东方省生物多样性跨界合作的平台，开展了大量跨界保护行动，对于中蒙两国跨界生物多样性保护，尤其对保护蒙古高原特有种黄羊的栖息、繁殖和迁徙等起到了非常重要的作用。

（5）编制了《呼伦贝尔草原放牧指导手册》，开展了退化草原最佳放牧模式示范，对草原生物多样性保护具有非常重要的指导意义。

（6）建立了沙化草原恢复治理示范点，并总结出8种模式，这些成果和模式已应用于2009年呼伦贝尔市667.7 km^{2}沙地治理工作中，为在呼伦贝尔市大面积开展治沙项目提供了技术支持。

（7）编写了《草原生物多样性监测技术手册》，提高了呼伦贝尔市生态监测站、达赉湖国家级自然保护区和辉河国家级自然保护区的生物多样性监测能力。

（8）通过电视、电台、报纸、政府宣传栏、大型活动等，将生物多样性保护知识传播给了大众，提高了大众的生物多样性保护意识。

2.4 国家生物多样性战略和行动计划的总体进展评估

自2010年颁布《中国生物多样性保护战略与行动计划》（2011—2030年）以来，中国实施该《战略与行动计划》开局良好，正朝着正确的方向推进。其中取得很大进展的行动有1项，即行动10"促进和协调生物遗传资源信息化建设"；取得较大进展的行动有15项；取得一定进展的行动有14项（附表1）。主要进展表现在以下三个方面。

2.4.1 基本建立了具有中国特色的生物多样性保护与管理体系

（1）保护和可持续利用生物多样性的法律法规日益完善。

（2）基本形成生物多样性保护工作机制，政府管理能力得到进一步提升。

（3）形成了类型比较齐全、布局比较合理、功能比较健全的自然保护区网络，还建立了大量风景名胜区、森林公园、自然保护小区、农业野生植物保护点、湿地公园、地质公园、海洋特别保护区、种质资源保护区。自然保护区保护了中国90%的陆地生态系统类型、85%的野生动物种群和65%的高等植物群落，涵盖了25%的原始天然林、50%以上的自然湿地和30%的典型荒漠地区。

（4）公众保护积极性和参与能力有较大提高。

（5）大专院校、科研院所的创新能力有较大提升。

（6）国际合作交流取得新进展。

2.4.2 生态破坏加剧的趋势有所减缓，部分区域生态系统得到恢复

（1）森林资源持续增长，森林面积较10年前增长了23.0%；森林蓄积量较10年前增长了21.8%。

（2）治理小流域1.2万条，完成水土流失综合防治面积27.0万km^2；实施封育保护面积72万km^2，其中45万km^2的生态环境得到了初步修复。

（3）一些国家重点保护野生动植物种群数量稳中有升，分布范围越来越大，生境质量不断改善。大熊猫数量从20世纪80年代的1 000多只增加到现在的1 590只，朱鹮数量从80年代的7只增加到目前的1 800多只，红豆杉、兰科植物、苏铁等保护植物种群不断扩大。

（4）主要污染物年排放量总体呈下降趋势。2000年以来，单位GDP污染物排放强度下降了55%以上。2004年以来，单位GDP能耗下降19.6%，单位GDP二氧化碳排放

强度下降 15.2%。

2.4.3 在保护生物多样性的同时地方经济社会也得到全面发展

在保护和恢复生态系统的同时，当地社区福祉也在改善。中国农村居民家庭人均纯收入 2011 年比 2000 年增加了 40.8%，贫困人口数量也大幅下降。

第3章
生物多样性保护融入部门和跨部门规划的情况

生物多样性保护与利用涉及众多部门。本章阐述在生物多样性保护融入行业规划方面，中国生物多样性保护国家委员会相关成员单位开展的工作和采取的主要措施。

3.1 国家发展和改革委员会

国家发展和改革委员会充分考虑了生态保护与建设对经济社会可持续发展的重要作用，以《中国生物多样性保护战略与行动计划》（2011—2030年）为行动指南，在退耕还林、退牧还草、石漠化、沙尘源区、生态关键区等优先区域，建立相应的保护政策和法规。

（1）以国家生态安全为依据构建国土空间开发战略格局

作为中国第一部全国性国土空间开发规划，《全国主体功能区规划》提出构建城市化地区、农产品主产区和生态功能区三大格局和优化开发、重点开发、限制开发及禁止开发四类开发模式，规划了以青藏高原生态屏障、黄土高原—川滇生态屏障、东北森林带、北方防沙带和南方丘陵山地带为主体的生态安全战略格局（表3-1），并制定了国家重点生态功能区、国家禁止开发区域两个名录和相关评价和规划图。

表3-1 中国"两屏三带"生态安全战略格局

区　域	生态保护工作重点
青藏高原生态屏障	保护好多样、独特的生态系统，发挥涵养大江大河水源和调节气候的作用
黄土高原—川滇生态屏障	加强水土流失防治和天然植被保护，发挥保障长江、黄河中下游地区生态安全的作用
东北森林带	保护好森林资源和生物多样性，发挥东北平原生态安全屏障的作用
北方防沙带	加强防护林建设、草原保护和防风固沙，对暂不具备治理条件的沙化土地实行封禁保护，发挥"三北"地区生态安全屏障的作用
南方丘陵山地带	加强植被修复和水土流失防治，发挥华南和西南地区生态安全屏障的作用

（2）积极推进生态补偿机制建设

2010年，国家发展和改革委员会与相关部委共同组织起草了《生态补偿条例》（草案）。该条例规定了生态补偿的原则、领域、对象、方式、标准等。目前，国家发展和改革委员会正在研究起草"关于建立完善生态补偿机制的若干意见"。

（3）将生态保护纳入地方经济发展与转型规划

为帮助地方创造生态保护与经济转型的良好政策环境，2010年国家发展和改革委员会和国家林业局会同有关部门发布了《大小兴安岭林区生态保护与经济转型规划》(2010—2020)，目前正在编制《黑龙江和内蒙古东部沿边开放开发规划》和《长白山林区生态保护与经济转型规划》等重要规划。

（4）推进重点地区生态建设综合治理

国家发展和改革委员会会同相关部委组织编制了《西部大开发"十二五"规划》。该规划明确了"十二五"期间中国西部地区实施重点生态工程的战略和行动计划。国家发展和改革委员会还发布了《西藏生态安全屏障保护与建设规划》(2008—2013)、《京津风沙源治理二期工程规划》(2013—2022)等重要规划，安排中央预算内资金支持天然林资源保护、石漠化综合治理、"三北"防护林建设等重点生态建设工程。

（5）将保护生物多样性纳入节能减排和应对气候变化行动中

国家发展和改革委员会相继发布了《"十二五"节能减排综合性工作方案》《节能减排"十二五"规划》《"十二五"控制温室气体排放工作方案》，通过调整产业结构和提高能效、发展低碳能源和优化能源结构、增加碳汇和加强生态保护，充分发挥应对气候变化与保护生态环境和生物多样性之间的协同效应。

3.2 教育部门

教育部门高度重视生物多样性方面的教育教学和人才培养，加强基础教育、科普活动、专业建设和人才培养。

（1）在基础教育与科普活动中普及生物多样性保护知识

为提高中小学生对生物多样性重要性的认识，教育部印发了《义务教育生物学课程标准》(2011年版)和《普通高中生物课程标准》(实验)，将生物多样性相关知识纳入其中。《义务教育生物学课程标准》(2011年版)设有"生物多样性"主题，明确提出了生物多样性学习内容和要求。《普通高中生物课程标准》(实验)要求学生"概述生物进化与生物多样性的形成"、"概述生物多样性保护的意义和措施"等，鼓励学生在课余时间参与生物多样性保护工作。各地和学校通过开展绿色社区、绿色学校、绿色家庭创建活动，举办知识竞赛、讲座、展览会、生态夏令营、征文、各类纪念活动，广泛宣传有关生物多样性的法律法规和科学知识。

（2）开展本科专业建设与人才培养

根据《普通高等学校本科专业目录(2012年)》，与生物多样性相关的专业类有"生物科学类"、"海洋科学类"、"自然保护与环境生态类"、"林学类"、"草学类"等。目前，全国共有近2 000所高等学校（约占全国高等学校总数的80%)设有生物多样性相关专业，其中本科院校约800所，高职（专科）院校约1 200所。全国共有298所高等学校设有生物科学专业，52所高等学校设有生态学专业，大多数综合性大学和师范院校都开设了生命科学和生态学课程。中国生物多样性相关专业的毕业人数逐年递增，2008年全国相关专业毕业人数约为10.6万人，到2012年相关专业毕业人数上升为11.7万人，比2008年增长了约10%。截至2012年，中国共培养相关专业人才总计约55.7万人。中国高校十分重视标本馆建设，目前，全国高校各类型标本馆共约180所，为生物多样性的教育教学与科学研究提供了平台。

（3）加强生物多样性相关学科建设

在现行《学位授予和人才培养学科目录》中，与生物多样性相关的学科领域涉及多个一级学科。仅"生物学"、"生态学"、"生物工程"、"林学"和"草学"5个一级学科，截至2012年，全国共设有博士学位授予权一级学科点172个，硕士学位授予权一级学科点225个；2008—2012年，这5个学科共培养相关博士学位研究生17 110人、硕士学位研究生63 634人。中国鼓励有关学位授予权单位根据经济社会发展需要和自身学科条件，按照有关规定，自主设置与生物多样性相关的二级学科，加强该领域的人才培养和科学研究工作。

（4）增加高层次人才培养和引进

近年来，中国开展了千人计划、长江学者奖励计划等一系列重大人才计划，引进、培养了一大批生物多样性研究领域的高水平学科带头人，带动相关国家重点建设学科赶超或保持国际先进水平。特别是自2008年开展了千人计划以来，全国高等学校共引进聘任生物多样性相关学科领域的"千人计划"专家118人，"长江学者"78人。教育部还通过"高层次创造性人才计划"为创新人才的成长搭建台阶、创造条件，提升创新能力和竞争实力。2013年，教育部支持了涉及生物多样性领域的"教育部创新团队"18个、"教育部新世纪优秀人才"100余人。

3.3 科技部门

《国家中长期科学和技术发展规划纲要》（2006—2020）在环境重点领域设有"生态脆弱区域生态系统功能的恢复重建"主题和"全球环境变化监测与对策"主题，在农业重点领域设有"农林生态安全与现代林业"主题，在面向国家重大战略需求的基础研究中设立了"人类活动对地球系统的影响机制"研究方向。这些主题和研究方向对生物多样性方面的科学研究进行了布置。

科技部在"国家科技支撑计划"、"国家重点基础研究发展规划"、"国家高技术发展计划"和其他一些专项资金中专门设立有关生物多样性保护与可持续利用的项目。"十一五"、"十二五"期间，国家科技支撑计划均对生物多样性和生态保护科技工作进行了重点部署，已先后安排了"西南生态安全屏幕（一期）构建技术与示范"、"生物多样性保护与濒危物种保育技术研究及示范"、"中国重要生物物种资源监测和保育关键技术与应用示范"等32个生态保护类的重大和主题项目，总投入经费13.7亿元人民币。国家重点基础研究发展规划设立了"中国—喜马拉雅地区生物多样性演变和保护研究"、"农业生物多样性控制病虫害和保护种质资源的原理与方法"等项目。国家高技术研究发展计划（"863"计划）中的一些项目也涉及了生物多样性保护与生物资源可持续利用的技术开发。国际科技合作专项计划设立了"洞庭湖流域生态系统管理关键技术与应用示范"等重点项目。国家在自然科技资源平台建设方面，安排了涉及动物、植物、微生物、种质资源等方面的资源调查与收集、信息平台建设、实物和信息共享工作。这些研究工作形成了一系列有价值、有影响的科研成果，为中国生物多样性保护提供了科技支撑。

2010年8—12月和2013年9—11月，科技部先后组织了两次"生态保护科技创新专家行"系列活动，邀请专家赴贵州毕节石漠化地区、青海三江源高寒退化草地、南疆沙区等20余个典型生态脆弱地区的生态保护科技示范区开展实地调研与座谈，凝练出一批具有推广应用前景的生态系统保护技术模式。

自"十一五"以来，科技部通过科技基础性工作专项支持教育部、农业部、中国科学院等中央部门所属的有关科研机构在生物多样性和气候变化等方面开展了科学考察与调查的工作。截至2013年12月底，科技部通过该专项立项近70个项目，经费总额近6亿元。调查区域主要包括泛喜马拉雅地区、青藏高原及新疆地区、海南岛及西沙群岛、罗霄山脉地区、南方丘陵山区、东南沿海主要丘陵平原生态区、热带岛屿和海岸带、华北地区、西北干旱区、东北温带针阔混交林区等；调查领域主要涵盖了湖泊、海岛、海洋、森林和特殊生境下的生物资源和种质资源，外来入侵物种等；先后完成了《中国动物志》《中国植物志》《中国植被图》《海洋物种名录》等的编研；建立和完善了"中国森林生物多样性监测网络"和"中国生态系统研究网络"等；编写了《植物、动物和微生物等主要类群的监测标准和规范》。

中国科学院等部门利用丰富的数字化信息，建立了国家标本资源共享平台（NSII）、亚洲生物多样性保护与信息网络（ABCDNet）和全球生物多样性信息网络（GBIF）中国节点；完成了中国3万多种高等植物的受威胁状况评估，发布了中国植物红色名录，并依据保护植物和特有植物的分布确定了中国植物保护的热点地区，为生物多样性保护和可持续利用提供了丰富的信息。中国科学院还启动了应对气候变化的碳收支认证及相关项目。

3.4 国土资源部门

国土资源部门在国土规划、土地利用规划及土地整治规划的制定与实践中，从战略层面高度重视生物多样性工作。

（1）全国国土规划中注重生物多样性保护

国土资源部已全面开展国土规划工作，旨在统筹国土开发、利用、保护与整治，优化国土空间开发格局。《全国国土规划纲要》提出，要严格保护森林、草原、湖泊、湿地和海岸带生态系统，强化各类国家级自然保护区和重要生态功能区的保护；要以自然保护区为主体，以各类国家公园、种质资源保护区、禁猎区、禁伐区、原生境保护小区（点）

等为补充，建立重要生物资源就地保护空间体系。

（2）各级土地利用规划重视生物多样性保护

国务院颁布的《全国土地利用总体规划纲要》（2006—2020）强调"统筹安排生活、生态和生产用地，优先保护自然生态空间"的指导原则。该纲要第五章明确指出，保护基础性生态用地，加大土地生态环境整治力度，因地制宜改善土地生态环境。2010年，国土资源部发布了《市（地）级土地利用总体规划编制规程》《县级土地利用总体规划编制规程》和《乡（镇）土地利用总体规划编制规程》，其中规定各级规划要加强对城乡建设用地的空间管制，将自然保护区核心区、森林公园、地质公园、列入省级以上保护名录的野生动植物自然栖息地、水源保护区的核心区等划入禁止建设区，严格禁止与主体功能不相符的建设活动。在国务院批准的《海南省土地利用总体规划》（2006—2020）中，生物多样性保护已被纳入土地利用总体规划；同时，海南省人民政府批复了《基于生物多样性保护的海南省陵水黎族自治县土地利用总体规划》和《基于生物多样性保护的海南省乐东黎族自治县土地利用总体规划》。

（3）全国土地整治规划强化生物多样性保护

2011年国务院颁布的《土地复垦条例》提出，土地复垦义务人应当遵守土地复垦标准和环境保护标准，保护土壤质量与生态环境，避免污染土壤和地下水。国务院颁布的《全国土地整治规划》（2011—2015）将生态环境建设与生物多样性保护作为土地整治的重要目标之一，"坚持保护优先，自然恢复为主，推进土地生态环境综合整治，提高退化土地生态系统的自我修复能力"。为全面改善矿区生态环境，国土资源部于2012年6月开始在全国范围内部署开展"矿山复绿"行动，通过采取工程、生物等措施，对重要自然保护区、景观区等区域采矿活动引起的矿山地质环境问题进行综合治理，该行动将有助于受损生态系统的修复。贵州省强化土地整治过程中生物多样性保护理念，2010年9月批复了基于生物多样性保护的《关岭自治县土地整理开发复垦专项规划》和《荔波县土地整理复垦开发专项规划》。

（4）建立国内外多方合作机制

国土资源部与商务部、德国环境、自然保护与核安全部、欧盟和联合国开发计划署共同实施了"土地利用总体规划和土地整理中的生物多样性保护"、"中德低碳土地利用"等项目。通过与比利时、德国、欧盟等国家和机构开展土地利用规划、土地整理、土地复垦等领域的生物多样性保护技术交流，提高了中方人员生物多样性保护知识，拓宽了国内外土地利用与生物多样性研究合作平台。

（5）加大生物多样性宣传力度

在2010年上海世博会期间，举办了"土地利用总体规划和土地整理中的生物多样性保护"主题讲座和系列宣传活动，提高公众对于土地管理中生物多样性保护的认知度；在"4·22地球日"、"6·25土地日"等节日、纪念日，向土地管理与科研人员、高等院校学生、农民普及生物多样性保护理念和知识，为全面开展土地整理与生物多样性保护提供技术服务。

3.5 住房和城乡建设部门

（1）完善相关法规政策、行业标准和发展规划

住房和城乡建设部已将"加大城市规划区及风景名胜区的生物多样性保护力度、应对气候变化"列入"十二五"工作计划，2010年印发了《关于进一步加强动物园管理的意见》，2013年印发了《全国动物园发展纲要》，制定了物种种群发展目标和措施，计划对圈养的国家Ⅰ级、Ⅱ级及《濒危野生动植物国际贸易公约》附录Ⅰ、附录Ⅱ中的动物实行全行业物种种群管理，并组织制定《城市植物园设计规范》、《城市动物园设计规范》、《动物园管理技术规范》，促进城市生物多样性保护工作步入法制化、规范化的轨道。

（2）将生物多样性保护纳入国家园林城市、生态园林城市考核体系

在"国家园林城市"创建活动中，住房和城乡建设部2010年印发的《国家园林城市标准》和2012年印发的《生态园林城市申报与定级评审办法和分级考核标准》都把生物多样性保护作为重要考核内容，将综合物种指数、本地木本植物指数、水体岸线自然化率、城市自然生态保护等指标纳入具体考核范围。通过创建活动，大多数城市都完成了生物物种资源普查，制定了《城市生物多样性保护规划》和实施措施，加强对地貌、水文、植被、物种的有效保护，城市规划区范围内的生物物种多样性日趋丰富。

（3）充分发挥植物园、动物园、湿地公园对物种保护与研究的基地作用

据不完全统计，住房和城乡建设系统建有植物园200个，迁地保护植物种类2万种。2012年国家林业局、住房和城乡建设部、中国科学院联合印发了《关于加强植物园植物物种资源迁地保护工作的指导意见》。建成动物园（动物展区）240多个，目前已培养了26位谱系保存人，建立了37个物种谱系，在珍稀濒危物种特别是在大熊猫和华南虎的迁地保护方面取得了较大成绩：大熊猫圈养种群到2011年年底达到332只个体，种群管理的主要任务已从个体数量的快速增长转向提高种群遗传质量；华南虎圈养种群在6只的基础上发展到目前超过100只，成为野外种群恢复的希望。针对城市生态系统受到较多人为干扰的现状，住房和城乡建设部印发了《国家城市湿地公园管理办法》（试行）和《城市湿地公园规划设计导则》，在全国设立了49个国家城市湿地公园，提高了城市湿地资源的保护管理水平，促进了城市规划区内的湿地物种资源保护。

（4）强化风景名胜区物种资源监测管理系统

风景名胜区是生物多样性保护的重要区域。住房和城乡建设部自2002年开始建立国家级风景名胜区监管信息系统，应用遥感和地理信息系统等技术，对国家级风景名胜区资源保护和规划实施情况进行监测，建立了遥感动态监测核查制度与城乡规划督察员制度，迄今已对208处国家级风景名胜区开展了监测。2012年和2013年，住房和城乡建

设部又集中组织开展了国家级风景名胜区保护管理执法检查。2013年开展了云南、贵州、四川、新疆四省（区）风景名胜区生态补偿试点工作。这些措施切实保护了风景名胜区内的生物物种资源和生态环境，为维护风景名胜区生物多样性发挥了重要作用。

（5）积极建立同国外机构和组织的合作关系

近年来，住房和城乡建设部与美国内政部、世界自然保护联盟、联合国教科文组织等机构和组织在风景名胜区和世界遗产资源保护方面建立了密切的合作与联系。近几年，以典型生态系统和珍稀物种栖息地标准列入《世界遗产名录》的遗产地日益增多，住房和城乡建设部加大了对世界遗产和风景名胜区生态系统、生物多样性的保护和研究力度。2012年，住房和城乡建设部同世界自然保护联盟合作开展了《中国绿色名录》试点及研究，与联合国教科文组织就四川大熊猫栖息地、荔波樟江、武夷山等世界自然遗产地的生物多样性保护与资源监测开展了持续的多层次的合作，并在四川都江堰—青城山世界遗产地开展并实施生物多样性保护行动计划。自中国加入《保护世界文化和自然遗产公约》以来，截至2013年年底，住房和城乡建设部已审核并成功申报世界遗产21处，得到了全球范围的关注和重视。

3.6 水利部门

近年来，水利部门认真贯彻《中国生物多样性保护战略与行动计划》（2011—2030年）的要求，结合部门工作实际，积极开展了生物多样性保护相关工作。

（1）完善了水利支撑生物多样性保护的法律、政策和规划

通过建立健全《中华人民共和国水法》配套制度体系，修订《中华人民共和国水土保持法》，发布《中共中央、国务院关于加快水利改革发展的决定》和《国务院关于实行最严格水资源管理制度的意见》，将生物多样性保护所涉及的水资源配置、生态用水保障、水资源保护、水土保持等工作法制化、规范化，进而将生物多样性保护相关要求上升为法律政策要求，明确了水行政主管部门、项目建设单位以及公众在生物多样性保护方面的责任与义务。水利部门在制定行业发展战略、规划或计划时，均在一定程度上考虑了生物多样性因素，如《全国水资源综合规划》（2010—2030）、《全国水资源保护规划》、《全国地下水利用与保护规划》、《水利发展规划》（2011—2015）、《七大流域综合规划》（修编）等，主要内容涉及生态用水水量与水质保障、水土流失预防、水土保持、生态补偿机制建设等方面。2011年国务院批复了《全国重要江河湖泊水功能区划》。该区划确定了全国4 493个重要水功能区的功能定位和水质管理目标，是全国水资源开发利用与保护、水污染防治和水环境综合治理的重要依据。

（2）保障生态环境用水，遏制了许多重要保护区的生态恶化趋势

近年来，水利部对黄河、塔里木河、黑河、石羊河进行综合治理和水资源科学调度。黄河已经实现连续14年不断流，塔里木河下游干涸20多年的台特马湖重新过流，黑河

下游的东居延海已经连续 9 年不干涸，石羊河民勤蔡旗断面下泄水量逐步增加。连续 9 年实施"引江济太"，将长江水调入太湖，实现了"以动治静、以清释污、以丰补枯、改善水质"的目标。对扎龙湿地、南四湖、白洋淀、衡水湖等湖泊和湿地实施生态补水，当地生态环境得到显著改善，维护了生态脆弱地区的水生态安全；开展了长江三峡水库优化调度，尝试实施兼顾生态需求的流域水库群联合调度。

（3）加大水土流失治理力度，极大改善了当地的生态环境

近年来，进一步加快水土流失治理步伐，在长江中上游、黄河中上游、珠江上游、丹江口库区及上游、首都水源区、京津风沙源区、晋陕蒙砒砂岩区、岩溶石漠化区、东北黑土区等重点地区大力开展水土流失综合治理工程。2009—2012 年，全国共治理小流域 1.2 万条，完成水土流失综合防治面积 27.0 万 km^2。继续推进水土流失封育保护。全国累计实施封育保护面积 72 万 km^2，其中有 45 万 km^2 生态环境得到了初步修复。特别是青海"三江源"、新疆内陆河流域、西藏等重点修复区，封育保护效果日益明显，生态功能得到有效保护。

（4）实施水生态系统保护与修复，部分地区的水生态环境质量得到极大改善

通过以城市为单元开展水生态系统保护与修复试点工作，无锡、武汉、桂林、哈尔滨等 14 个城市采取了水系连通、截污导流、河湖清淤、岸线整治与修复、水源保护等水生态保护与修复的综合措施，取得了良好成效，水生态环境质量得到提升。武汉市围绕河湖水系连通建设，实现 16 个中心城区湖泊水质的提档升级；桂林市围绕生态水库建设和科学调度，保证了漓江枯季流量，有效保护了生物多样性；无锡市围绕湖泊水环境的综合整治，使水功能区达标率由 2005 年的 11.8% 提升到 46.8%；丽水市围绕小流域综合治理，使瓯江水土流失面积减少 $567km^2$，干流水质达标率提高到 98.7%；莱州市围绕地下水超采治理，海水入侵面积由 $261km^2$ 减少至 $228.5km^2$。

3.7 农业部门

农业生物多样性是生物多样性的重要组成部分。中国政府一贯重视农业生物多样性的保护与可持续利用，并制定相关法规和规划予以实施。

（1）加强组织领导

农业部加强了农业野生植物保护领导小组、农业野生植物保护专家审定委员会、国家畜禽遗传资源委员会、农业部水生野生动植物保护办公室、农业部濒危水生野生动植物种科学委员会、长江流域渔业资源管理委员会、黄河流域渔业资源管理委员会、珠江流域渔业资源管理委员会、全国外来入侵生物防治协作组、外来物种管理办公室和农业部外来入侵生物预防与控制研究中心，提高农业生物多样性保护管理水平。成立了国家农作物种质委员会，负责协调全国农作物种质资源包括农业野生植物的管理工作，研究提出国家农业种质资源包括农业野生植物发展战略和方针政策，指导编制农作物种质资

源包括农业野生植物中长期发展规划等。

（2）完善法规和标准体系

在原有法律法规的基础上，农业部颁布了《中华人民共和国畜禽遗传资源进出境和对外合作研究利用审批办法》《畜禽遗传资源保种场保护区和基因库管理办法》《畜禽新品种配套系审定和畜禽遗传资源鉴定办法》《家畜遗传材料生产许可办法》《从境外首次引进畜禽遗传资源技术要求(试行)》等一系列法规，健全了畜禽遗传资源保护的法律体系。2010年国家发展和改革委员会与财政部出台了《关于同意收取草原植被恢复费有关问题的通知》，明确了对草原矿藏勘察开采和进行工程建设征用或使用草原，应交纳草原植被恢复费；2012年国家颁布了《最高人民法院关于审理破坏草原资源刑事案件应用法律若干问题的解释》，明确了破坏草原资源犯罪行为的定罪量刑标准。农业部组织制定了《农业野生植物原生境保护点监测预警技术规程》和《农业野生植物异位保存技术规程 第一部分：总则》等4项农业野生植物保护行业标准；颁布了《水产种质资源保护区管理暂行办法》《水生生物增殖放流管理规定》《建设项目对水生生物国家级自然保护区影响专题评价管理规范》《农业部关于做好海洋伏季休渔管理工作的通知》；发布了《农业重大有害生物及外来入侵生物突发事件应急预案》和《国家重点管理外来入侵物种名录（第一批）》，陆续颁布了《微甘菊综合防治技术规程》、《福寿螺综合防治技术规程》等17项外来入侵物种防治行业标准，发布了40项农业重大外来入侵物种应急防控技术指南。这进一步健全了农业生物多样性保护的法规体系。

（3）把生物多样性保护纳入相关规划和计划中

国务院颁布的《全国现代农业发展规划》（2011—2015）明确加强农业资源和生态环境保护，坚持基本草原保护制度，加大水生生物资源养护力度，强化水生生态修复和建设，加强畜禽遗传资源和农业野生植物资源保护的目标。在草原生物多样性方面，农业部发布了《全国草原保护建设利用总体规划》《全国畜牧业发展第十二个五年规划》（2011—2015）等规划和《农业部办公厅　财政部办公厅关于进一步推进草原生态保护补助奖励机制落实工作的通知》《中央财政草原生态保护补助奖励资金绩效评价办法》等管理办法，陆续启动了退牧还草、京津风沙源草原治理、游牧民定居、草原防火等一系列草原保护建设工程项目。《全国畜牧业发展第十二个五年规划》（2011—2015）提出，"继续实施畜禽良种工程建设，扶持畜禽遗传资源保护场、保护区和基因库的基础设施建设，健全国家畜禽遗传资源保护体系"。在水生生物多样性方面，农业部发布了《全国渔业发展第十二个五年规划》（2011—2015），明确了积极发展环境友好的增殖渔业，改善水域生态环境，强化休渔禁渔制度，开展珍稀物种放流，加强水生生物自然保护区建设，保护水生生物多样性等工作内容和目标。《农业科技发展"十二五"规划》（2011—2015）要求进一步强化农业生物种质资源、农业基因资源的搜集、保护、鉴定及育种材料的改良和创制。《全国种植业发展第十二个五年规划》（2011—2015）提出，要完善国家种质资源保存与利用体系。这些规划的颁布实施，有力地推动了全国农业生物多样性的保护与可持续利用。

（4）加强农业野生植物资源保护

一是开展了农业野生植物资源调查和监测。查清了全国172个农业野生植物物种的分布，建立了国家级农业野生植物资源管理信息系统。二是推进农业野生植物保护点建设与监测，对42种濒危野生物种进行保护，在广西、海南、云南、河南、吉林、黑龙江、新疆和宁夏选择野生稻、野生大豆、小麦野生近缘植物等15个原生境保护点开展跟踪监测。三是加强农业野生植物资源迁地保护。开展了主要农作物野生植物抢救性收集，新建立了一批国家种质资源圃，已收集保存主要农作物野生近缘植物3万多份；进一步完善了由种质圃、长期库、复份库、中期库和试管苗库组成的农作物种质资源迁地保护体系，为农作物野生植物多样性保护和利用提供了可靠保证。

（5）强化草原生态系统保护与恢复

农业部开展了第二次全国草原遥感快速调查。健全了草原监督管理机构，到2012年年底，全国县级以上草原监督管理机构共有844个，其中国家级1个、省级23个、地级126个、县级694个。2003—2009年，共查处各类草原破坏违法行为案件9万余起；2010—2012年，每年查处草原违法案件2万余起。中国初步建立了布局较为合理、类型较为齐全、分布范围较广泛、代表性较强的草原保护区网络框架。为了保护草原生态系统，先后启动了退耕还林工程、退牧还草工程、京津风沙源工程、青海"三江源"生态保护和建设工程、西南岩溶地区草地治理、草原防灾救灾工程等多项工程。按照2012年农业部监测结果，草原生态保护建设工程成效显著。与非工程区相比，草原植被盖度提高11%，草丛高度提高43.1%，鲜草产量提高50.7%，草原利用状况有较大改善，2012年全国268个牧区县超载率较2011年下降34.5%～36.2%。总体上，局部草原生态状况在改善，工程区和保护区内的生物多样性在恢复。但是整体看，草原大部分仍处于超载过牧状态，草原退化、沙化、盐渍化、石漠化尚未得到有效控制，草原生态保护建设任务依然繁重。

（6）积极开展农业外来入侵物种防治

一是开展外来入侵物种调查和监测预警。近年来，农业部采取点面结合的方法，连续开展了全国外来入侵物种普查，重点对紫茎泽兰、薇甘菊、银胶菊、长芒苋、刺苍耳等22种危险性农业入侵物种进行了全面调查和跟踪监控。二是开展农业外来入侵生物的集中灭除。近5年来，农业部分别在云南、湖南、湖北、四川、贵州、江西、吉林等地开展十余次薇甘菊、福寿螺、紫茎泽兰、水花生、刺萼龙葵集中灭除活动，有效控制了重大恶性外来入侵物种的扩散和蔓延。三是开展综合防控技术研究。开发出低容量喷雾、静电超低量喷雾等高效施药技术，研究利用牧草、灌木、农作物等替代入侵植物的生态调控技术。筛选出椰心叶甲天敌——椰心叶甲啮小蜂和椰扁甲姬小蜂，控制效果达到85%以上，在海南建立4个椰心叶甲天敌工厂，寄生蜂日生产规模达到200万头，防治面积达1 000 km^2。开展外来入侵物种变废为宝、化害为利技术攻关取得突破性进展。

（7）依法开展水生生物资源的保护

一是加强水生生物自然保护区建设。全国已建立各类水生野生动植物及水域生态系统类型自然保护区 200 多处，初步形成了布局较为合理、类型较为齐全的水生生物保护区网络。二是加强水产种质资源保护区建设。截至 2013 年年底，划定了 428 个国家级水产种质资源保护区，对保护上百种国家重点保护经济水生动植物和地方珍稀特有水生物种及其栖息繁衍场所发挥着重要作用。三是加大水生生物资源增殖放流。在适于渔业资源增殖的水域，开展增殖放流活动，并逐步扩大增殖品种、数量和范围。一些重要渔业资源品种的种群数量有所恢复，在渤海和黄海北部部分海域，曾经消失的中国对虾、海蜇、梭子蟹等秋季渔汛又重新形成。四是加强海洋牧场建设。各地积极开展了以人工鱼礁为载体，以底播增殖和海藻种植为手段，以增殖放流为补充的海洋牧场建设。各地结合减船工作，利用报废渔船等废旧物资，积极降低人工鱼礁（巢）建设成本。全国已累计投放各类人工鱼礁礁体近 200 万 m^3，积极营造重要区域海洋生物的栖息、繁衍场所。

（8）推进农业转基因生物安全评价和管理

详见本书第 2 章 2.3.9 节。

案例 3.1　作物野生近缘植物保护

2007—2013 年，在全球环境基金的资助下，联合国开发计划署和中国农业部联合执行了"作物野生近缘植物保护与可持续利用"项目。在 8 个省（自治区）选定 8 个代表不同社会经济状况的项目点，分别对野生稻、野生大豆和小麦野生近缘植物实施保护。项目制定的策略是：以政策法规为先导，通过约束人的行为减少对作物野生近缘植物及其栖息地的破坏；以生计替代为核心，切实帮助农牧民解决生计问题，降低农牧民对作物野生近缘植物及其栖息地的依赖程度；以资金激励为后盾，引导农牧民发展生物多样性友好型家庭经济；以提高意识为纽带，鼓励农牧民主动参与作物野生近缘植物的保护活动。

项目实施的结果表明，与 2008 年的基线数据相比，各项目点的资源状况指数均有所提高，总体呈现上升的趋势，取得了预期的保护效果；各项目点的威胁因素缩减指数（TRA）均在 80% 以上，说明对作物野生近缘植物的威胁均有减缓，保护活动具有较好的可持续性（见下图）。更为重要的是，各项目点居民的人均收入逐年递增，且增幅均高于条件相似的邻近村落，项目在保护作物野生近缘植物资源的同时，促进了当地经济的可持续发展。

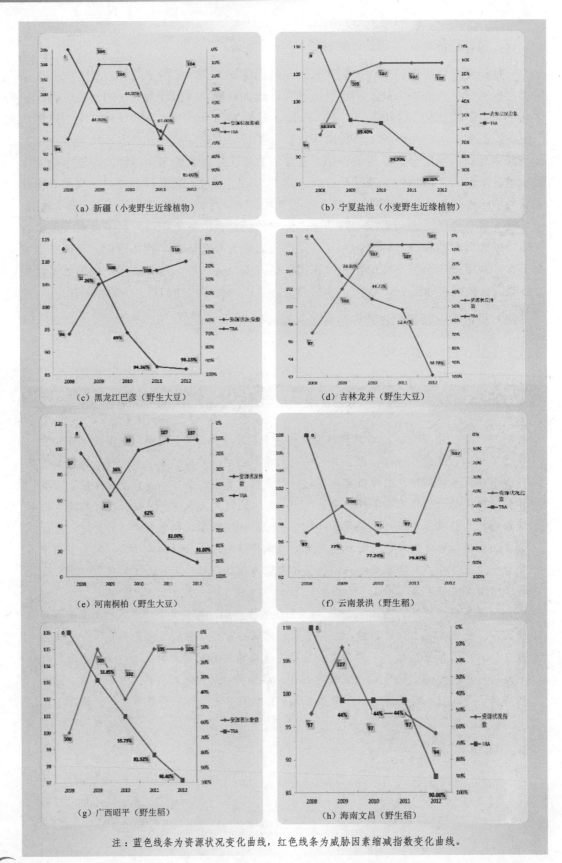

注：蓝色线条为资源状况变化曲线，红色线条为威胁因素缩减指数变化曲线。

3.8 商务部门

(1) 在多边贸易体系内促进《生物多样性公约》的效力

中国积极推进在世界贸易组织知识产权理事会中讨论《与贸易有关的知识产权协定》（《TRIPs协定》）与《生物多样性公约》的关系，与其他发展中成员国一起主张根据《生物多样性公约》修改 TRIPs。《TRIPs 与公共健康多哈宣言》第 19 条要求世贸组织知识产权理事会"审查《TRIPs 协定》与《生物多样性公约》之间的关系，审查对传统知识和民俗的保护……"鉴于此，发展中成员国主张《TRIPs 协定》应当与《生物多样性公约》相协调，提出应符合国家主权、知情同意、利益分享三原则，并提出对于依赖基因资源完成的发明应规定在专利申请过程中披露基因资源来源的强制性义务，确保基因资源拥有者的知情权和获得报酬权。中国与巴西、印度、欧盟、瑞士以及支持基因资源来源强制性披露的其他成员于 2008 年 5 月提交了 W52 号联合提案，要求同步推进地理标志多边注册体系、地理标志保护范围扩大与基因资源信息披露三个议题的谈判。2011 年 4 月，中国与巴西、印度等 W52 成员向 WTO 提交了《关于促进〈TRIPs 协定〉与〈生物多样性公约〉相互支持的决议草案》（TN/C/W/59），建议在《TRIPs 协定》中增加第 29 条"披露基因资源的来源和 / 或相关传统知识"，使遗传资源的来源和相关传统知识等信息披露成为专利申请的实质要件。

(2) 将生物多样性因素纳入商务部出台的政策、法规、指南等文件中

自 2009 年 3 月以来，商务部联合其他部委共同下发了多个有关生物多样性保护的指南或指导意见，对中国企业在国内外经营过程中的环境保护和生物多样性保护提出了要求，鼓励企业在国际贸易和投资中考虑生物多样性问题。2009 年 3 月，国家林业局、商务部联合颁布了《中国企业境外森林可持续经营利用指南》。该指南要求采取科学合理的采伐方式和作业措施，尽量减少森林采伐对生物多样性的影响。2013 年 2 月，商务部和环境保护部联合颁布了《对外投资合作环境保护指南》，规定企业应优先采取就地、就近保护等措施，减少对当地生物多样性的不利影响。

(3) 在对外贸易谈判中考虑生物多样性问题

2009 年 4 月签署的《中国—秘鲁自由贸易协定》第 145 条"遗传资源、传统知识和民间文艺"中明确提出，"缔约双方承认并且重申 1992 年 6 月 5 日通过的《生物多样性公约》确立的原则和规定，并鼓励建立《TRIPs 协定》与《生物多样性公约》之间相互支持关系的努力。"

(4) 积极开展生物多样性领域的国际合作

商务部利用多边和双边合作渠道，组织实施了多个生物多样性保护项目，如"中国和挪威生物多样性和气候变化"项目。2012 年 11 月，商务部和联合国开发计划署、工发组织、贸发会议、环境规划署等在广州联合主办了"加强南南合作，促进绿色发展"图片展。图片展共有 120 多幅图片，内容涵盖节能环保、应对气候变化、保护生物多样性和能力建设等。商务部多次举办了"生物多样性跨国界保护研究及管理研修班",邀请发展中国家的管理及国际合作官员来华交流,向其介绍中国生物多样性跨国界联合保护机制、保护成果及经验。

(5) 加大生物多样性保护的新闻报道和宣传

自2009年以来,商务部、驻外商务机构和地方商务主管部门收集并报道了大量有关生物多样性保护的新闻和动态,介绍全球有关生物多样性的政策、法规及发达国家对发展中国家的生物多样性保护援助项目,既为中国企业和产品走出去奠定基础,也可借鉴国外的经验和做法为我所用,提高公众对生物多样性保护的认知和意识。

3.9 海关

海关作为国家进出境监督管理机关,始终高度重视生物物种资源的进出境管理。

(1) 加强生物物种进出境实际监管

海关认真履行《生物多样性公约》,积极落实国务院批准发布的《全国生物物种资源保护与利用规划纲要》,依法认真验核《濒危物种允许进(出)口证明书》《非〈进出口野生动植物种商品目录〉物种证明》《血液进出口批件》《人类遗传资源材料出口/出境证明》《合法捕捞产品通关证明》等进出境监管证件,依据各主管部门发布的各项法规、规章进行生物物种资源的进出境管理。同时,海关积极配合相关主管部门,为科研用或以保护为目的的生物物种进出境提供通关便利。

(2) 参与生物安全法律体系建设

多年以来,海关始终致力于推动生物多样性进出境法律、法规体系建设,并根据《濒危野生动植物进出口管理条例》,及时会同有关部门制定、调整并公布相关进出口管理目录,包括《进出口野生动植物种商品目录》《两用物项和技术进出口管理目录》。同时,结合工作实际,海关对《古生物化石保护条例实施办法》的颁布及《中华人民共和国生物遗传资源管理条例》《外来入侵物种环境监督管理办法》的制定提出具体意见,完善生物安全立法。

(3) 不断强化现场关员能力培训

海关系统坚持定期开展现场关员执法能力培训,努力提高海关关员执法水平。培训内容主要包括:《生物遗传资源进出口管理目录》相关内容;生物遗传资源保护法律法规;海关查验人员相关专业知识。

(4) 进出境环节有关案件查缉情况

2009年1月—2012年12月,海关缉私部门共查获走私珍贵动物、珍稀植物及其制品刑事案件406起;查获走私珍贵动物、珍稀植物及其制品共计381t,案值53.8亿元;查获珍贵动物、珍稀植物及其制品行政违法违规案件3 573起,案值1.3亿元。特别是

2012年，海关总署在全国范围内开展打击走私的"国门之盾"行动中，组织了重点查缉行动，查获了一批濒危物种重特大走私案件。2012年5月，《濒危野生动植物种国际贸易公约》秘书处向中国海关与国家林业局授予"CITES秘书长表彰证书"。2013年，中国又牵头发起了包括亚洲、非洲22国参与的打击濒危物种走私的"眼镜蛇行动"，中国海关共查处走私濒危物种案件71起，处理嫌疑人85人。中国海关查获案件数占行动期间全球查获案件总数的三分之一。

3.10 工商行政管理部门

工商行政管理部门认真贯彻落实《中华人民共和国野生动物保护法》等法律法规，加大市场监管执法力度，严厉打击非法收购和经营国家重点保护野生动植物及其产品等违法违规行为。

（1）加强组织领导，周密安排部署

工商总局在各项市场监管中，将生物多样性保护工作纳入重点工作以及各类规范性文件中。在每年的节日市场专项整治和检查中，工商总局多次下发专门通知，要求各级工商机关对经营利用野生动植物及其产品的违法违规行为进行整治。2012年12月，工商总局下发了《关于进一步加强野生动物保护工作的紧急通知》，要求各地工商机关集中执法力量，严厉查处非法销售、收购国家重点保护野生动物及其产品的违法违规行为。各地工商机关按照工商总局的部署，迅速组织开展了为期一个月的专项整治行动。在专项整治行动期间,江苏省工商局共检查集贸市场6 485个（次）,检查经营户7.8万户（次），取缔无照经营户58户，查处违法案件9件，案值8.6万元，罚没金额1.8万元。

（2）加强日常监管执法，切实维护市场秩序

一是严格市场准入管理,规范经营资格。严格按照《中华人民共和国野生动物保护法》的规定，要求从事出售、收购、贩运、驯养繁殖国家重点保护野生动物的单位，在申请办理营业执照之前，必须取得有关主管部门或者其授权单位的批准文件或者有关许可证。二是严格规范经营行为，强化基层监管执法。各地工商机关在野生动植物执法工作中严格落实基层工商所日常巡查和属地监管责任制，落实网格化责任区，切实做到任务到岗、责任到人。自2009年3月以来，云南省工商局共检查野生动物及其产品市场3 068个，检查经营野生动物及其产品的餐饮企业3.2万户，检查繁殖、驯养野生动物的企业264家，查处取缔无证照经营利用野生动物及其产品的企业120户，查缴各类野生动物3 300只。

（3）加大案件查办力度，严厉查处违法违规行为

各地工商机关深入贯彻落实《中华人民共和国野生动物保护法》等法律法规，严厉打击涉及野生动植物的违法违规行为。自2009年3月以来，广东省工商局在开展保护野生动植物资源专项行动中，共检查各类市场3.5万个，各类经营户15.5万户；查处违法违规案件154件，案值12万元，罚没金额15万元。贵州省工商局共检查各类市场2 094

个，各类经营户 2.2 万户，查获野生动物产品 857 kg，收缴野生动物 5 455 只（条）。黑龙江省工商局共查处有关案件 100 多件，收缴各类野生动物 3 万多只，案值 300 多万元。青海省工商局组织开展了非法销售湟鱼专项整治行动，与经营户签订《销售湟鱼及其产品责任书》3 000 多份，查处违法案件 200 多件，罚没金额 20 多万元。

（4）开展执法培训，加强宣传教育

一是加大培训力度。各地工商机关通过举办培训班、大讲堂、发放培训材料等方式，加大对执法人员的教育培训力度。新疆维吾尔自治区、吉林省和黑龙江省工商局多次邀请有关专家对执法人员进行培训，增强了依法行政能力。二是加强宣传教育。各地工商部门采取多种有效形式，利用"国际生物多样性日"、"世界环境日"、"全国法制宣传日"、"国际消费者权益保护日"等，大力宣传野生动植物保护的法律法规，广泛普及生物多样性科学知识，增强全社会的法律意识和生态保护意识。例如，辽宁省工商局与省林业厅联合召开了野生鸟类资源专项行动新闻发布会，组织开展了"爱鸟护鸟春季行动"，发放宣传单 4 万余份。

3.11 质量监督检验检疫部门

（1）注重宏观规划

质量监督检验检疫部门在制定本行业发展战略和规划时把生物多样性和物种资源保护工作纳入其中。2011 年，国家质量监督检验检疫总局颁布了《质量监督检验检疫事业发展"十二五"规划》，从法律法规、体制机制、技术措施等方面规划了生物多样性保护和出入境物种资源检验检疫工作，明确提出要初步构建"物种资源检验检疫体系"和"物种资源截获评价体系"。为全面落实《中国生物多样性保护战略与行动计划》（2011—2030 年），国家质量监督检验检疫总局专门编制了《出入境物种资源检验检疫配套行动计划》。

（2）加强组织领导

为加强对本系统和行业关于生物多样性和物种资源保护工作的领导，国家质量监督检验检疫总局在内部成立物种资源监管处和旅邮检处，在中国检验检疫科学研究院成立物种资源检验鉴定中心的基础上，2012 年又成立了出入境生物多样性和物种资源保护领导小组，科学领导和指导本部门生物多样性和物种资源保护工作。

（3）稳步科学指导

为指导出入境生物物种资源检验检疫工作，国家质量监督检验检疫总局出台了《关于加强出入境生物物种资源检验检疫工作的指导意见》，要求全系统认清形势、明确使命，清醒认识生物物种资源保护面临的艰巨任务和严峻挑战，增加责任感和使命感，从全局和战略高度全面加强包括濒危野生动植物种在内的出入境生物物种资源检验检疫工作。要求全系统加大投入，夯实基础；充实队伍，提升能力；强化协作，形成合力；注重宣传，营造氛围；强化考核，落实责任，努力提升生物物种资源检验检疫水平和能力。

（4）认真开展调查

为深入掌握进出口贸易中物种资源出入境状况，国家质量监督检验检疫总局自2008年以来组织开展了进出口贸易中生物物种资源调查工作，基本掌握了中国当前生物物种资源出入境的种类和流向，调查表明，出境生物物种资源的批次和种类分别相当于入境的两倍，物种资源流失现象比较严重。

（5）积极开展试点

为积累物种资源口岸查验经验，为工作全面铺开做好准备，国家质量监督检验检疫总局与有关部门配合，在山东烟台、新疆等海、陆、空口岸开展了物种资源查验试点，取得丰硕成果，为物种资源口岸查验体系建设奠定了基础。

（6）严防外来物种入侵

出入境检验检疫部门持续加强对外来有害生物的检测与监测，2008—2012年截获的有害生物种类和批次持续大幅度增长。口岸有害生物截获种类由2008年的2 856种增加到2012年的4 331种，年均增长10%以上；次数由22.9万次增加到57.9万次，年均增长30%以上。国家质量监督检验检疫总局专门出台《关于进一步加强旅客携带物、邮寄物检疫工作的指导意见》，加大对旅客携带物、邮寄物检疫工作指导。截至2012年年底，全国61个国际邮件互换局（交换站），已有31个设立邮检办事机构。"十一五"以来，出入境检验检疫部门开展了大量与外来入侵物种管控相关的科研和标准研究，制定了多项标准。以植物检疫相关标准为例，2006年至今制定发布行业标准352项、国家标准104项、国际标准2项，初步构建了中国进出境植物检疫标准体系。

（7）开展转基因生物检测

国家质量监督检验检疫总局对进境转基因动植物、微生物及其产品和食品实行申报制度。2008—2013年，国家质量监督检验检疫总局成立了进出境转基因生物检验监测技术中心和进出境转基因重点实验室，全面提升了系统内30余家实验室的转基因检测能力，建成了较为完备的技术体系和标准体系。据不完全统计，近几年已对3亿多t的大豆、玉米、油菜籽、棉籽，600万t大米制品和其他农产品开展转基因检测。有关检验检疫机构从美国进口玉米和制品中检出未经中国政府批准的MIR162转基因品系，从爱尔兰混合马饲料和台湾非转基因大豆粕中检出转基因品系，相关货物均做退货处理。

3.12 林业部门

（1）继续实施林业重点生态工程

近十年来，天然林资源保护工程完成造林任务8.3万 km^2，现有104.9万 km^2 森林得到有效保护,森林面积净增加 10 万 km^2,森林覆盖率增加 3.8 个百分点,森林蓄积量增加 7.3 亿 m^3。退耕还林工程共完成营造林任务 21.8 万 km^2,工程区森林覆盖率平均提高 3 个百分点。"三北"防护林及长江流域等防护林工程共完成建设任务 8.3 万 km^2。其中,"三北"防护林工程区森林覆盖率提高至 12.4%,区域生态状况明显好转。2000—2010 年,京津风沙源治理工程累计完成治理总面积 8.9 万 km^2,治理工程区森林覆盖率提高 4.1 个百分点,植物群落发生良性演替。岩溶地区石漠化综合治理工程累计完成 3 万 km^2 的石漠化治理任务,工程区林草植被综合覆盖度比治理前提高了 15%。实施全国湿地保护与恢复工程。2006—2010 年,全国完成了 205 个湿地保护和恢复示范工程,恢复湿地 $800km^2$。

(2) 完善自然保护区的规范化管理

为贯彻国务院办公厅《关于做好自然保护区管理有关工作的通知》,2011 年国家林业局颁布了《关于进一步加强林业系统自然保护区管理工作的通知》,对自然保护区的总体规划、土地权属、机构人员、编制经费、管理体制等提出了要求。一些省区市也相继制定了贯彻落实的意见措施。"一区一法"制度得到进一步落实。甘肃、福建等省先后批准了两处国家级自然保护区的专项管理办法,内蒙古汗马国家级自然保护区也在积极争取地方人大立法。进一步推进自然保护区规范化管理,宁夏回族自治区出台了《自治区林业局关于进一步加强林业系统自然保护区林地管理的通知》,强化自然保护区土地管理措施;广西壮族自治区全力推进 12 处地方级自然保护区的确界工作,上报政府申批核准,并提出在未来三年内实现地方级保护区总体规划编制完成率达 100% 的目标;贵州省对全省林业系统的各级自然保护区组织了全面调查清理核实,建立了全省林业自然保护区名录和主要信息数据库。

(3) 制定并实施野生动植物保护与自然保护区发展规划

2012 年,国家林业局颁布了《全国野生动植物保护及自然保护区建设"十二五"发展规划》。该规划提出,"十二五"时期中国将优先保护 60 余种野生动物,优先保护 120 种野生植物,优先建设 51 个国家级自然保护区。2012 年,国家林业局与国家发展和改革委员会联合印发《全国极小种群野生植物拯救保护工程规划》(2011—2015)。通过该规划的实施,加强对极小种群物种的拯救保护,推动和带动野生植物保护工作。另外,国家林业局还制定了《全国极小种群野生动物拯救保护规划(草案)》。

(4) 探索自然保护区管理机制和生态补偿机制

各地对自然保护区管理机制进行了探索。广东省将林业系统国家级、省级自然保护区确定为公益一类事业单位,海南省将委托市县管理的省级自然保护区收归省林业局直接管理,湖北省于 2011 年将 5 个国家级自然保护区人员经费和编制都纳入省政府财政预算和编制,广西壮族自治区于 2011 年将 9 个国家级自然保护区列为参公管理的事业单位,为自然保护区长远发展提供了有力保障。浙江省林业厅和财政厅联合下达了《关于组织开展省级以上森林类型自然保护区集体林租赁工作的通知》,对省级以上保护区的核心区和缓冲区的集体林实行国家租赁,租赁价为 5.0 万元 / (a·km^2),以进一步维护林权所有者合法权益。针对由于保护野生动物造成贫困地区人民财产受到损失的情况，从 2008 年起开展了野生动物损害补偿试点工作。安徽、甘肃和青海等省对野生动物造成的人身伤害和财产损失出台了补偿办法。2012 年对试点省区野生动物损害补偿补助试点进行了监督检查，并初步制定了《国家重点保护野生动物损害补偿中央财政补助资金定额标准（试行）》。

（5）强化外来林业有害生物防治

已建成国家级中心测报点 1 000 处、省级测报点 1 200 多处，科学布局了 2.8 万个国家、省、市、县四级测报站点，基本形成全国林业有害生物监测预警网络体系。全国共建立林业有害生物防治检疫机构 3 117 个，检疫检查站 858 个，防治队伍趋于稳定。防治基础设施设备建设明显加强，初步建成以监测预警、检疫御灾、防灾减灾、服务保障为核心的防控体系。

（6）加强人工种群野化与野生动植物种群恢复

国家林业局加大了濒危野生动物拯救繁育力度，大熊猫圈养种群数量达到 312 只，朱鹮人工种群已达 600 多只，扬子鳄、虎、金丝猴、藏羚羊等 50 多种野生动物繁育种群持续扩大，成功实施了朱鹮、野马、麋鹿、梅花鹿、瑶山鳄蜥、扬子鳄、塔里木马鹿、黄腹角雉 8 种濒危野生动物的放归自然，且实现了自然繁殖，逐步建立起新的野外种群。迁地保护了东北、西北、西南地区 1 000 多种珍稀、濒危或特有植物，建立了野生植物种质资源保育基地 400 余处，成立了苏铁、兰科植物、木兰科植物、棕榈植物种质资源保护中心（基地）。针对松茸、雪莲、珙桐、肉苁蓉、红豆杉、珍稀兰科植物等 10 种（类）市场需求较大的珍稀野生植物，扶持开展人工培育技术研究和种源建设。建立珍稀野生植物培育基地 280 处，使千余种野生植物建立了稳定的人工种群；对五唇兰、杏黄兜兰等中国特有的濒危兰科植物以及德保苏铁、华盖木、西畴青冈等极度濒危物种开展了回归自然的救护行动。

（7）积极推进栽培植物、养殖和驯养动物及野生近缘物种的遗传多样性保护、研究和开发利用

各级林业部门积极实施以利用野外资源为主向利用人工繁育资源为主的战略转变。对如中医药和蟒皮乐器等传统产业中资源消耗或需求量较大的野生动植物，大力发展养殖业和种植业，引导、推进野生动植物人工繁（培）育技术，积极争取对野生动植物驯养繁（培）育企业免收企业所得税，对人工驯养繁殖野生动物及其产品免收野生动物资源保护管理费。同时大力推行专用标识管理、标准化管理等手段，强化监管力度，严厉打击非法经营利用、走私野生动植物及其产品的行为。这不仅有效缓解了野外资源保护压力，还有效促进了当地特色产业的发展和农民增收。

（8）积极开展生物多样性调查和监测

国家林业局稳步推进森林、湿地、荒漠三大生态系统及生物多样性调查和监测工作。

> **案例 3.2　红豆杉的人工培育与产业化**
>
> 　　红豆杉为国家 I 级重点保护植物，世界自然保护联盟（IUCN）易危（VU）物种，已被列入 CITES 附录 II，在野外十分稀有，具有重要的药用价值。江苏红豆杉生物科技有限公司于 1997 年开始研究红豆杉种子发育和人工培育、种植，并在红豆杉栽培技术领域取得显著成果，攻克了红豆杉的快繁技术，解决了红豆杉种苗严重短缺的问题。国家林业局于 2008 年批准江苏红豆杉生物科技有限公司建设的红豆杉高科技生态产业园为中国首个红豆杉科技示范区。2010 年该公司在红豆杉产业方面的销售收入达到 20 多亿元，形成了红豆杉的培植、盆景、成树、制药、保健养生的庞大产业链。
>
> 　　按照产学研发展模式，浙江海正药业股份有限公司在东北林业大学的指导下，以"公司+高校+基地+农户"的模式在浙江富阳建立了大型基地，开展红豆杉种植与提取的现代化产业链建设，至 2010 年已实现销售收入 10 亿元，成为中国林业科技创新平台。

2009—2013 年开展了第八次全国森林资源清查，掌握了森林覆盖率、森林蓄积量两项约束性指标及森林类型多样性的最新状况。2009 年完成的第四次全国荒漠化和沙化土地监测表明：截至 2009 年年底，中国荒漠化土地面积为 262.4 万 km^2，与前五年相比，荒漠化土地面积年均减少 2 491 km^2，荒漠化呈整体得到初步遏制、局部仍在扩展的局面。荒漠化地区平均植被盖度由 2004 年的 17.0% 提高到 2009 年的 17.6%，重点保护治理区植物种类明显增加，植被群落稳定性增强。

　　（9）划定沙化土地封禁保护区，全面保护沙区生物多样性

　　对于暂不具备治理条件以及因保护生态的需要不宜开发利用的连片沙化土地，划定沙化土地封禁保护区，实行严格的封禁保护，严格禁止滥开垦、滥放牧、滥樵采、滥用水资源等行为，严格管控开发建设活动，保护荒漠植被，逐步形成稳定的天然荒漠生态系统。

　　（10）推进林业转基因生物安全和森林遗传资源管理

　　根据国家林业局《开展林木转基因工程活动审批管理办法》和《转基因森林植物及其产品安全性评价技术规程》，对转基因林木的中间试验、环境释放和生产性试验，进行严格的安全评审。对许可的转基因林木进行生物安全性监测。加强林业生物遗传资源管理，开展中国特有林业生物遗传资源调查编目工作，对林业生物遗传资源及相关传统知识获取和惠益分享进行试点。

3.13　知识产权部门

　　国家知识产权局历来重视与履约相关的遗传资源、传统知识和民间文艺的知识产权保护工作。

（1）积极参与推动建立遗传资源、传统知识保护制度的国际磋商

国家知识产权局一直牵头代表中国参加世界知识产权组织知识产权与遗传资源、传统知识和民间文艺政府间委员会会议的国际谈判，并参加世界贸易组织 TRIPs 理事会会议等其他多边场合涉及遗传资源、传统知识保护的国际讨论，在国际场合与广大发展中国家共同提出在专利申请中强制性披露遗传资源来源的主张。

国家知识产权局积极参加中国与新西兰、秘鲁、哥斯达黎加、瑞士等国的自由贸易协定谈判。在 2008 年、2009 年、2010 年和 2013 年先后生效的中国—新西兰、中国—秘鲁、中国—哥斯达黎加、中国—瑞士自由贸易协定中，均纳入了涉及遗传资源、传统知识保护的相关条款。

国家知识产权局在遗传资源和传统知识保护领域，加强与各国知识产权部门的交流合作，近年来多次举办面向亚洲、非洲和拉丁美洲发展中国家以及东盟国家的知识产权涉外培训班，介绍中国遗传资源、传统知识和民间文艺的知识产权保护状况以及相关国际进展，加强彼此的了解和交流，为相关国际合作创造良好条件。

（2）积极参加涉及生物多样性的立法、执法以及宣传培训等工作

国家知识产权局努力推进涉及生物多样性的知识产权立法研究和政策制定工作，2011 年与中医药局等部门联合发布了《关于加强中医药知识产权工作的指导意见》，努力推动建立有关中医药产业发展的遗传资源、传统知识保护制度。在专利审查实践中贯彻执行《中华人民共和国专利法》及实施细则中的遗传资源来源披露条款，要求申请人在相关专利申请中披露遗传资源来源，并通过课题研究对相关制度的实施效果进行了评估。

国家知识产权局先后于 2011 年、2012 年和 2013 年在湖北、四川、甘肃等地举办了传统知识和遗传资源知识产权保护研讨会、培训班，邀请相关领域专家介绍国际、国内相关保护制度和发展动态，与环保、文化、农业、林业、中医药等部门加强交流，并结合地方产业发展需求，总结地方实践经验，扩大普及宣传，推动相关研究。

3.14 旅游部门

旅游业已成为国民经济的重要战略性支柱产业，也成为对生物多样性影响最大的人类活动之一。旅游业在生物多样性保护中扮演着重要的角色。

（1）将生物多样性保护纳入旅游业发展战略和规划

2009 年 12 月，国务院颁布的《国务院关于加快发展旅游业的意见》明确提出"推进节能环保"等要求。2010 年发布的《中国旅游业"十二五"发展规划纲要》设立专门章节"保护资源环境，实现可持续发展"。近年来国家旅游局主持编制跨区域旅游战略规划，如《东北地区旅游业发展规划》等也都把保护生态环境和生物多样性作为重要内容。

(2) 行业标准和规范充分兼顾到《生物多样性公约》的要求

《旅游景区质量等级评定与划分》对生物多样性保护有明确的规定，成为景区等级评定的核心指标之一。2012年国家旅游局和环境保护部共同制定了《国家生态旅游示范区建设与运营规范》（GB/T 26362—2010）国家标准，并配套出台了《国家生态旅游示范区管理规程》和《国家生态旅游示范区建设与运营规范评分实施细则》，为生态旅游发展提供了依据。2013年，国家旅游局联合环保部确定了38个国家生态旅游示范区。

(3) 开展"生态旅游年"主题活动，有效推动《生物多样性公约》的履行

国家旅游局将2009年确定为"中国生态旅游年"，并将主题年口号确定为"走进绿色旅游、感受生态文明"。"生态旅游年"的开展，充分发挥了生态旅游区的环境教育功能，提高了公众生态保护意识。特别是通过社区参与为当地居民提供就业机会，促进地方经济发展，使他们自觉地成为生态保护的拥护者和实践者。

(4) 结合生物多样性优先保护区域，实现地理生态空间单元整体保护性开发

结合已划定的35个生物多样性优先保护区域，2012年国家旅游局相继启动了《秦岭山区旅游发展规划纲要》（2013—2020）、《大武陵山区旅游发展规划纲要》（2013—2020）、《大别山区旅游发展战略纲要》（2013—2020）等战略性规划，力图突破地域行政界线，积极推进"山系"、"水系"等完整地理生态单元的整体旅游开发，统筹旅游开发中的生态环境保护工作。

(5) 鼓励行业创新，探索旅游开发与生物多样性保护协调发展的模式

国内诸多知名生态型旅游区，如安徽黄山旅游区、福建武夷山旅游区、四川九寨沟旅游区、云南普达措旅游区等，结合旅游区自身特点，大胆创新，形成了旅游开发与生物多样性保护协调发展的良好模式。

3.15 海洋部门

中国各级海洋行政主管部门把海洋生物多样性保护要求纳入涉海相关战略和计划，采取多种保护措施，取得明显成效。

(1) 完善海洋生物多样性保护法律法规

中国初步形成了以《中华人民共和国海洋环境保护法》为中心、配套条例和地方各级海洋环境保护行政法规为辅助的海洋环境法律体系。近年来，中国颁布了《中华人民共和国海岛保护法》、《海洋特别保护区管理办法》等法律法规，天津、河北、浙江、广东、海南等省陆续出台了地方海洋环境保护法规。这些法律法规的颁布实施进一步完善了海洋生态环境保护法律法规体系。

案例 3.3 福建厦门五缘湾滨海湿地生态修复项目

2005—2007年，福建厦门开展了五缘湾生态修复工程，内容包括海堤开口、内湾疏浚清淤、海湾低水位水坝、海湾沿岸护岸、湿地公园建设等工程。生态修复实施后，退塘还海，水文条件、景观、环境质量等都得到大幅度改善，生物多样性也逐渐得到恢复。目前，五缘湾不仅成为厦门市的景观与休闲区，而且是良好的自然生态科普宣传基地，为开展厦门国际海洋周、世界地球日活动以及观鸟等活动提供了平台，有力地提高了公众的海洋生物多样性保护意识。

厦门五缘湾滨海湿地生态修复成效图（左侧为修复前；右侧为修复后）

(2) 将海洋生物多样性保护要求融入行业战略与计划

2012 年国务院批准了《国家海洋事业发展"十二五"规划》。该规划提出了"到 2020 年，陆源污染得到有效治理、近海生态环境恶化趋势得到根本扭转、海洋生物多样性下降趋势得到基本遏制"的主要目标，提出了加强海洋生物多样性保护、推进海洋生态系统修复、强化海洋生态监测和生态灾害管理的重要任务。由国务院批准发布的《全国海洋经济发展"十二五"规划》《全国海洋功能区划》（2011—2020）和《全国海岛保护规划》（2011—2020）都把海洋生物多样性放到十分突出的地位，并对海洋生物多样性保护提出了明确的目标和要求。中国沿海各级地方均高度重视海洋生态保护和建设，制定的区域涉海发展规划也都把生物多样性保护摆到十分突出的地位，将保护和恢复生物多样性作为重要目标和任务，实施了一批入海污染物控制与治理、海洋生物多样性保护与修复等工程。

(3) 开展海洋生物多样性调查和监测

2006—2008 年，中国实施了近海近岸海洋生物调查。通过调查基本摸清了中国海洋生物的"家底"，出版了《中国海洋物种和图集》，全面、系统阐述了中国海洋生物的种类及其分布特征。为掌握中国海洋生物多样性的变化状况，自 2004 年起，中国在近岸海域部分生态脆弱区和敏感区建立了 18 个海洋生态监控区，监测总面积达 5.2 万 km^2，涵盖海湾、河口、滨海湿地、珊瑚礁、红树林和海草床等典型海洋生态系统。2008 年以来，中国每年均开展国家级海洋自然保护区和海洋特别保护区的常规性监测，基本掌握了海洋保护区生物多样性动态。

(4) 加强海洋保护区网络建设

国家海洋局出台了《海洋特别保护区管理办法》，建立了海洋特别保护区评审委员会制度，修订了《海洋特别保护区功能分区和总体规划编制技术导则》。近两年，国家海洋局利用中央分成海域使用金批准实施了 10 个保护区能力建设项目，累计投资 1 亿多元。目前，大多数国家级海洋保护区已建立管理机构，落实了一定的人员编制和管理运行经费，强化了保护区执法力量。2008 年以来，海洋保护区数量尤其是国家级海洋保护区数量有较大幅度的增长，新建了多个国家级海洋自然保护区和海洋特别保护区。截至 2012 年年底，全国共建有各级、各类海洋保护区 240 多处以上，总面积达到 8.7 万 km^2，占中国主张管辖海域的近 3%。

(5) 开展海洋生态保护与修复

2012 年，国家海洋局印发了《海洋生态文明示范区建设管理暂行办法》和《海洋生态文明示范区建设指标体系（试行）》。目前，完成了山东、浙江、福建、广东等首批国家级海洋生态文明示范区的申报创建工作。国家海洋局正在探索建立海洋生态红线制度，其中重要河口、重要滨海湿地、海洋保护区、重要渔业区域等海洋生物多样性保护区域是海洋生态红线的重点，已在渤海率先启动海洋生态红线制度建设，其中山东省已建立起海洋生态红线制度，超过 40% 的渤海管辖海域将得到严格保护。国家海洋局从

2010年开始，利用中央分成海域使用金，投入约44.3亿元，先后支持了180个海岸带修复、海岛保护与修复和海洋生态修复项目，修复红树林、滩涂等退化湿地面积达2 800多 km^2。

（6）应对气候变化对生物多样性的影响

国家海洋局专门成立了应对气候变化领导小组，对海洋领域应对气候变化工作进行规划，具体工作已逐步开展实施。定期开展厄尔尼诺/拉尼娜等海洋与气候变化研究与形势预测工作，定期开展海水水温、海平面、海水入侵以及土壤盐渍化等与气候变化密切相关的监测工作。国家海洋局加强了海洋领域应对气候变化的科学研究，建立滨海湿地碳固定/碳埋藏能力的计算方法，开发与集成滨海湿地碳固定和碳封存技术。

3.16 中医药管理部门

中药资源是中医药发展的核心物质基础。中医药管理部门十分重视中药资源的可持续利用。

（1）在国家法律法规和行业规划中考虑生物多样性因素

《中华人民共和国药典》对中药资源的保护规定如下：对于贵重中药材，取消野生物种的药用标准；对于资源匮乏的中药材，在允许的情况下，药材的药用部分由全草改为地上部分，以保留地下部分使其可继续生长；一些中药增加人工栽培品种作为新的基源，以减少对原有野生物种资源的采挖量。国家中医药管理局印发的《中医药事业发展"十二五"规划》，确定了野生中药资源培育、研究开发和合理利用能力不断提高的发展目标，提出了"开展全国中药资源普查，加快种质资源库建设，加强野生中药资源培育基地建设，强化对重要、资源有限的野生中药原材料的宏观调控"的任务。《中医药创新发展规划纲要》（2006—2020）要求，开展中药材珍稀濒危品种保护、繁育和替代品等的研究，建立中药材种质库等，完善中药材资源保护与可持续利用的关键技术。《国务院关于扶持和促进中医药事业发展的若干意见》提出："加强对中药资源的保护、研究开发和合理利用。保护药用野生动植物资源，加快种质资源库建设，在药用野生动植物资源集中分布区建设保护区，建立一批繁育基地，加强珍稀濒危品种保护、繁育和替代品研究，促进资源恢复与增长。"

（2）中药资源普查试点工作

中国启动了中药资源普查试点工作。2011年起分批在25个省（直辖市、自治区）的698个县开展了中药资源普查试点工作。主要内容包括：①摸清中药资源种类和分布及563种重点中药资源（药材）的蕴藏量等家底情况；②与中药资源相关的传统知识调查，重点调查具有地方特色的用药知识、民族的用药知识和经验；③建设16个中药材种苗繁育基地和2个种质资源库，对繁育有困难的中药资源进行人工繁育技术研究并建立种质资源库；④建设国家基本药物中药原料资源动态监测和信息服务体系。

(3) 积极开展中药资源的收集和保护

建立了药用植物种质资源离体保护技术体系，收集药用植物离体种质近 3 万份，涉及 3 599 个物种；首次成功建立了中国第一座国家药用植物种质资源库。创建了药用植物种质迁地保护技术体系，实现 5 282 种药用植物迁地保护，迁地保护药用物种数量居世界首位。

(4) 开展中药材种植、减轻对野生资源的压力

在多个部委的共同推动和扶持下，开发中药材种植新技术，探索中药材规模化种植模式，中药材规范化、规模化种植取得初步进展，在实现增产增收的同时减轻了对野生资源的压力。例如，当归、甘草、大黄、金银花等大众品种中药材连片种植面积达到 66.7 km^2 以上。

3.17 扶贫开发

中国生物多样性丰富的地区，往往也是贫困地区。中国在减贫工作中，十分重视生物多样性保护。

(1) 在国家扶贫规划中充分考虑生物多样性保护

2011 年，中国政府颁布实施了《中国农村扶贫开发纲要》（2011—2020）。该纲要明确提出，坚持扶贫开发与生态建设、环境保护相结合，充分发挥贫困地区资源优势，发展环境友好型产业，提倡健康科学生活方式，促进经济社会发展与人口资源环境相协调。在连片特困地区区域发展与扶贫攻坚规划中，优先考虑贫困地区的生态建设和生物多样性保护。

(2) 在贫困地区继续实施退耕还林、退牧还草、水土保持、天然林保护、防护林体系建设和石漠化、荒漠化治理等重点生态工程

逐步建立并完善生态补偿机制。加大对重点生态功能区的生态补偿力度。加强草原保护和建设，加强自然保护区建设和管理，大力支持退牧还草工程。采取禁牧、休牧、轮牧等措施，恢复天然草原植被和生态功能。

(3) 因地制宜发展清洁能源

加快贫困地区可再生能源开发，因地制宜发展小水电、太阳能、风能、生物质能，推广应用沼气、节能灶、固体成型燃料、秸秆气化集中供气站等能源建设项目，带动改水、改厨、改厕、改圈和秸秆综合利用。

(4) 更加重视贫困地区人力资源开发，缓解人与资源的矛盾

以促进扶贫对象稳定就业为核心，对农村贫困家庭未继续升学的应届初、高中毕业生参加劳动预备制培训，给予一定的生活费补贴；对农村贫困家庭新成长劳动力接受中

等职业教育给予生活费、交通费等特殊补贴。同时,,对农村贫困劳动力开展实用技术培训。近几年来,通过扶贫培训实现转移就业的贫困地区劳动力每年均超过100万人,带动了400多万人摆脱贫困,有效缓解贫困地区人地矛盾,促进了这些地区的生物多样性保护。

(5)坚持自愿原则,对生存条件恶劣地区扶贫对象实行异地扶贫搬迁

减轻对自然条件极度恶劣地区的生态压力。同时,引导其他移民搬迁项目优先在符合条件的贫困地区实施,加强与异地扶贫搬迁项目的衔接,共同促进改善贫困群众的生产生活环境。加强统筹协调,切实解决搬迁群众在生产生活等方面的困难和问题,确保搬得出、稳得住、能发展、可致富。目前,国家正在制定异地扶贫搬迁规划,力求通过减轻这些地区的人口承载力,积极改善生物多样性保护的外部条件。

> **案例 3.4　贵州省毕节市开发扶贫与生物多样性保护协调发展**
>
> 从1988年开始,为缓解毕节市生存与发展、人口与资源之间日益尖锐的矛盾,贵州省在毕节市建立了以"开发扶贫、生态建设、人口控制"为三大主题的试验区。20多年来,作为全国唯一的地区级"开发扶贫、生态建设"试验区,贵州省毕节试验区从经济社会发展滞后、群众生活贫困、生态环境极差的现实条件出发,坚持开发与扶贫并举、生态恢复与建设并进、人口数量控制与质量提高并重,跳出了"越生越垦—越垦越穷—越穷越生"的怪圈,走出了一条开发扶贫、生态建设的新路,取得了经济社会发展的重大成就。1988—2011年,农民人均年收入从182元跃升至4 300元,森林覆盖率从15.0%增加到41.5%。

3.18　履行其他相关公约

3.18.1　《联合国防治荒漠化公约》

(1)防治荒漠化规划充分考虑了生物多样性因素

《全国防沙治沙规划》(2011—2020)提出了"预防为主,综合治理"的基本原则,提出了"到2020年使全国一半以上可治理的沙化土地得到治理,沙区生态状况进一步改善"的目标。规划期内,完成沙化土地治理任务20万 km^2,其中,2011—2015年为10万 km^2,2016—2020年为10万 km^2。

(2)完善了防沙治沙扶持政策

国家出台了包括集体林权制度改革、森林生态效益补偿、林业贷款贴息、造林补贴、草原生态保护补助等一系列支持沙区生态建设和产业发展的政策措施。各地结合当地实际,在投资、税收、金融等方面完善了防沙治沙优惠政策,极大地调动了企业、个人等各种社会主体参与防沙治沙的积极性,初步形成了全社会参与、多元化投资防沙治沙的

新格局。

(3) 推进了防沙治沙重点工程建设

"十一五"以来，中国继续实施京津风沙源治理、"三北"防护林体系建设、退耕还林、退牧还草、草原保护、小流域综合治理等一系列重点生态工程，还相继启动了新疆塔里木盆地防沙治沙、石羊河流域防沙治沙及生态恢复、西藏生态安全屏障保护与建设等区域性防沙治沙工程，对沙化重点地区和薄弱环节进行集中治理，推动了全国沙区生态状况的持续好转。监测结果显示，"十一五"期间，全国沙化土地年均净减少 $1\ 717\ km^2$，五年间中度、重度和极重度沙化土地面积共减少 3.6 万 km^2，沙化程度减轻。局部地区的水土流失得到有效控制，土壤侵蚀模数大幅度下降，年入黄泥沙减少 3 亿多 t。

(4) 提高了防沙治沙支撑保障能力

一是进一步提升科技支撑能力。国家林业局专门成立了荒漠化研究所，强化防沙治沙科研技术力量。"沙漠化发生规律及其综合防治模式研究"、"中国北方沙漠化过程及其防治"等科研成果荣获国家科技进步奖励，一批防沙治沙科研成果和适用技术得到推广应用。二是制定和完善技术标准。制定颁发了《防沙治沙技术规范》、《沙化土地监测技术规程》、《京津风沙源治理工程技术标准》等一批防沙治沙技术标准。三是强化荒漠化沙化监测和沙尘暴应急处置。完成了第四次全国荒漠化沙化监测，建立了重大沙尘暴灾害应急体系，形成了以遥感监测和地面监测为主、信息员测报为辅的沙尘暴灾害监测体系。

(5) 强化了防沙治沙部门协作机制

从中央到地方基本成立了专门的防沙治沙组织协调和领导机构，加强对防沙治沙工作的组织、领导和协调工作。"十一五"以来，各有关部门充分发挥各自的职能作用，形成了各负其责、密切配合、协同作战、齐抓共管的工作机制。

(6) 落实了防沙治沙目标责任制

按照《中华人民共和国防沙治沙法》的要求，受国务院委托，国家林业局与防治任务较重的北方 12 个省级政府和新疆生产建设兵团签订了"十一五"防沙治沙目标责任书。2009 年，国家林业局会同有关部门制订的《省级政府防沙治沙目标责任制考核办法》经国务院批准后实施。防沙治沙目标责任制的建立和实施，在中国防沙治沙史上第一次真正实现了中央政府对省级政府防沙治沙目标责任进行问责，切实提高了地方各级政府和有关部门防沙治沙的责任意识，进一步推动了全国防沙治沙工作。目前，"十二五"防沙治沙目标责任书已签订。

(7) 鼓励发展沙区特色产业

为切实推动沙区特色产业发展，国家林业局制定了《国家林业局关于进一步加快发展沙产业的意见》，确定了加快发展沙产业的指导思想、原则和目标，提出了沙产业发展的总体布局和重点领域，明确了促进和扶持沙产业发展的政策措施和保障措施。各地结合实际，制定了一系列发展沙区特色产业的措施，在有效治理和严格保护的基础上，积极引导各种实体充分利用沙区的优势资源，发展特色优势产业，增加沙区农民收入，促进沙区经济发展。

> **案例 3.5　甘肃省定西市安定区水土保持重点治理工程**
>
> 甘肃省定西市安定区地处黄土高原，全区总人口 43 万人，面积 3 638 km²，水土流失面积占土地面积的 90% 以上。中国相关部门依据生态系统方法，对该地区开展了以小流域为单元的水土流失综合治理工程。
>
> （1）在缓坡耕地连片修梯田，陡坡耕地弃耕改种多年生牧草，荒山荒坡挖反坡种草种树，沟道打骨干坝拦水淤地，支毛沟打谷坊，庄前屋后打水窖、修涝池，形成了被群众誉之为"山顶戴帽子，山腰系带子，山脚穿靴子"的综合治理体系。
>
> （2）全区 90 多条小流域、1 620 多 km² 的水土流失面积得到了有效控制。原先跑水、跑土、跑肥的"三跑田"变成了保水、保土、保肥的"三保田"，林草覆盖率由原来的 8% 上升到 43%。
>
> （3）在治理过程中，注重利用生物多样性维持生态系统的结构和功能。例如，坡地改为梯田后，大面积推广马铃薯，开展农业产业化经营；实施雨水集流工程，建成集流场 349 万 m²，打水窖 5 万多眼，解决了 20 多万人和 30 多万头家畜的饮水问题；发展畜牧业，使全区畜牧业产值由 1982 年的 878 万元猛增到 9 846 万元。通过以上各种措施，农业生产条件极大好转。近 10 年来，全区粮食产量连年保持在 10 万 t 以上，人民生活水平普遍好转，已经摘掉了贫困的帽子。
>
>
>
> 甘肃省定西市安定区水土保持综合治理效果

3.18.2 《联合国气候变化框架公约》

中国在履行《联合国气候变化框架公约》的过程中，充分考虑气候变化、生态系统与生物多样性保护之间的依存关系，并采取了一系列减缓和适应气候变化的措施。在减缓气候变化方面，中国政府承诺 2020 年单位国内生产总值温室气体排放比 2005 年下降 40%～45% 的行动目标，并进一步明确 2015 年单位 GDP 二氧化碳排放比 2010 年下降 17% 的约束性指标。其主要做法包括以下几项。

（1）调整产业结构和提高能效

2011 年国家发展和改革委员会修改并发布了《产业结构调整指导目录（2011 年版）》，淘汰高投入、高消耗、高污染、低效益的建设项目，鼓励环境友好和资源综合利用的生态恢复、技术产品开发和基础建设项目。

(2) 发展低碳能源和优化能源结构

通过政策引导和资金投入，推进煤炭的清洁化利用，加强煤层气、页岩气等清洁能源的开发，积极支持风能、太阳能、地热能、生物质能等新型可再生能源的发展。

(3) 增加碳汇和加强生态保护

继续实施重点生态工程，加强农田水利设施建设和实行保护性耕作，加强湿地保护，大幅提高森林、草原、湿地与农业碳汇，在保护生物多样性的同时进一步增强碳汇对温室气体的吸收能力。

(4) 开展低碳发展试验试点

积极开展低碳省区和低碳城市、碳排放权交易等多种形式的试点工作，充分发挥应对气候变化与节能环保、新能源发展、生态建设等方面的协同效应。

除有效控制温室气体排放外，《国民经济和社会发展第十二个五年规划纲要》同时强调要增强适应气候变化的能力，包括提高农业、林业、水资源等重点领域和沿海、生态脆弱地区适应气候变化水平。主要行动包括：

1) 积极构建生态系统对气候变化响应监测评价工作，强化对重点地区海平面变化、海水入侵、土壤盐渍化和海岸侵蚀的监测评价，扎实推进全国林业碳汇计量监测体系建设，开展林业适应气候变化研究，积极开展包括黄河、珠江、辽河等流域在内的重点区域生物多样性与气候变化现状调查。

2) 进一步完善极端天气和气候事件监测预警系统建设，出台气象灾害防御规划，扎实推进极端天气气候条件下的应对极端事件能力建设。

3.18.3《关于特别是作为水禽栖息地的国际重要湿地公约》（以下简称《湿地公约》）

中国是世界上湿地资源最为丰富的国家之一。中国建立了专门的湿地保护管理机构，健全湿地保护管理体系，全国湿地保护面积大幅增加，许多重要的自然湿地得到抢救性保护，重要区域的湿地生态系统得到有效恢复。主要做法包括如下几项。

(1) 推动湿地保护立法和规划

一系列有关湿地生物多样性保护的法律法规先后颁布实施，其中《中华人民共和国森林法》《中华人民共和国野生动物保护法》《中华人民共和国水法》《中华人民共和国环境保护法》《中华人民共和国海洋环境保护法》《中华人民共和国渔业法》等法律法规及实施条例，对湿地保护和利用发挥了积极作用。17个省（区）也出台了省级湿地保护条例。为加强湿地保护和管理，国务院还颁布了《全国湿地保护工程规划》（2002—2030）、《全国湿地保护工程实施规划》（2011—2015）。

(2) 完善湿地保护政策

中国采取多种措施，使湿地得到更为有效的保护。"十一五"期间中央累计投入14

亿元，地方配套超过 17 亿元，完成了 205 个湿地保护和恢复示范工程，恢复湿地近 800 km^2。截至目前，中国已建湿地自然保护区 577 处、湿地公园 468 处、国际重要湿地 46 处，使约 43.5% 的湿地得到较为有效的保护，保护体系在维护湿地生态系统健康方面发挥着骨干作用。

（3）夯实湿地保护基础工作

2009 年起组织开展了第二次全国湿地资源调查，利用"3S"技术，参照《湿地公约》标准，对面积在 0.08 km^2 以上的湿地进行了全面调查，掌握了中国湿地类型、面积、野生动植物、保护管理状况、面临的威胁和问题等。调查成果已于 2014 年 1 月向社会发布。

（4）建立湿地生态补偿、生态补水等长效机制

2009 年中央一号文件明确提出了"启动湿地生态效益补偿试点"，同年中央林业工作会议再次提出要建立湿地生态效益补偿机制。同时，将湿地生态用水纳入流域水资源利用规划，既保证湿地生态用水，又发挥湿地在补充地下水、蓄洪防旱等方面的生态功能，使湿地生态系统进入良性循环轨道；重点对面临严重缺水威胁的国家重要湿地进行生态补水。

中国湿地保护虽然取得了显著成效，局部地区湿地生态状况有了明显改善，但是整体上全国湿地仍面临干旱缺水、开垦围垦、泥沙淤积、水体污染和生物资源过度利用等严重威胁，湿地功能退化的趋势尚未得到根本遏制，湿地仍然是最脆弱、最容易遭到侵占和破坏的生态系统。

3.18.4 《濒危野生动植物种国际贸易公约》

中国严格执行《濒危野生动植物种国际贸易公约》（CITES）的决定和特有的"非致危性判定（NDF）"机制，确保相关濒危物种的出口不会危及其野生种群的生存，同时严厉打击非法野生动植物贸易。中华人民共和国濒危物种进出口管理办公室会同有关部门共同编制并实施了《2010—2013 中国履行 CITES 第十五届缔约国大会有效决议决定行动方案》，确定了虎、鲨鱼、象牙、赛加羚羊、犀牛、蛇类、苏眉、热带木材、海上引进、非致危性判定 10 项重点议题以及海参、鳄鱼制品统一标记、执法、个人携带和家庭财产、网上贸易、来源和目的代码 6 项一般性议题，明确了履约活动的方向和重点。2011 年，中国成立了以林业、农业、海关、公安、工商、检验检疫为主体的多部门联合执法协调工作组，提高了执法效率，遏制了犯罪势头。针对大宗木材物种及远洋性商用水生物种，濒危物种进出口管理办公室向国内相关主管部门传达了 CITES 的关切和可能采取的管制措施，提请相关主管部门重视商用资源的保护问题。在国内的日常履约过程中，由于 CITES 已经建立了一套进出口管制许可证系统，因此濒危物种进出口管理办公室采用《进出口濒危野生动植物种商品目录》与海关联合确定管制范围，在强制性的许可证系统之外，有效利用行政许可及物种的技术性管控需求，建立了一套行之有效的"物种证明"管理体系。但由于公约附录管制范围与生物多样性保护要求不完全重叠，且部分特有资源尚未纳入公约附录，相关物种的管理措施尚未落实，其中问题比较明显的包括新种或未定

名种较多的昆虫，分布地狭小、种群数量小的两栖爬行类等。

濒危物种进出口管理办公室加强了与 CITES 秘书处的合作。2010 年 9 月，濒危物种进出口管理办公室与 CITES 秘书处及《联合国迁徙物种保护公约》秘书处在中国新疆乌鲁木齐市召开了"赛加羚羊保护与可持续利用研讨会"，邀请哈萨克斯坦、俄罗斯、乌兹别克斯坦及蒙古等赛加羚羊分布国的保护与管理机构与会，中国中药协会也参加了会议并作为关键利益攸关方与分布国管理机构进行了讨论。这不仅体现了环境公约的协同增效，同时也有利于各利益攸关方融入保护、敦促行业践行可持续发展战略。2011 年 4 月和 2012 年 5 月，濒危物种进出口管理办公室与 CITES 秘书处在广州分别召开了"亚洲蛇类管理研讨会"和"CITES 濒危物种电子许可系统开发研讨会"，对深入履行 CITES 中的新问题进行了研讨。

中国与相关缔约方开展了交流与合作。中国与美国共同梳理龟鳖类贸易管理，并根据其野生资源极度濒危的现状，联合制定了野生种群贸易零限额、促进龟鳖野生种群监控和复壮、促进养殖业进一步反哺野生种群的政策，这得到了 CITES 第十六届缔约方大会的认可。2012 年 9 月，中国政府在杭州为非洲国家代表举办了培训班，起到了很好的效果，加强了与非洲国家的合作。中国进一步深化了与俄罗斯、印度、蒙古、越南、老挝、印尼、泰国等周边国家的履约及执法合作。

第 4 章
2020 年生物多样性目标的实施进展以及对千年发展目标的贡献

4.1 2020 年生物多样性目标的评估指标

采用"压力—状况—惠益—响应"模型，设计了 2020 年生物多样性目标评估指标体系。指标体系的设计遵循了以下原则：①代表生物多样性的各个方面；②真实、及时地反映生物多样性的变化；③容易被决策者、公众和管理人员理解，有广泛的认可度；④能精确测量，数据采集成本较低，尽量利用现有数据；⑤能表征政策变革所产生的变化。该指标体系包括 17 个一级指标、42 个二级指标（见表 4-1）。

表 4-1 中国有关 2020 年生物多样性目标的评估指标

一级指标	二级指标及其含义
生物多样性状况	
1. 生态系统宏观结构	（1）指森林、湿地、草地、沙漠等生态系统的面积及其比例的变化。可采用遥感数据计算。还可对天然林面积进行单独分析。天然林又称自然林，包括自然形成与人工促进天然更新或萌生所形成的森林
2. 生态系统健康状况	（2）活立木总蓄积量：指一定区域范围土地上所生长着的全部树木的蓄积量之和。还可分析天然林蓄积量。可采用林业清查数据。 （3）森林生态系统年均净初级生产力。净初级生产力是指绿色植物在单位时间、单位面积上由光合作用所产生的有机物质总量中扣除自养呼吸后的剩余部分。其作为表征植被活动的关键变量，在全球碳平衡中扮演着重要作用。可采用遥感数据计算。 （4）**天然草原鲜草总产量**。可采用农业部草原监测数据。 （5）海洋营养指数：指海洋渔获物的平均营养级，反映海洋食物链的长短，进而反映海洋生态系统的抗干扰能力和完整性。可采用 FAO 数据计算
3. 物种多样性	（6）红色名录指数：指受威胁物种在濒危等级、种群数量等方面的变化，表示特定生物类群濒危等级的总体变化。可分别对相关类群进行计算。 （7）**内陆水域特有鱼类种数**，可反映内陆水域鱼类多样性的变化。可采用中国科学院水生生物研究所监测数据计算
4. 遗传资源	（8）地方品种资源，可反映传统遗传资源的保护状况
生态系统服务	
5. 生态系统提供的服务	（9）**产品提供功能**：指生态系统提供给人类的可以在市场中进行交换的产品或服务。主要包括粮食和畜牧产品等。可采用遥感数据计算。 （10）**调节功能**：包括水源涵养功能、固沙功能、土壤保持功能。可采用遥感数据计算。 （11）**支持功能**：生态系统具有为野生动植物物种提供生境的作用，保障野生动植物繁衍生息的功能。可采用遥感数据计算
6. 直接依赖于当地生态系统产品和服务的社区健康和福祉的变化	（12）**农村居民家庭人均纯收入**。 （13）**贫困人口数量**。 以上指标可采用国家统计局统计数据和国家林业局林业重点生态工程监测数据

一级指标	二级指标及其含义
压力	
7. 环境污染	（14）工业废水中化学需氧量、废气中二氧化硫和烟尘、工业固体废物等的年排放量，表明环境污染对生物多样性构成的威胁。可采用环境统计数据。 （15）单位GDP污染物排放强度：指单位国内生产总值污染物排放量。可采用环境统计数据。 （16）单位GDP二氧化碳排放强度：指单位国内生产总值二氧化碳排放量。可采用统计数据。 （17）农用化学品施用量：说明农业活动对生物多样性的影响程度，高的氮投入和氮失衡常对生物多样性构成重大威胁。可采用农业统计数据。 （18）单位农业增加值农用化学品施用量。可采用农业统计数据计算
8. 气候变化对生物多样性的影响	（19）指气候变化对生态系统的结构和功能、对物种和遗传资源的分布与增长所造成的影响
9. 外来入侵物种危害程度	（20）每20年新发现的外来入侵物种种数。可采用调查数据计算。 （21）口岸截获有害生物的批次和物种种数。可采用口岸查验数据
响应	
10. 就地保护体系	（22）自然保护区数量和面积覆盖率：面积覆盖率指的是自然保护区面积占陆地国土面积的百分比，反映生物多样性就地保护状况。可采用环境统计数据。 （23）保护小区的数量与面积。可采用国家林业局统计数据。 （24）风景名胜区的数量与面积。可采用住房和城乡建设部统计数据。 （25）森林公园的数量与面积。可采用国家林业局统计数据。 （26）湿地公园的数量与面积。可采用国家林业局统计数据。 （27）国家级水产种质资源保护区的数量和面积。可采用农业部统计数据。 （28）海洋特别保护区的数量和面积。可采用国家海洋局统计数据
11. 政策和规划的实施	（29）国家重点生态工程的实施情况和发布省级生物多样性保护战略与行动计划的数量
12. 生境保护与恢复	（30）重点生态工程区森林蓄积量。 （31）重点生态工程区木材产量。可反映木材产量调减情况。 （32）重点生态工程区水土流失情况。 以上指标可采用国家林业局林业重点生态工程监测数据
13. 污染控制	（33）烟气脱硫机组装机容量及其占全部火电机组的比例。 （34）全国城市污水处理能力。 （35）固体废物处理情况。 以上指标可采用环境统计数据
14. 资源综合利用	（36）可再生资源利用量：太阳热水器、太阳灶的数量。 （37）处理农业废弃物工程年产量。 （38）处理农业废弃物工程总池容。 （39）生活污水净化沼气池村级处理系统总池容。 以上指标可采用农业统计数据
15. 公众意识	（40）不同年份通过Google或百度检索到有关中国生物多样性的条目
16. 与生物多样性保护与持续利用有关的知识	（41）有关生物多样性保护论文的数量。可采用相关文献数据库计算
17. 生物多样性保护相关资金的投入	（42）天然林资源保护工程、野生动植物保护及自然保护区建设工程、湿地保护工程的资金投入。可采用林业统计数据

注：黑体字的二级指标指的是与第四次国家报告相比本报告新采用的指标。
2020年目标与指标的对应关系，请见附表2。

4.2 2020年生物多样性目标评估指标的数据分析

4.2.1 生物多样性状况

（1）生态系统宏观结构

根据全国生态环境十年变化（2000—2010）遥感调查与评估项目的结果，2000—2010年，中国森林、湿地、城镇生态系统面积有所增加，灌丛、草地、荒漠、农田生态系统面积减少；森林生态系统面积增加了约2.9万 km²，湿地生态系统面积增加了400 km²，城镇面积增加了5.5万 km²，灌丛和草地生态系统面积分别减少了约1.2万 km²和1.6万 km²，荒漠生态系统面积减少了4 400 km²，农田减少了4.8万 km²（见图4-1）。城市化区、退耕还林集中区、农业扩展区等部分地区生态系统格局变化剧烈。20世纪80年代末以来，中国天然林面积持续增加（见图4-2）。

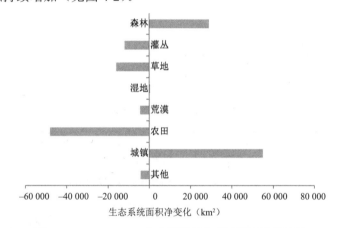

图4-1　2000—2010年中国陆域生态系统面积净变化

数据来源：全国生态环境十年变化（2000—2010）遥感调查与评估项目，欧阳志云等提供。

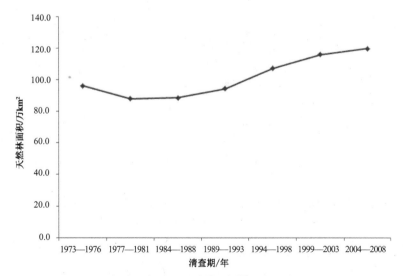

图4-2　中国不同时期天然林面积的变化

数据来源：《中国林业统计年鉴》。

（2）生态系统健康状况

20 世纪 80 年代以来，中国的森林蓄积量持续增加。第七次清查与第六次清查相比，全国活立木总蓄积净增 11.3 亿 m^3，森林蓄积净增 11.2 亿 m^3，天然林蓄积净增 6.8 亿 m^3（见图 4-3）。2000—2010 年，中国森林生态系统年均净初级生产力总体上呈增加趋势（见图 4-4）。尽管中国森林资源总量持续增长，森林生态系统功能有所恢复，但仍然存在总量不足、质量不高、分布不均衡的问题。

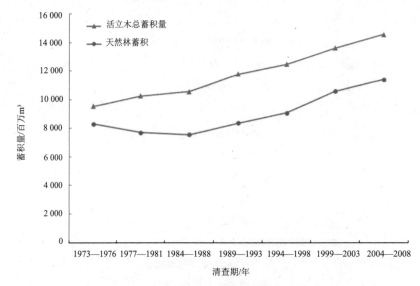

图 4-3　中国不同时期森林蓄积量的变化

数据来源：《中国林业统计年鉴》。

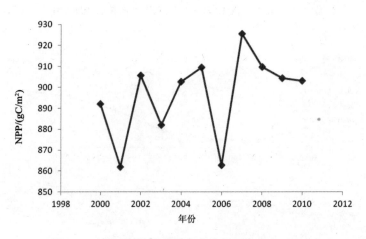

图 4-4　中国森林生态系统年均净初级生产力（NPP）

数据来源：全国生态环境十年变化（2000—2010）遥感调查与评估项目，欧阳志云等提供。

中国不断加强草原生态系统的保护和恢复，草原生态系统发生积极变化。2005—2012 年，全国草原鲜草总产量基本呈不断增加的趋势（见图 4-5）。这表明局部草原生态系统状况在改善。但当前全国大部分草原仍处于超载过牧状态，草原退化、沙化、盐碱化、石漠化现象依然严重，生态环境形势依然严峻。

由于过度捕捞，从20世纪80年代初到90年代中期，中国海洋营养指数持续下降，低于同期全球平均水平，海洋生态系统严重退化。1997年至今，中国海洋营养指数开始呈平稳上升趋势（见图4-6）。这可能是伏季休渔政策的实施对海洋渔业资源的养护起到了积极作用。但中国海洋营养指数仍处于较低的水平，生态功能低下，海洋生物多样性保护仍然任重道远。

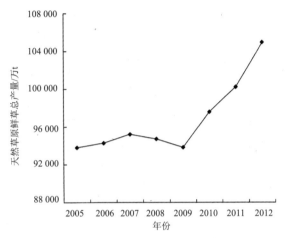

图4-5　中国天然草原鲜草总产量

资料来源：《全国草原监测报告》。

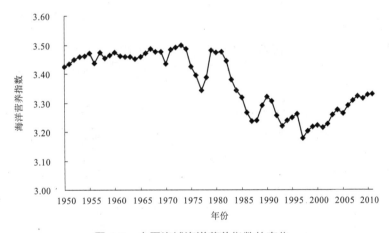

图4-6　中国海域海洋营养指数的变化

（3）物种多样性

红色名录指数（RLI）是指特定生物类群濒危等级的总体变化，RLI为0时指所有物种都灭绝；RLI为1时指所有物种都不受威胁，不需要保护。1998—2004年，中国淡水鱼类的RLI下降；1996—2008年，中国兽类的RLI下降；1988—2012年，根据Equal-steps方法计算的鸟类RLI略有下降，但根据Extinction-risk方法计算的RLI先略有好转又呈下降趋势（见图4-7）。由于生境的退化和消失，兽类和鱼类受威胁程度在加剧；鸟类整体上受威胁程度在加剧，尽管高濒危物种的保护状况得到一定程度的改善。

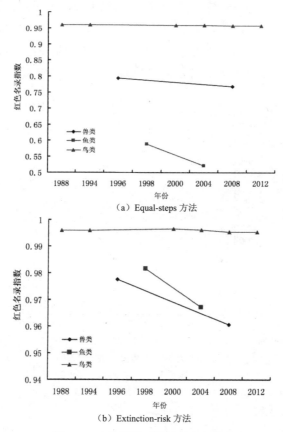

图 4-7　中国脊椎动物红色名录指数

数据来源：鸟类，Birdlife International, http://www.birdlife.org/，$n=1\,208$；兽类，IUCN 红色名录数据库，http://www.iucnredlist.org/，$n=99$；鱼类，《中国濒危动物红皮书》和《中国物种红色名录》，$n=81$。

中国开展了部分内陆水域鱼类的监测。1997—2009 年，长江上游特有鱼类种数总体呈下降趋势（Liu and Gao, 2012）（见图4-8），表明长江流域生物多样性仍在下降。

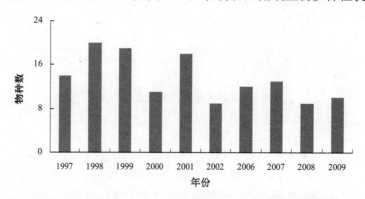

图 4-8　1997—2009 年长江上游木洞江段长江上游特有鱼类种数

资料来源：Liu and Gao, 2012。

（4）遗传资源

据估计，中国遗传资源的丧失十分严重，但由于数据的限制，这里只能以案例来说明。中国种植的农作物以粮食作物为主，而粮食作物又以水稻、小麦、玉米三大作物为主。20 世纪 50 年代，中国各地农民种植水稻地方品种达 46 000 多个，至 2006 年，全国种植水稻品种仅 1 000 多个，且基本为育成品种。20 世纪 50 年代中国种植的玉米地方品种达 10 000 多个，到目前生产上已基本没有地方品种。50 年代初期每年种植小麦地方品种约 4 000 个，截至 2000 年减少至不到 400 个，目前基本都是育成品种。根据第二次全国畜禽遗传资源调查的结果，全国有 15 个地方畜禽品种资源未发现，超过一半以上的地方品种的群体数量呈下降趋势（国家畜禽遗传资源委员会，2011）。

4.2.2 生态系统服务

（1）生态系统提供的服务

根据全国生态环境十年变化（2000—2010）遥感调查与评估项目的结果，2000—2010 年，中国陆域生态系统粮食和畜牧产品等产品提供功能持续增加，从 2000 年的 2014.7 万亿 kcal 增加到 2010 年的 2 805.2 万亿 kcal，10 年间增加了 39.2%。

就生态系统的调节功能而言，中国陆域生态系统水源涵养功能从 2000 年的 122.6×10^{10} m³ 增加到 2010 年的 123.5×10^{10} m³，10 年间增加了 0.7%；防风固沙功能从 2000 年的 121.2×10^{8} t 增加到 2010 年的 137.5×10^{8} t，10 年间增加了 13.4%；土壤保持量从 2000 年的 1966.5×10^{8} t 增加到 2010 年的 1979.6×10^{8} t，10 年间提高了 0.7%。

就生态系统的支持功能而言，采用生境质量指数来评估生物多样性维持功能 10 年变化。2000—2010 年，"低"等级生境质量生态系统面积占较大比例，且"高"和"较高"等级生境质量的生态系统面积持续下降（见图 4-9）。这说明，尽管中国实施了重点生态工程，但由于生境质量的改善是一个缓慢长期的过程，生物多样性维持功能仍在不断下降。

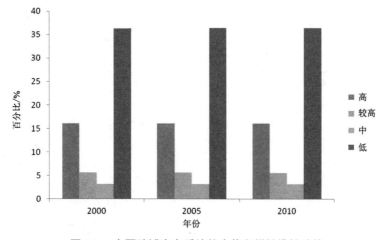

图 4-9　中国陆域生态系统的生物多样性维持功能

数据来源：全国生态环境十年变化（2000—2010）遥感调查与评估项目，欧阳志云等提供。

（2）直接依赖于当地生态系统产品和服务的社区健康和福祉的变化

中国农村居民家庭人均纯收入增加较快，2011年比2000年增加了40.8%（见图4-10），这在一定程度上得益于生态系统产品提供功能的增加。对于减贫工作，以天然林资源保护工程和退耕还林工程区的数据反映全国的情况。天然林资源保护工程和退耕还林工程样本县贫困人口数量呈下降趋势。天然林资源保护工程样本县贫困人口由1997年的395万人下降至2011年的183万人，退耕还林工程由1998年的830万人下降至2008年的570万人（见图4-11）。在保护和恢复森林生态系统的同时，直接依赖于当地生态系统产品和服务的社区福祉也在改善。

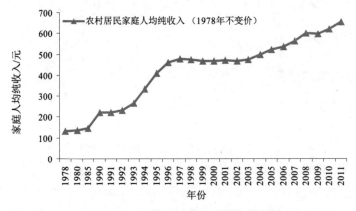

图 4-10　中国农村居民家庭人均纯收入

数据来源：《中国统计年鉴》。

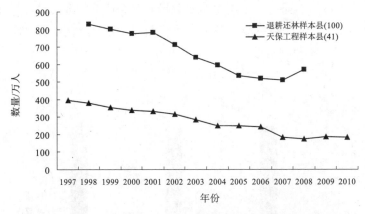

图 4-11　退耕还林工程样本县、天然林资源保护工程样本县贫困人口

数据来源：《国家林业重点工程社会经济效益监测报告》。

4.2.3　压力

（1）环境污染

2006年以来，中国工业废水中化学需氧量（COD），废气中二氧化硫（SO_2）、烟尘、工业粉尘，工业固体废物的排放量呈下降趋势（见图4-12）。近10年，尽管经济高速增长，但单位 GDP 污染物排放量大幅下降了 55% 以上（见图 4-13）。2004 年以来，单位 GDP 二氧化碳排放量下降 15.2%（见图 4-14）。1991—2011 年，农用化学品施用量增长了 1 倍以上（见图 4-15），但 2003 年后单位农业增加值农用化学品施用量持续下降（见图 4-16），说明农用化学品的使用效率在提高。

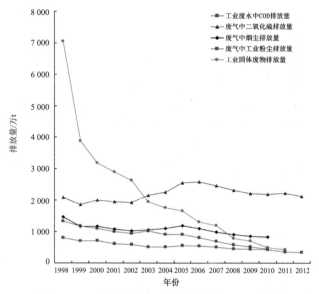

图 4-12　工业废水中 COD、废气中 SO_2、废气中烟尘、
废气中工业粉尘、工业固体废物的年排放量

数据来源：《全国环境统计公报》和《中国环境统计年鉴》。

中国废水排放量仍在增加，单位 GDP 污染物排放量仍然很高。地表水总体仍为轻度污染，长江、黄河、珠江、浙闽片河流、西南诸河等十大流域的国控断面中，劣Ⅴ类水质的断面比例仍达 10.2%；在监测的 60 个湖泊（水库）中，25% 处于富营养化状态；中国管辖海域水环境状况总体较好，但近岸海域海水污染依然严重。环境污染对生物多样性构成严峻威胁。

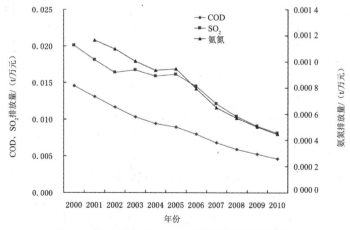

图 4-13　全国单位 GDP 污染物排放量的变化

数据来源：《中国环境统计年鉴》和《中国统计年鉴》。

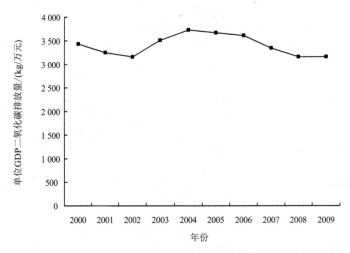

图 4-14　单位 GDP 二氧化碳排放量的变化

数据来源：《中国环境统计年鉴》和《中国统计年鉴》。

（2）气候变化

1951—2009 年，中国陆地表面平均温度上升了 1.38℃，变暖速率为每 10 年 0.23℃。气候变化对中国的生态系统和物种产生了可辨识的影响。气候变化使中国草地退化加剧，内陆湿地功能下降（第二次气候变化国家评估报告编写委员会，2011）。近几十年来，呼伦湖地区气候暖干化趋势明显，湖面萎缩，环湖草场退化、土地沙化，植被盖度下降，对栖息物种造成了很大威胁（赵慧颖等，2008）。

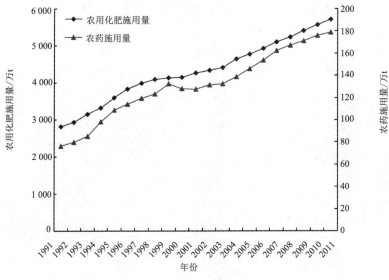

图 4-15　农用化学品施用量

数据来源：《中国农业统计资料》。

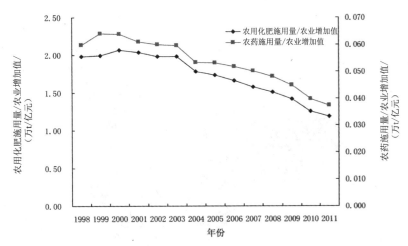

图 4-16　单位农业增加值农用化学品施用量
（农业增加值使用居民消费价格指数进行调整，基准年为 1978 年）
数据来源：《中国统计年鉴》（2012 年）。

气候变化使生物物候、分布和迁移发生改变。东北、华北及长江下游等地区木本植物的春季物候（展叶期、始花期）提前，而秦岭以南包括西南东部、长江中游等地区的物候期推迟（郑景云等，2003）。1980 年以来，一些鸟类如青海大杜鹃（祁如英等，2008）、鲁西南四声杜鹃（张翠英等，2011）的自然物候提早，始鸣期提前、绝鸣期推迟。气候变化导致一些地区林线海拔升高。在气候变暖背景下，中国有 120 种鸟类的分布区向北或向西发生了扩展。气候变化还使一些物种在原栖息地消失。青海湖地区气候呈现暖干化趋势，气候变化和人为活动的影响使该地区物种组成改变，尤其是鸟类组成有很大变化。与 20 世纪中期相比，豆雁、灰头鹀、白头鹞、鹌鹑、白背矶鸫等 26 种鸟从湖区消失（马瑞俊等，2006）。

气候变化使有害生物的分布范围改变，危害加剧。例如，气候变暖使加拿大一枝黄花（吴春霞等，2008）和马尾松毛虫（国家林业局森林病虫害防治总站，2013）等分布范围扩大。

（3）外来入侵物种

外来入侵物种种数在中国的分布大致分三个台阶，由沿海向内陆逐步减少。最多的是沿海省份及云南；中部地区及一些与之相邻的东部和西部省份次之（见图 4-17）（徐海根，强胜，2011）。对 396 种有明确入侵时间记载的外来入侵物种的分析表明，新出现的外来入侵物种种数总体呈逐步上升的趋势，1950 年后的 60 年间，新出现 212 种外来入侵物种，占外来入侵物种种数的 53.5%（见图 4-18）（徐海根，强胜，2011）。

随着对外开放的深入和国际贸易的高速发展，口岸截获植物疫情呈大幅增长趋势，2012 年截获有害生物的种类是 1999 年的 18.9 倍，批次更达 230.2 倍（见图 4-19），这给中国农林业生产和生态安全构成严重威胁。

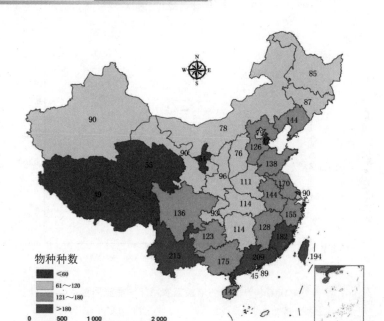

图 4-17　外来入侵物种种数的分布（徐海根，张胜，2011）

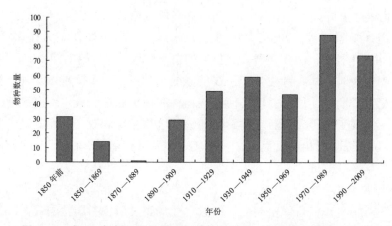

图 4-18　每 20 年新发现的外来入侵物种种数（徐海根，张胜，2011）

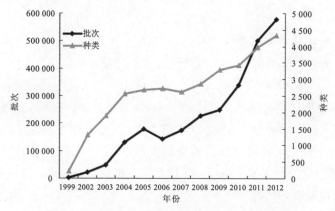

图 4-19　全国口岸截获有害生物的批次和种类

数据来源：国家质量监督检验检疫总局门户网站，www.aqsiq.gov.cn。

4.2.4 响应

(1) 就地保护体系

近年来,中国自然保护区的数量和面积保持稳定(见图4-20),自然保护区面积达到全国陆域面积的14.8%,初步形成了类型较为齐全、布局较为合理、功能比较健全的自然保护区网络,但在保护区的生态代表性、管理有效性、科学研究和生态补偿等方面有待完善;国家级风景名胜区数量和面积保持增长(见图4-21),逐步实现由注重视觉景观保护向视觉景观、文化遗产、生物多样性等方面综合保护的转变,但一些地方仍然存在破坏风景名胜区资源的问题;森林公园数量和面积快速增加(见图4-22),初步形成森林风景资源保护和利用的管理体系,但尚有大量珍贵的森林风景资源没有得到有效保护,一些地方破坏森林风景资源的现象突出;全国林业自然保护小区数量基本稳定,但面积自2008年以来持续下降(见图4-23);国家级水产种质资源保护区和海洋特别保护区的数量与面积持续增加(见图4-24和图4-25)。

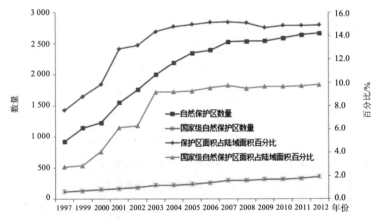

图 4-20　全国自然保护区数量和面积百分比

数据来源:《中国环境状况公报》和《中国环境统计年鉴》。

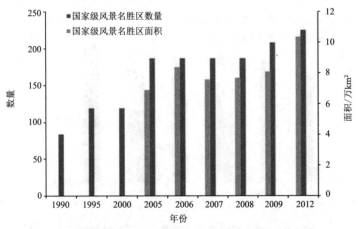

图 4-21　全国国家级风景名胜区数量和面积

数据来源:《中国风景名胜区事业发展公报》。

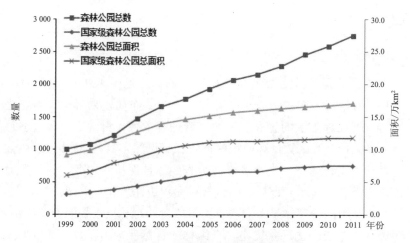

图 4-22　全国森林公园数量和面积

数据来源：《中国林业统计年鉴》。

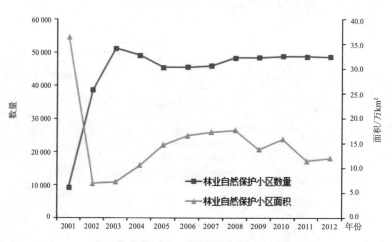

图 4-23　全国林业自然保护小区数量和面积

数据来源：《中国林业统计年鉴》。

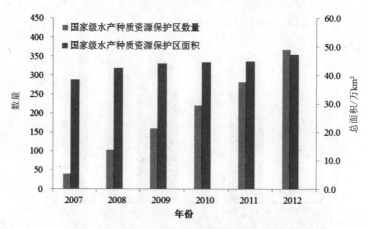

图 4-24　国家级水产种质资源保护区的数量和面积

数据来源：农业部。

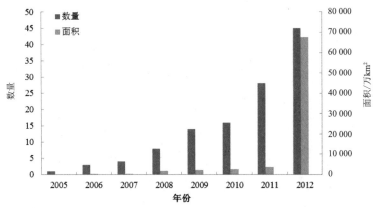

图 4-25　国家级海洋特别保护区的数量和面积

数据来源：国家海洋局。

（2）政策和规划的实施

中国实施了一系列有利于保护生物多样性的政策和规划，先后启动了天然林资源保护、退耕还林、退牧还草、"三北"防护林建设、湿地保护与恢复、水土流失综合治理等重点生态工程。这些重点生态工程的实施，促进了退化生态系统和野生物种生境的恢复，有效保护了生物多样性。各省（自治区、直辖市）都在制定本地区生物多样性保护战略与行动计划，目前已发布 7 个省级生物多样性保护战略与行动计划。

（3）生境保护和恢复

自 2001 年，中国林业重点工程建设取得巨大成就，森林保护和恢复效果良好。天然林资源保护工程、退耕还林工程、京津风沙源治理工程样本县与样本企业森林蓄积量自 1999 年起呈上升趋势（见图 4-26、图 4-27）。天然林资源保护工程样本企业木材产量持续下降，由 1997 年的 624.3 万 m^3 降至 2011 年的 179.5 万 m^3（见图 4-28），反映出木材产量调减的显著成绩。天然林资源保护工程和退耕还林工程样本县水土流失面积总体呈下降趋势（见图 4-29）。总体看来，林业重点工程在生态保护方面起到了极其重要的作用。

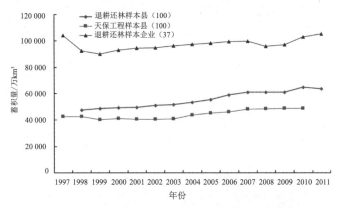

图 4-26　退耕还林工程、天然林资源保护工程样本县和
天然林资源保护工程样本企业森林蓄积量

注：图例括号中数字表示样本数。

数据来源：《国家林业重点工程社会经济效益监测报告》。

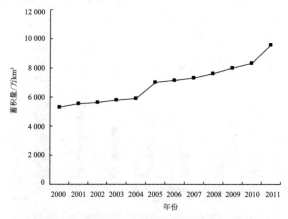

图 4-27　京津风沙源治理工程样本县森林蓄积量

资料来源：《国家林业重点工程社会经济效益监测报告》。

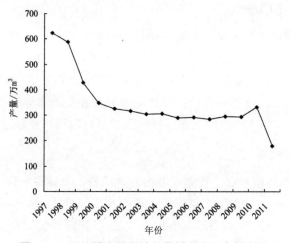

图 4-28　天然林资源保护工程样本企业木材产量

资料来源：《国家林业重点工程社会经济效益监测报告》。

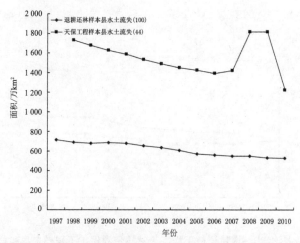

图 4-29　退耕还林工程样本县、天然林资源保护工程样本县水土流失

注：图例中数字表示样本数。

资料来源：《国家林业重点工程社会经济效益监测报告》。

（4）污染控制

污染物减排成效显著，全国烟气脱硫机组装机容量及其占全部火电机组的比例（见图 4-30）、全国城市污水处理率（见图 4-31）、工业固体废物综合利用量（见图 4-32）均持续大幅增长，但近两年的工业固体废物综合利用率有所下降。

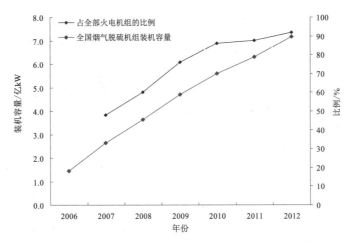

图 4-30　全国烟气脱硫机组装机容量及其占全部火电机组容量的比例

数据来源：《中国环境状况公报》和《环境统计年报》。

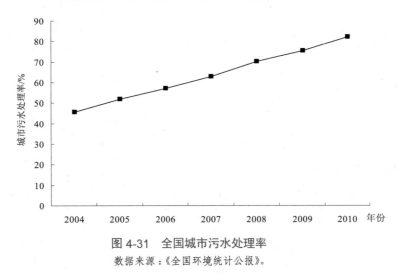

图 4-31　全国城市污水处理率

数据来源：《全国环境统计公报》。

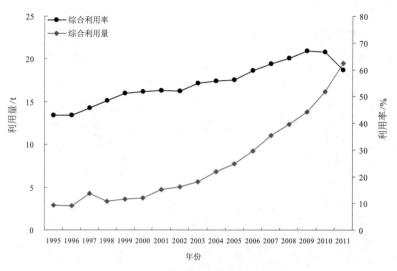

图 4-32　全国固体废物利用情况

数据来源：《全国环境统计公报》。

（5）资源综合利用

全国太阳能热水器和太阳灶的数量呈逐年递增趋势（见图 4-33），2011 年太阳能热水器的数量达到 6 232 万 m^3，与 1997 年相比增加了近 10 倍。2011 年太阳能灶的数量是 1997 年的近 10 倍。全国处理农业废弃物工程年产气量和处理农业废弃物工程总池容呈逐年递增趋势（见图 4-34），两项指标 2011 年分别是 1997 年的 60 倍和 77 倍。

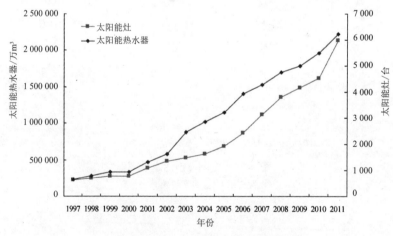

图 4-33　全国太阳能热水器和太阳灶的数量

数据来源：《中国农业统计资料》。

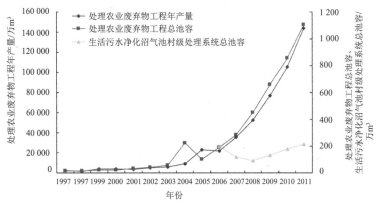

图 4-34　全国农业废弃物利用情况

数据来源：《中国农业统计资料》。

（6）公众意识

利用网络高级检索，查询不同年份关键词为中国生物多样性的信息条目，结果表明有关"生物多样性"的条目呈增加趋势（见图 4-35、图 4-36）。这说明生物多样性越来越多地被公众所关注。

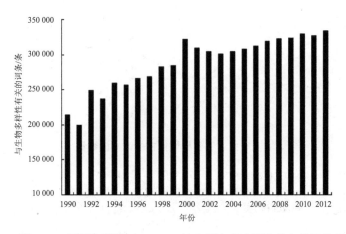

图 4-35　不同年份通过 Google 检索到有关中国生物多样性的条目

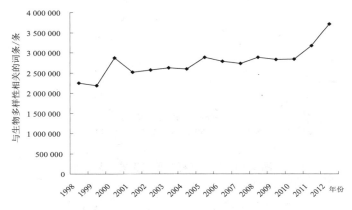

图 4-36　不同年份通过百度检索到有关中国生物多样性的条目

(7) 与生物多样性保护与持续利用有关的知识

通过中文维普数据库查询 1990—2012 年每年发表的有关生物多样性的论文，通过外文数据库（EBSCO 和 ISI WEB OF SCIENCE）查询 1990—2012 年每年发表的有关中国生物多样性的论文。结果表明，有关生物多样性保护的论文呈逐年递增趋势（见图 4-37）。

(8) 生物多样性保护相关资金的投入

近年来，中国生物多样性保护相关资金投入大幅度增加。以天然林资源保护工程、野生动植物保护与自然保护区建设工程、湿地保护工程为例，资金投入从 2001 年的 97.0 亿元增加到 2011 年的 217.7 亿元，年均增长 13.7%，对生物多样性保护提供了资金支持（见图 4-38）。

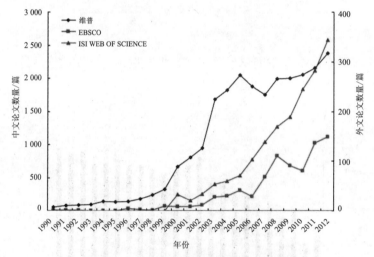

图 4-37　有关生物多样性保护的论文的数量

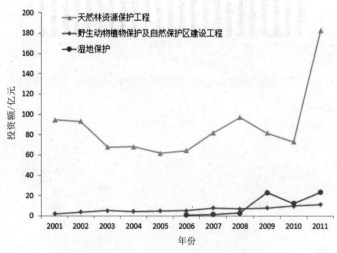

图 4-38　生物多样性保护相关资金

数据来源：《中国林业统计年鉴》。

综上所述，近年来中国政府加大了生物多样性保护力度，通过完善保护政策、加强保护体系建设、恢复退化生态系统、控制环境污染、强化科学技术研究、推动公众参与、增加资金投入等措施，使生态破坏加剧的趋势有所减缓，部分区域生态系统功能得到恢复，一些重点保护物种种群有所增长，生物多样性保护取得积极进展。但总体上，野生生物生境面积萎缩和功能退化，一些重点流域、海域污染严重，少数地区外来入侵物种呈现扩大蔓延之势，单一品种大规模种植和气候变化对生物多样性的不利影响日益凸显，全国生物多样性下降的趋势尚未得到根本遏制，生物多样性保护形势依然严峻。

4.3 中国实现全球《生物多样性战略计划》和2020年生物多样性目标的总体进展评估

采用生物多样性评估指标，对2020年目标的实施进展进行了评估（附表2）。对于目标2、目标16和目标18，由于缺乏相应的国家指标，无法进行评估。目标1、目标3、目标4、目标5、目标7、目标8、目标10、目标11、目标14、目标15、目标17、目标19、目标20的相关评估指标均有不同程度的改善，表明这些目标的实施正沿着正确的轨道推进，特别是目标3（鼓励措施）、目标5（减少生境退化和丧失）、目标8（控制环境污染）、目标11（强化保护区系统和有效管理）、目标14（恢复和保障重要生态系统服务）、目标15（增强生态系统的复原力和碳储量）进展较大。但目标5中的草原生态系统保护，目标6（可持续渔业）、目标9（防治外来入侵物种）、目标12（保护受威胁物种）、目标13（保护遗传资源）的相关评估指标大多呈现恶化的趋势，表明虽然已开展了大量工作，但尚需要采取更加有效的策略和措施来实施这些目标。总体上，中国在实现全球《生物多样性战略计划》和2020年目标方面，正沿着正确的轨道推进并取得积极进展，今后应下更大的决心，采取更加有效的措施，投入更多的资源，努力实现2020年目标。

同时，今后应进一步开发生物多样性价值、可持续消费、生态退化、农林渔业对生物多样性的影响、气候变化对生物多样性的影响、保护区的生态代表性和管理有效性、遗传资源和相关传统知识的获取与惠益分享等方面的指标，更加重视生态系统特别是草原生态系统和海洋生态系统的整体保护，重视濒危物种和遗传资源的保护，重视外来入侵物种的防控。

4.4 对实现千年发展目标的贡献

2010年9月，中国政府发布了《中国实施千年发展目标进展情况报告》（2010年版）。报告指出，中国在落实千年发展目标方面取得积极进展，已提前实现"将贫困与饥饿人口减半"、"普及初级教育"及"降低儿童死亡率"三项目标及"环境可持续力"目标下的"安全饮用水"子目标，其他目标亦有望如期实现。正如本书第1章所述，生物多样性是人类赖以生存的条件，是经济社会可持续发展的物质基础，中国政府在保护生物多样性的同时，对实现千年发展目标也有十分重要的贡献。当然，中国在全面实现千年发展目标方面也面临挑战，特别是在实现可持续发展领域，包括城乡区域发展不平衡、环境压力较大、生物多样性丧失等。中国将继续深入贯彻落实科学发展观，全面深化改革，加快调整经济结构和转变增长方式，努力建设生态文明和美丽中国，促进经济社会又好又快发展，为全面实现千年发展目标做出不懈努力。

4.5 在执行《生物多样性公约》方面取得的经验

经过长期的探索实践，中国初步走出了一条适合其基本国情的生物多样性保护之路，为推动全球生物多样性保护事业积累了宝贵经验。

（1）坚持政府主导，实行国家扶持与公众参与相结合

作为一项社会公益事业，做好生物多样性保护，必须以国家投入为主导。中国各级政府把生物多样性保护纳入经济社会发展规划，加强组织领导，加大投入力度。同时，完善政策和措施，充分调动社会各界力量参与生物多样性保护，提高公众的保护意识和参与程度，形成全社会关心、支持生物多样性保护的良好局面。例如在神农架林区通过建立政府、保护区、民间社会组织和所有关键相关者的保护联盟，推进生物多样性保护。

（2）坚持部门联动，实行"环保部门统一协调、各部门齐抓共管"的工作机制

进一步完善了中国履行《生物多样性公约》工作协调组和生物物种资源保护部际联席会议制度，新建了由环保部牵头、25个部门参加的中国生物多样性保护国家委员会。这三个工作机制在中国生物多样性保护中发挥了重要作用。实践证明，"环保部门统一协调、各部门齐抓共管"是符合中国国情的工作机制。

（3）坚持保护优先，实行保护与合理利用相结合

在经济社会发展中优先考虑生物多样性保护，遵循自然规律，坚持"在发展中保护，在保护中发展"的方针，对重要生态系统、生物物种及遗传资源实施有效保护。在有效保护的前提下，合理开发利用生物资源，增加群众收入，改善民生福祉。

（4）坚持工程带动，实行重点突破和面上推进相结合

通过实施重点生态工程，在森林、草原、湿地等重要生态系统和生态脆弱区域、敏感区域开展生物多样性保护相关示范项目。同时，完善政策和法规，颁布和实施战略与行动计划，推动生物多样性保护在全国的全面展开。

（5）坚持创新驱动，实行科技创新和管理创新相结合

加强生物多样性专业教育和人才培养，着力培育生物多样性保护和利用技术的创新能力，开发具有独立知识产权的生物多样性保护技术、模式和知识体系。同时，通过完善政策、制度、标准体系和监测预警体系，促进生物多样性保护管理科学化、规范化，提高生物多样性保护管理水平。

（6）坚持开放合作，实行国际履约和国内履约相结合

始终认真履行国际义务，积极参与国际谈判和多边体系的建设，积极开展国际合作与交流，引进国际上先进的技术和管理理念，实施好生物多样性保护示范项目。同时，努力推动国内生物多样性保护与管理上水平、上台阶，充分发挥地方各级政府作为生物多样性责任主体的作用，使生物多样性保护的目标和任务纳入各级政府和各部门的相关规划、计划之中。

第5章
生物多样性保护面临的主要问题与优先行动

在中国政府的正确领导下,在全社会的支持下,中国生物多样性保护工作取得了巨大成就。但应当清醒地看到,中国生物多样性保护形势依然严峻。《战略与行动计划》中有近一半的优先行动的实施存在较多困难,生物多样性下降的趋势尚未得到根本遏制,保护工作仍然任重道远。

5.1 面临的主要问题

(1) 法制和体制有待进一步完善

中国现有生物多样性保护法律还不完善。一些法律法规已不适应当前社会经济发展的要求,如《中华人民共和国环境保护法》、《中华人民共和国野生动物保护法》、《中华人民共和国野生植物保护条例》和《中华人民共和国自然保护区条例》等亟须修订完善。一些领域如遗传资源获取与惠益分享、湿地保护、外来入侵物种防治等尚没有专门的法规。部分法律法规处罚力度低,违法成本低,难以起到震慑效果。

(2) 保护意识有待进一步提高

全社会生物多样性保护意识和风险意识尚需进一步提高。一些群众对保护生物多样性的重要性认识不足,保护意识比较淡薄,参与程度低。地方政府片面追求经济发展,一旦经济发展与生物多样性保护相冲突时,往往以牺牲生物多样性为代价推动经济发展。企业参与生物多样性保护的积极性不高。决策者和管理人员的生物多样性保护知识缺乏。社会监督力量和能力也不足。

(3) 保护与开发利用的矛盾突出

随着城镇化和工业化进程的加快,生物多样性保护面临着严峻的威胁。例如,牧区人口成倍增长,北方干旱草原区人口密度达到 11.2 人 $/km^2$,为国际公认的干旱草原区生态容量 5 人 $/km^2$ 的 2.2 倍。野生中药材资源需求量大,一些物种由于被长期过度利用,导致野生资源量下降。渔业捕捞作业方式仍处于粗放式和掠夺式阶段,特别是电鱼等非法作业方式对渔业资源和水域生态环境造成严重的损害。中国生物多样性热点地区大多处于偏远、经济欠发达的地区,这些地区保护与发展经济之间的矛盾将长期存在,保护的压力有可能进一步加大。

(4) 保护经费投入不足

尽管中国政府已投入巨额资金用于生物多样性保护，但由于中国国土辽阔、生物多样性丰富、保护工作量大，资金缺口很大，特别是在生物多样性调查和监测、自然保护区建设和管理、生物多样性恢复等方面能力十分薄弱、资金严重缺乏。执法条件差，执法场所尚不能满足一线工作实际需要，缺乏必要的设备和条件。

(5) 技术研究相对滞后

由于长期以来投入不足，专业人才和技术储备欠缺。生物多样性本底、保护实用技术和模式等相关领域研究十分薄弱，许多新问题、新技术有待深入探索，特别是生物多样性调查和监测、生物多样性就地保护、生物多样性恢复等方面没有得到应有重视。

5.2 优先行动

(1) 完善生物多样性保护法律法规，加大执法监督力度

修订《中华人民共和国环境保护法》、《中华人民共和国野生动物保护法》、《中华人民共和国野生植物保护条例》和《中华人民共和国自然保护区条例》；制定《湿地保护条例》《外来入侵物种管理条例》《遗传资源管理条例》等法律法规。健全自然资源资产产权制度和用途管制制度，实行最严格的源头保护制度、损害赔偿制度和生态环境损害责任终身追究制度；尽快建立生态补偿机制，把生物多样性保护纳入生态补偿政策，特别是对生物多样性保护优先区域加快建立相应的补偿机制，在资金和制度上给予支持。加强执法能力建设，加大对破坏生物多样性违法活动的打击力度，加大对物种资源出入境的执法检查力度。

(2) 促进公众参与，提高公众保护意识

开展多种形式的生物多样性保护宣传教育活动，充分发挥各社团组织和企业的作用，不断提高全民保护意识。探索建立社会监督生物多样性保护的机制和政策。积极发展公民科学，推动公众参与生物多样性保护活动，形成全社会共同推进生物多样性保护和可持续利用的氛围。

(3) 落实《全国主体功能区规划》和《中国生物多样性保护战略与行动计划》

落实《全国主体功能区规划》，建立国土空间开发保护制度，优化国土空间开发格局，提出针对各主体功能区的生物多样性保护政策措施；划定生态保护红线，确保国土生态安全。切实执行《中国生物多样性保护战略与行动计划》，强化对生物多样性保护优先区域的监管，将生物多样性保护纳入国家、部门和地方相关规划。加强对开发建设活动的环境管理，把生物多样性影响评价纳入大型工程环评、区域环评和规划环评之中，并落实生态恢复责任。建立评估监督机制，促进各项规划的有效实施。

(4) 进一步完善就地保护网络, 加大保护力度

优化自然保护区空间结构, 科学构建生物多样性保护网络体系, 建立国家公园体制。加强自然保护区、风景名胜区、森林公园、湿地公园、水产种质资源保护区等管理机构能力建设。继续实施天然林资源保护、退耕还林、退牧还草、"三北"防护林及长江流域等防护林建设、京津风沙源治理、岩溶地区石漠化综合治理、湿地保护与恢复、自然保护区建设、水土流失综合治理等重点生态工程, 启动生物多样性保护重大工程, 建立生物多样性保护与减贫相结合的机制。

(5) 提高应对新威胁和新挑战的能力

抓紧建立外来入侵物种的预警和监测体系, 采取预防措施, 对有意引进的外来物种进行规范的风险评估, 并落实风险管理措施, 组织开展对重大危害外来物种的灭杀工作。开展对转基因生物风险评估和环境影响检测的基础性研究, 开展环境监测和检测的技术开发, 完善相关技术标准和规范。

(6) 加强机制与机构能力建设, 提高管理水平

加强中国生物多样性保护国家委员会的统筹协调能力, 继续发挥中国履行《生物多样性公约》工作协调组和生物物种资源保护部际联席会议的作用。进一步加强各有关部门生物多样性保护相关机构的能力建设, 尤其要加强对地方生物多样性保护工作的支持力度, 促进管理能力不断提高。

(7) 加大资金投入

拓宽投入渠道, 加大国家和地方资金投入, 引导社会、信贷、国际资金参与生物多样性保护, 形成多元化投入机制。整合生物多样性保护现有分散资金, 提高使用效率。加大各级财政对生物多样性保护能力建设、基础科学研究和生态补偿的支持力度。

(8) 建立生物多样性调查、监测和发布制度

定期开展生物多样性调查, 建立生物多样性监测预警体系, 及时掌握生物多样性的动态变化, 发布生物多样性红色名录, 有针对性地实施对重要物种和生态系统的有效监控。

(9) 加强人才培养和科学研究

加强生物多样性保护人才培养。进一步加大科研攻关力度, 着力解决生物多样性形成机制、丧失途径、保护模式、恢复模式、价值评估、生态补偿等方面的技术问题, 加强生物遗传资源的收集、保存和开发力度, 为生物多样性保护和管理提供有力的科技支撑。

(10) 加强国际合作与交流

认真履行公约义务, 积极参与国际多边体系建设, 广泛开展国际交流与合作, 引进国外先进技术和经验, 促进中国生物多样性保护能力和水平的提高。

附　录

附录 1　有关缔约方和国家报告编写的情况

A. 缔约方

缔约方	中国
国家联络处	
机构全称	环境保护部国际合作司
联系人姓名和职称	张洁清　处长
通信地址	中国北京市西直门内南小街 115 号
电话	+86-10-66556520
传真	+86-10-66556513
电子邮件	zhang.jieqing@mep.gov.cn
国家报告联系人（若与上面不同）	
机构全称	环境保护部自然生态保护司（生物多样性保护办公室）生物多样性保护处
联系人姓名和职称	张文国　处长
通信地址	中国北京市西直门内南小街 115 号
电话	+86-10-66556309
传真	+86-10-66556329
电子邮件	zhang.wenguo@mep.gov.cn
提交	
负责提交国家报告的官员签字	张洁清
提交日期	2014 年 3 月 31 日

B. 国家报告编写过程

1.制定编制大纲，组成编制专家组

2013 年 1—2 月，环境保护部制定了第五次国家报告编制工作方案和编制大纲，并发函邀请中国履行《生物多样性公约》工作协调组成员单位推荐相关专家。各成员单位均按要求推荐了专家。环境保护部经过资格审查确定了编写专家。

2.召开编制专家组会议，启动项目，明确任务分工

2013 年 3 月 8 日，在北京召开了编制专家组第一次会议，正式启动编制工作，进一步完善了编制大纲，明确了时间进度和专家任务分工。

3. 资料调研、整理和起草部门分报告

2013年4月，各部门推荐的专家按照编制大纲的要求，收集整理本部门的材料和数据，5—8月分别提交了部门分报告。

4. 召开专家研讨会，研讨2020年目标评估指标

2013年6月9日，环境保护部在北京召开了2020年目标评估指标研讨会，讨论并完善专家组提出的2020年目标评估指标，并进一步邀请相关单位的专家为2020年目标评估提供数据。

5. 起草第五次国家报告初稿

编制专家组在充分吸收和整理部门分报告的基础上，形成了第五次国家报告初稿。

6. 召开3次专家研讨会，讨论国家报告初稿

2013年11月4日，环境保护部在北京召开了编制专家组第二次会议，讨论国家报告初稿，提出修改意见。同时，11月12日和12月24日，分别召开了专家研讨会和编制专家组第三次会议，对2020年目标评估指标及国家报告初稿进行了研讨。在充分吸收各方面意见的基础上，形成第五次国家报告征求意见稿。

7. 向履约协调组成员单位征求意见

2014年1月初，环境保护部向中国履行《生物多样性公约》工作协调组成员单位等征求意见。各部门对国家报告进行了认真研究，提出了很好的意见。在此基础上编制专家组进一步修改了国家报告，形成了国家报告报批稿，报环境保护部审批。

8. 第五次国家报告的审批、翻译与提交

2014年3月7日，第五次国家报告得到环境保护部批准。2014年1—3月，第五次国家报告中文版被翻译成英文。2014年3月，第五次国家报告中英文版被提交秘书处。

参与编写的各方：

环境保护部、国家发展和改革委员会、教育部、科技部、财政部、国土资源部、住房和城乡建设部、水利部、农业部、商务部、国家新闻出版广电总局、海关总署、国家工商行政管理总局、国家质量监督检验检疫总局、国家林业局、国家知识产权局、国家旅游局、国家海洋局、中国科学院、国家中医药管理局、国务院扶贫办公室、国家濒危物种进出口管理办公室、中国湿地公约履约办公室、国家气候变化对策协调小组办公室、国家林业局防治荒漠化管理中心、中国科学院地理与资源科学研究所、植物研究所、动物研究所、微生物研究所、中国农业科学院作物科学研究所、农业环境与可持续发展研究所、北京畜牧兽医研究所、农业部农业生态与资源保护总站、中国水产科学研究院、中国林业科学研究院、中国检验检疫科学研究院、北京林业大学、华中农业大学、国家海洋环境监测中心、第三海洋研究所、中国中医科学院、环境保护部对外合作中心、环境保护部南京环境科学研究所、中国科学院生态环境研究中心。

感谢全球环境基金（GEF）对本报告编制工作的支持。

附录2　中国履行《生物多样性公约》第五次国家报告参与编写人员名单

报告最终审核： 周生贤　环境保护部部长
　　　　　　　　李干杰　环境保护部副部长
报告编写组织： 庄国泰　环境保护部自然生态保护司司长
　　　　　　　　柏成寿　环境保护部自然生态保护司副司长
报 告 统 稿： 徐海根　环境保护部南京环境科学研究所 研究员/副所长
　　　　　　　　张文国　环境保护部自然生态保护司生物多样性保护处处长
报告英文翻译： 蔡立杰

报告编制专家组名单：

序号	推荐部门	姓名	单位	职称/职务
1	环境保护部	徐海根	环境保护部南京环境科学研究所	研究员/副所长/专家组组长
2	环境保护部	欧阳志云	中国科学院生态环境研究中心	研究员/书记
3	环境保护部	朱留财	环境保护部对外合作中心	研究员
4	环境保护部	王智	环境保护部南京环境科学研究所	副研究员/副主任
5	环境保护部	丁晖	环境保护部南京环境科学研究所	副研究员
6	环境保护部	徐卫华	中国科学院生态环境研究中心	副研究员
7	环境保护部	吴军	环境保护部南京环境科学研究所	副研究员
8	国家发展和改革委员会	陈怡	国家应对气候变化战略研究和国际合作中心	博士
9	国家发展和改革委员会	朱建华	中国林业科学研究院	副研究员
10	国家发展和改革委员会	岳天祥	中国科学院地理科学与资源研究所	研究员
11	教育部	张志翔	北京林业大学	教授
12	国土资源部	王军	国土资源部土地整治中心	研究员
13	住房和城乡建设部	包满珠	华中农业大学园艺林学院	教授/院长
14	水利部	王建平	水利部发展研究中心	高工/副处长
15	农业部	杨庆文	中国农业科学院作物科学研究所	研究员
16	农业部	张国良	中国农科院农业环境与可持续发展研究所	研究员
17	农业部	师荣光	农业部农业生态与资源保护总站	副研究员
18	农业部	樊恩源	中国水产科学研究院	研究员
19	农业部	马月辉	中国农业科学院北京畜牧兽医研究所	研究员
20	农业部	卢欣石	北京林业大学	教授
21	农业部	王志兴	中国农业科学院生物技术研究所	研究员
22	商务部	李丽	对外经济贸易大学	助理研究员/主任
23	海关总署	周亚春	海关总署监管司	处长
24	国家新闻出版广电总局	王京	宣传司	
25	国家工商行政管理总局	白谨毅	国家工商行政管理总局市场司	副处长

序号	推荐部门	姓名	单位	职称/职务
26	国家质量监督检验检疫总局	李明福	中国检验检疫科学研究院	研究员
27	国家林业局	刘增力	国家林业局	副处长
28	国家林业局	吴波	中国林业科学研究院	研究员
29	国家林业局	李迪强	中国林业科学研究院	研究员
30	国家林业局	徐基良	北京林业大学	副教授
31	国家林业局	张明祥	北京林业大学	教授
32	国家知识产权局	张清奎	国家知识产权局医药生物发明审查部	处长
33	国家旅游局	席建超	中国科学院地理科学与资源研究所	副研究员
34	国家海洋局	潘增弟	国家海洋局东海分局	研究员/副巡视员
35	国家海洋局	马明辉	国家海洋环境监测中心	研究员/室主任
36	国家海洋局	陈彬	国家海洋局第三海洋研究所	研究员/室主任
37	中国科学院	马克平	中国科学院植物研究所	研究员
38	中国科学院	解焱	中国科学院动物研究所	副研究员
39	中国科学院	郭良栋	中国科学院微生物研究所	研究员
40	国家中医药管理局	黄璐琦	中国中医科学院	研究员/副院长
41	国务院扶贫办	张良	国务院扶贫办开发指导司	处长
42	国家濒危物种进出口管理办公室	吕晓平	国家濒危物种进出口管理办公室	处长
43		崔鹏	环境保护部南京环境科学研究所	副研究员
44		陈炼	环境保护部南京环境科学研究所	副研究员
45		曹铭昌	环境保护部南京环境科学研究所	副研究员
46		卢晓强	环境保护部南京环境科学研究所	博士
47		刘立	环境保护部南京环境科学研究所	助理研究员

第五次国家报告编制专家组第一次会议参会人员名单
（2013年3月8日，北京）

姓名	单位	职称/职务
柏成寿	环境保护部自然生态保护司	副司长
张文国	环境保护部自然生态保护司生物多样性保护处	处长
卢晓强	环境保护部自然生态保护司生物多样性保护处	博士
徐海根	环境保护部南京环境科学研究所	研究员/副所长
朱留财	环境保护部对外合作中心	研究员
王爱华	环境保护部对外合作中心	项目官员
陈怡	国家应对气候变化战略研究和国际合作中心	博士
岳天祥	中国科学院地理科学与资源研究所	研究员
张志翔	北京林业大学	教授
王军	国土资源部土地整治中心	研究员
王建平	水利部发展研究中心	高工/副处长
杨庆文	中国农业科学院作物科学研究所	研究员

附录 2　中国履行《生物多样性公约》第五次国家报告参与编写人员名单

姓　名	单　位	职称 / 职务
师荣光	农业部农业生态与资源保护总站	副研究员
马月辉	中国农业科学院北京畜牧兽医研究所	研究员
卢欣石	北京林业大学	教授
王志兴	中国农业科学院生物技术研究所	研究员
李丽	对外经济贸易大学	助理研究员 / 主任
李宁	海关总署监管司	主任科员
张广领	国家工商行政管理总局市场司	干部
李明福	中国检验检疫科学研究院	研究员
刘增力	国家林业局	副处长
吴波	中国林业科学研究院	研究员
李迪强	中国林业科学研究院	研究员
徐基良	北京林业大学	副教授
张明祥	北京林业大学	教授
席建超	中国科学院地理科学与资源研究所	副研究员
潘增弟	国家海洋局东海分局	研究员 / 副巡视员
兰冬东	国家海洋环境监测中心	助理研究员
陈彬	国家海洋局第三海洋研究所	研究员 / 室主任
罗茂芳	国际生物多样性计划中国委员会	工程师
解焱	中国科学院动物研究所	副研究员
郭良栋	中国科学院微生物研究所	研究员
陈美兰	中国中医科学院	副研究员
张良	国务院扶贫办公室开发指导司	处长
吕晓平	国家濒危物种进出口管理办公室	处长
吴军	环境保护部南京环境科学研究所	副研究员

2020 年生物多样性目标评估指标体系研讨会参会人员名单
（2013 年 6 月 9 日，北京）

姓　名	单　位	职称 / 职务
张文国	环境保护部自然生态保护司生物多样性保护处	处长
卢晓强	环境保护部自然生态保护司生物多样性保护处	博士
徐海根	环境保护部南京环境科学研究所	研究员 / 副所长
刘纪远	中国科学院地理科学与资源研究所	研究员
杨庆文	中国农业科学院作物科学研究所	研究员
樊恩源	中国水产科学研究院	研究员
李明福	中国检验检疫科学研究院	研究员
李迪强	中国林业科学研究院	研究员
张清奎	国家知识产权局	部长
解焱	中国科学院动物研究所	副研究员
潘增弟	国家海洋局东海分局	研究员 / 副巡视员

姓 名	单 位	职称/职务
陈彬	国家海洋局第三海洋研究所	研究员/室主任
徐卫华	中国科学院生态环境研究中心	副研究员
范泽孟	中国科学院地理科学与资源研究所	研究员
兰冬东	国家海洋环境监测中心	助理研究员
丁晖	环境保护部南京环境科学研究所	副研究员
曹铭昌	环境保护部南京环境科学研究所	副研究员
陈炼	环境保护部南京环境科学研究所	副研究员
吴军	环境保护部南京环境科学研究所	副研究员

第五次国家报告编制专家组第二次会议参会人员名单
（2013年11月4日，北京）

姓 名	单 位	职称/职务
张文国	环境保护部自然生态保护司生物多样性保护处	处长
卢晓强	环境保护部自然生态保护司生物多样性保护处	博士
朱留财	环境保护部对外合作中心	研究员
丁晖	环境保护部南京环境科学研究所	副研究员
陈怡	国家应对气候变化战略研究和国际合作中心	博士
张志翔	北京林业大学	教授
王军	国土资源部土地整治中心	研究员
王建平	水利部发展研究中心	高工/副处长
杨庆文	中国农业科学院作物科学研究所	研究员
张国良	中国农科院农业环境与可持续发展研究所	研究员
师荣光	农业部农业生态与资源保护总站	副研究员
马月辉	中国农业科学院北京畜牧兽医研究所	研究员
卢欣石	北京林业大学	教授
王志兴	中国农业科学院生物技术研究所	研究员
李丽	对外经济贸易大学	助研/主任
李宁	海关总署监管司	主任科员
张广领	工商行政管理总局市场司	干部
李明福	中国检验检疫科学研究院	研究员
徐基良	北京林业大学	副教授
张明祥	北京林业大学	教授
吴波	中国林业科学研究院	研究员
李迪强	中国林业科学研究院	研究员
席建超	中国科学院地理科学与资源研究所	副研究员
马克平	中国科学院植物研究所	研究员
郭良栋	中国科学院微生物研究所	研究员
徐卫华	中国科学院生态环境研究中心	副研究员
兰冬东	国家海洋环境监测中心	助理研究员

姓　名	单　位	职称/职务
陈彬	国家海洋局第三海洋研究所	研究员/室主任
陈美兰	中国中医科学院	副研究员
张良	国务院扶贫办公室开发指导司	调研员
吕晓平	国家濒危物种进出口管理办公室	处长
陈炼	环境保护部南京环境科学研究所	副研究员
崔鹏	环境保护部南京环境科学研究所	副研究员
刘立	环境保护部南京环境科学研究所	助理研究员

第五次国家报告专家研讨会参会人员名单
（2013年11月12日，北京）

姓　名	单　位	职务/职称
张文国	环境保护部自然生态保护司生物多样性保护处	处长
卢晓强	环境保护部自然生态保护司生物多样性保护处	博士
徐海根	环境保护部南京环境科学研究所	副所长/研究员
马克平	中国科学院植物研究所	研究员
王新	环境保护部对外合作中心	处长/研究员
欧阳志云	中国科学院生态环境研究中心	研究员
崔国发	北京林业大学	教授
唐小平	国家林业局勘探设计院	研究员
曹铭昌	环境保护部南京环境科学研究所	副研究员
陈炼	环境保护部南京环境科学研究所	副研究员

第五次国家报告编制专家组第三次会议参会人员名单
（2013年12月24日，北京）

姓　名	单　位	职称/职务
张文国	环境保护部自然生态保护司生物多样性保护处	处长
卢晓强	环境保护部自然生态保护司生物多样性保护处	博士
徐海根	环境保护部南京环境科学研究所	副所长/研究员
朱留财	环境保护部对外合作中心	研究员
王新	环境保护部对外合作中心	处长/研究员
王建平	水利部发展研究中心	高工/副处长
王军	国土资源部土地整治中心	研究员
杨庆文	中国农业科学院作物科学研究所	研究员
张国良	中国农业科学院农业环境与可持续发展研究所	研究员
樊恩源	中国水产科学研究院	研究员
卢欣石	北京林业大学	教授
王志兴	中国农业科学院生物技术研究所	研究员
李丽	对外经济贸易大学	助理研究员/主任

姓　名	单　位	职称/职务
李宁	海关总署监管司	主任科员
张广领	国家工商行政管理总局市场司	干部
李明福	中国检验检疫科学研究院	研究员
刘增力	国家林业局	副处长
席建超	中国科学院地理科学与资源研究所	副研究员
马克平	中国科学院植物研究所	研究员
解焱	中国科学院动物研究所	副研究员
王新歌	中国科学院地理科学与资源研究所	硕士
兰冬东	国家海洋环境监测中心	助理研究员
陈美兰	中国中医科学院	副研究员
吕晓平	国家濒危物种进出口管理办公室	处长
丁晖	环境保护部南京环境科学研究所	副研究员
崔鹏	环境保护部南京环境科学研究所	副研究员
刘立	环境保护部南京环境科学研究所	助理研究员

附表 1　国家生物多样性战略与行动计划的进展评估

总体目标：
1. 到 2015 年，力争使重点区域生物多样性下降的趋势得到有效遏制。
2. 到 2020 年，努力使生物多样性的丧失与流失得到基本控制。
3. 到 2030 年，使生物多样性得到切实保护。

优先领域	规定的行动	已开展的行动	进展评估
一、完善生物多样性保护与可持续利用的政策与法律体系	**行动 1**　制定促进生物多样性保护和可持续利用的政策	• 已制定并落实有利于生物多样性保护的鼓励措施（详见本书第 2 章 2.3.10 节）。 • 但在价格、税收、信贷、贸易、土地利用和政府采购等方面的政策有待完善	●
	行动 2　完善生物多样性保护与可持续利用的法律法规体系	• 初步建立了生物多样性保护的法律法规体系（详见本书第 2 章 2.3.1 节）。 • 近年来开展了大量有关保护区、湿地保护、遗传资源管理、外来入侵物种防治、转基因生物安全管理等方面的立法研究，但这些法律法规的出台存在较大的难度	●
	行动 3　建立健全生物多样性保护和管理机构，完善跨部门协调机制	• 分别于 1993 年和 2003 年成立了中国履行《生物多样性公约》工作协调组和生物物种资源保护部际联席会议制度，为组织实施好 2010 国际生物多样性年的相关活动，中国政府成立了"2010 国际生物多样性年中国国家委员会"，2011 年，国务院批准将"2010 国际生物多样性年中国国家委员会"更名为"中国生物多样性保护国家委员会"。这三个由环保部牵头、各部门参与的协调机制在推进中国生物多样性保护工作中发挥着重要作用。 • 大部分省（自治区、直辖市）人民政府加强了环保、农业、林业、海洋等涉及生物多样性保护的机构建设，并成立了跨部门的协调机制	●
二、将生物多样性保护纳入部门和区域规划、计划，促进持续利用	**行动 4**　将生物多样性纳入部门和区域规划、计划	• 发展和改革委员会、教育、科技、国土、水利、农业、扶贫等部门在制定本部门规划时都考虑了生物多样性保护的要求（详见本书第 3 章）。 • 各省（自治区、直辖市）均在制定本地区生物多样性保护战略与行动计划，辽宁省发布了辽河流域生物多样性保护战略与行动计划，其中已有 7 个省份发布了省级战略与行动计划的实施有待进一步加强	●

优先领域	规定的行动	已开展的行动	进展评估
二、将生物多样性保护纳入部门和区域规划、促进持续利用	行动5 保障生物多样性的可持续利用	• 2000年以来，组织开展了生态省、市、县创建活动。目前，已有15个省（区、市）开展生态省建设，13个省颁布生态省建设规划纲要，1 000多个县（市、区）开展生态县建设。"十一五"以来，命名了38个国家生态县（市、区），建成1 559个国家生态乡镇和238个国家级生态村。 • 生物多样性影响评价试点工作有待拓展，各行业中有利于生物多样性保护的生产反消费方式待有结和推广。	
	行动6 减少环境污染对生物多样性的影响	• 中国政府将主要污染物排放总量显著减少作为经济社会发展的约束性指标，着力解决突出环境问题。2000—2010年，主要污染物年均浓度总体呈下降趋势。近10年来，单位GDP污染物排放强度大幅下降了55%以上。2004年以来，单位GDP二氧化碳排放强度下降了15.2%。 • 颁布并实施《重点流域水污染防治规划》（2011—2015），深入推进江河湖泊污染治理。全国七大水系好于III类水质比例由2005年的41%提高到2012年的64%，劣V类水质比例由27%下降到12.3%；加强地下水污染防治，积极落实《全国地下水污染防治规划》，编制《华北平原地下水污染防治工作方案》。 • 颁布并实施《重点区域大气污染防治"十二五"规划》，发布新修订的《环境空气质量标准》。2012年，中央财政补助10.9亿元，支持《重点区域大气污染防治"十二五"规划》中15个重点城市实施燃煤锅炉综合整治工程，共改造燃煤锅炉28 997蒸t，其中除生设施改造15 406蒸t，清洁能源替代13 591蒸t。 • 截至2012年年底，《全国危险废物和医疗废物集中处置设施建设规划》确定的57个危险废物集中处置设施建设项目中，已基本建成36个，形成危险废物集中处置能力143万t/a；271个医疗废物集中处置设施建设项目中，已基本建成231个，形成医疗废物集中处置能力42.8万t/a。 • 截至2013年年底，中央财政共安排农村环保专项资金195亿元，4.6万个村庄，8 700多万农村人口受益。2012年，中央财政安排资金13.3亿元用于建设294.6万户农村无害化卫生厕所，超额完成年度工作目标。 • 全国历史遗留铬渣约670万t，多数堆存达二十年、甚至五十多年。从2005年年底启动治理工作，截至2012年年底，历史遗留铬渣基本处置完毕。其中，2012年一年处置铬渣230万t，相当于前6年年平均处置量的3倍。 • 但单位GDP污染物排放强度仍然很高，废水排放量仍在增加。	

附表1 国家生物多样性战略与行动计划的进展评估

优先领域	规定的行动	已开展的行动	进展评估
三、开展生物多样性调查、评估与监测	**行动7** 开展生物物种资源和生态系统本底调查	• 已完成七次全国森林资源清查、第二次全国湿地资源调查、第二次全国畜禽遗传资源调查、海洋生物多样性综合专项调查、西南物种和资源调查和编目、云南及周边地区农业地区农业资源调查、全国生态环境十年（2000—2010）变化遥感调查与评估。 • 正在进行第二次全国陆生野生动物资源调查、第二次全国重点保护野生植物资源调查、第四次大熊猫资源调查、中药资源普查试点工作。 • 通过以上调查，建立了国家生物多样性信息管理系统和国家资源标本资源共享平台等信息共享网络体系，但调查的范围和深度有待扩展，制度有待完善	●
	行动8 开展生物遗传资源和相关传统知识的调查与编目	• 截至2012年12月，农作物收集品总量已达42.3万份，比2007年增加了约3万份。为了妥善保存收集的遗传资源，中国政府加强了国家长期库、1座国家复份库、10座国家种质圃、32个国家种质圃、2个试管苗库）进行扩建和设施改造，另一方面新建了7个国家级种质圃。 • 已初步建立以保种场为主、保护区和基因库和辅的畜禽遗传资源保种体系。截至2011年，有效保护了100多个畜禽品种资源。中国的畜禽成纤维细胞库已经成为世界上最大的畜禽体细胞库，细胞库中包括95个地方畜禽品种共计5.8万份细胞。 • 在少数民族地区开展了与生物遗传资源相关的传统知识调查，建立了相关数据库	●
	行动9 开展生物多样性监测和预警	• 已制定典型生态系统和重要生物类群监测技术指南（草案），但这些标准尚未发布，标准化和规范化工作有待加强。 • 初步设计了全国生物多样性监测网络，建立了中国森林生物多样性监测网络，初步建立了鸟类和两栖动物的示范监测网络，正在开展生物多样性监测示范监测工作，但尚未建立覆盖面广、代表性高的全国长期生物多样性监测网络。 • 开展了生物多样性预测预警模型的研究，但尚未建立预测预警技术体系和应急响应机制	●
	行动10 促进生物遗传资源信息化建设和协调生物遗传资源信息共享	建立了以下三个主要的遗传资源数据库系统共享，实现信息共享： • 国家植物种质资源共享平台（http://icgr.caas.net.cn/pt/）。该平台涵盖农作物、多年生和无性繁殖作物、林木（含竹藤花卉）、药用植物、热带作物、重要野生植物及牧草种质资源。 • 家养动物种质资源平台（http://www.cdad-is.org.cn/）。该平台涵盖猪、牛、羊、家禽等遗传资源。 • 微生物资源平台（http://www.cdcm.net/indexAction.action）。该平台拥有16.2万株菌种的信息，菌种资源占中国微生物资源量的40%～45%	●

优先领域	规定的行动	已开展的行动	进展评估
三、开展生物多样性调查、评估与监测	行动11 开展生物多样性综合评估	• 2012年，完成了全国生物多样性评价工作。该项工作以县域为单元，束植物和3 865种野生脊椎动物的县域分布数据，基本掌握了全国陆域生物多样性现状、空间分布特征及主要威胁因素，识别了全国生物多样性热点地区，发现全国存在较大的保护空缺，并发布了《中国生物多样性本底评估报告》。首次系统采集了全国34 039种野生维管 • 2011年，启动了"全国生态环境十年（2000—2010）变化遥感调查与评估项目"。项目的总体目标是全面掌握十年来全国生态系统分布、格局、质量、生态服务功能等变化特点和演变规律。 • 已发布《生物多样性资源经济价值评价技术导则》，正在制定生态系统服务功能评价技术指南，并开展了大量有关生物多样性经济价值评估的试点工作。 • 2013年9月，发布了《中国生物多样性红色名录 高等植物卷》，正在制定《中国生物多样性红色名录 脊椎动物卷》。	●
四、加强生物多样性就地保护	行动12 统筹实施和完善全国自然保护区规划	• 1999年中国颁布了《中国自然保护区发展规划纲要》（1996—2010），2003年批准了《全国湿地保护工程规划》（2004—2030），目前正在开展"全国自然保护区发展规划"编制工作。 • 加强了生物多样性保护优先区域内的自然保护区建设，优化保护区的空间布局，提高整体保护能力。 • 2006年以来，中国与俄罗斯设立了专门的政府间跨界自然保护区与生物多样性保护工作组，每年定期召开会议，《中俄关于兴凯湖自然保护区协定》等六次工作会议。双方签订了《中俄黑龙江流域跨界自然保护区与俄罗斯巴斯达克、大赫黑契尔、兴安斯基和博龙基等保护区分别签订了协议。2013年，中俄签订野生东北虎保护协议，加快跨境生态通道建设，建立东北虎跨境自然保护区。 • 2009年，中老双方建立了第一个联合保护区域——"中国西双版纳尚勇－老挝南塔南木哈联合保护区"。2012年年初，中老双方划定"中国动腊腊曼庄－老挝丰沙里一老挝南木哈联合保护区域"。	●
	行动13 加强生物多样性保护优先区域的保护	• 在《中国生物多样性保护战略与行动计划》中，中国划定了35个生物多样性保护优先区域，加强了生物多样性保护优先区域内的自然保护区建设，优化保护区的空间布局，提高整体保护能力。 • 正在研究制定优先区域的保护规划、政策、制度和相关措施。	●

附表1　国家生物多样性战略与行动计划的进展评估

优先领域	规定的行动	已开展的行动	进展评估
四、加强生物多样性就地保护	行动14 开展自然保护区规范化建设，提高保护区管理质量	• 发布了《国家级自然保护区总体规划大纲》《自然保护区总体规划技术规程》和《国家级自然保护区规范化建设和管理导则（试行）》《自然保护区生态旅游规划技术规程》。基于这些规范和标准，2008年起，开展了国家级自然保护区管理评估。到2012年，完成了所有国家级自然保护区的评估。 • 加强了国家级自然保护区的规范化建设。截至2013年年底，全国共建立国家级自然保护区407个，面积约94.0万 km²，占陆地国土面积的9.8%。 • 强化管理设施，完善管理措施，强化监督措施。占全国自然保护区总面积的64.3%。 • 环保、林业、农业等主管部门多次举办自然保护区管理培训班，针对自然保护区政策法规、规范化管理、规划编制、能力建设项目设计、开发活动监管、管理信息系统建设、资源本底调查等内容进行培训。 • 2007—2012年，环境保护部等部门多次开展了自然保护区专项执法检查，防止不合理的开发建设活动对自然保护区的冲击和破坏	●
	行动15 加强自然保护区外生物多样性的保护	• 继续实施了天然林资源保护、退耕还林还草、防护林体系建设、湿地保护与恢复等重点生态工程（详见本书第2章2.3.6节），强化草原生态系统保护与恢复（详见本书第3章3.7节），加强了海洋生物多样性保护（详见本书第3章3.14节），取得明显成效。 • 2012年，中国启动了全国极小种群野生植物拯救保护工程，对120种极小种群野生植物开展为期5年的拯救保护行动。该行动的实施，将有效改善最濒危的珍稀植物的生存状态	●
	行动16 加强畜禽遗传资源保种场和保护区建设	• 初步建立了以保种场为主，保护区和基因库为辅的畜禽遗传资源保种体系。截至2012年8月，共建成150个国家级保种场、保护区和基因库，抢救了五指山猪、矮脚鸡、晋江马等一批濒临灭绝的畜禽品种，有效保护了100多个重点品种。 • 公布了《国家级畜禽遗传资源保护名录》，对138个珍贵、稀有、濒危的畜禽品种实施重点保护	●
五、科学开展生物多样性迁地保护体系建设	行动17 科学合理地开展物种迁地保护体系建设	• 目前已建有各级各类植物园200个，收集保存了占中国植物区系2/3的2万个物种；建立了野生植物种质资源保育基地400多处，成立了苏铁等植物种质资源保护中心和兰科植物种质资源库，分别收集保存苏铁类、兰科类植物240余种和500余种。 • 在云南昆明建立了"西南野生生物种质资源库"。截至2013年4月，该质资源库已收集和保存10 096种76 864份植物种子。 • 据不完全统计，中国建立了240多个动物园（含动物展区）、250处质动物地方品种资源场和国家畜禽重点保种场，保存各种家养动物138个品种了各具特色的养动物地方品种资源场和家养动物138个品种。在全国各地建立	●

优先领域	规定的行动	已开展的行动	进展评估
五、科学开展生物多样性迁地保护	行动 18 建立和完善生物遗传资源保存体系	见本表行动 8、行动 10、行动 17 的相关说明	●
	行动 19 加强人工种群野化与野生种群恢复	• 2012 年 3 月，发布了《全国极小种群野生植物拯救保护工程规划》（2011—2015），把 120 种极小种群野生植物作为工程一期拯救保护对象。其中国家 I 级保护植物 36 种，国家 II 级保护植物 26 种，省级重点保护植物 58 种。目前正在制定极小种群野生植物拯救保护工程规划。 • 加强珍稀野生植物的人工培育研究和种源建设，针对松萝、雪莲、珙桐、红豆杉、珍稀兰科植物等 10 种（类）市场需求较大的珍稀野生植物，扶持开展人工培育技术研究和种源建设；加强濒危动物繁育和保护技术研究，大熊猫、朱鹮、扬子鳄、虎、金丝猴、藏羚羊等 50 多种野生动物种群持续扩大。 • 对玉唇兰、杏黄兜兰等特有濒危兰科植物以及极度濒危物种开展了回归自然的前期准备和试验；成功实施了朱鹮、野马、麋鹿、梅花鹿、瑶山鳄蜥、扬子鳄、塔里木马鹿、黄腹角雉 8 种濒危野生动物的放归自然，且实现了自然繁殖，逐步建立起新的野外种群	●
	行动 20 加强生物遗传资源的开发利用与创新研究	详见本书第 2 章 2.3.5 节的相关说明	●
六、促进生物遗传资源及相关传统知识的合理利用与惠益共享	行动 21 建立生物资源及相关传统知识保护、获取和惠益共享的制度和机制	• 为规范遗传资源的开发活动，2008 年 12 月修订的《中华人民共和国专利法》增加了遗传资源披露制度，并明确遗传资源的获取或利用违反有关法律、法规规定的，不授予专利权。 • 正在研究制定生物遗传资源相关传统知识获取与惠益共享的政策，制度及相关信息交换机制	●
	行动 22 建立生物遗传资源出入境管理和检验体系	• 国家质量监督检验检疫总局建立了生物物种出境审批制度，2013 年颁布了《关于加强出入境生物物种资源检验检疫工作的指导意见》，防止中国特有和珍稀的物种资源外流。 • 对物种资源检疫查验体系进行布局，建立了两个物种资源检测鉴定研究中心，六个重点实验室，开展物种资源查验试点。 • 开展了大量科研项目，通过项目的实施，在动植物及人类资源、植物物种资源出入境生物物种资源检验检疫科研方法面取得一定进展。 • 2012 年年底召开了"出入境生物物种资源技术应用培训暨出入境生物物种调查培训班"，为一线工作人员提供专业知识培训。 开了"物种资源现场查验鉴定培训班"，生物遗传资源出入境管理名录尚未颁布，生物遗传资源快速检测鉴定方法、查验能力和条件还较薄弱	●

附表1 国家生物多样性战略与行动计划的进展评估

优先领域	规定的行动	已开展的行动	进展评估
七、加强外来入侵物种和转基因生物安全管理	行动23 提高对外来入侵物种的早期预警、应急与监测能力	详见本书第2章2.3.8节的相关说明	◐
	行动24 建立和完善转基因生物安全评价、检测和监测技术体系与平台	详见本书第2章2.3.9节的相关说明	●
八、提高应对气候变化能力	行动25 制定生物多样性保护应对气候变化的行动计划	● 开展了气候变化对生物多样性的影响研究，正在开发气候变化对生物多样性影响的监测技术，正在制订生物多样性保护应对气候变化的行动计划。 ● 今后有待进一步查明气候变化对生物多样性的有利和不利影响，进一步研究制定相关适应措施	◐
	行动26 评估生物燃料生产对生物多样性的影响	● 开展了个别能源植物种植对生物多样性影响的研究，但尚未建立生物燃料生产的环境安全管理体系	◐
九、加强生物多样性保护科学研究和领域人才培养	行动27 加强生物多样性保护科学研究	详见本书第2章2.3.11节的相关说明，但对生物多样性保护科学研究、基础设施和成果推广的投入有待加强	◐
	行动28 加强生物多样性保护领域的人才培养	详见本书第3章3.2节的相关说明。虽然对生物多样性保护领域人才的培养取得一定的进展，但应进一步加大人才队伍培养的力度，特别是要加大对生物分类人才培养的力度，造就一批领军人才	◐
十、建立生物多样性参与机制与公众参与伙伴关系	行动29 建立公众参与机制	详见本书第2章2.3.12的相关说明	◐
	行动30 推动生物多样性保护伙伴关系	在国家和省级层面建立了较为有效的生物多样性保护伙伴关系（详见本表行动3的相关说明），但有关国际组织、地方社区和市政府组织参与的伙伴关系尚有待加强	◐

注：● 全部实现；◕ 有很大进展；◐ 有较大进展；◑ 有一定进展；○ 没有进展

附表2　中国实现全球《生物多样性战略计划》(2011—2020)和2020年生物多样性目标的进展评估

全球目标	国家目标	国家行动	所取得的成果	国家指标	总体评估及变化趋势
战略目标A：将生物多样性纳入整个政府和社会的主流解决生物多样性丧失的根本原因					
目标1：最迟到2020年，人们认识到生物多样性的价值，并认识到采取何种措施来保护和可持续利用生物多样性	扎实开展环境宣传活动，普及环境保护知识，增强全民环境意识。到2030年，保护生物多样性成为公众的自觉行动	• 在中小学课堂教学中讲授生物多样性相关知识。 • 在高校开展相关专业教育。 • 利用电视、网络、报刊、广播等媒体，以及通过举办培训班、大讲堂、发放培训材料等方式，开展生物多样性保护宣传活动	生物多样性相关知识已被纳入中国中小学课堂教育中。截至2012年，中国1 908所普通高校共培养生物多样性相关专业人才55.6万余人。在开展2010国际生物多样性年中国行动宣传活动中，发放各类国家层面组织各类大型宣传活动40项，宣传活动37万余件，通过各类媒体宣传影响受众8.0亿人次；在地方层面，举办大型宣传保护专题片25部。各地还积极动员自然保护区、动物园、植物园、环境保护宣教与科研机构以及报刊、网络等媒体共发放宣传材料约2万余件，面向大中小学生、公众开展了系列宣传，公众的参与热情高涨，生物多样性保护意识有了明显的提高，生物多样性保护获得广泛认同	不同年份通过Google或百度检索到有关中国生物多样性的条目	呈增加趋势
目标2：最迟到2020年，生物多样性的价值已被纳入国家和地方发展及扶贫战略及规划进程，并正在被酌情纳入国民经济核算体系和报告体系	把资源消耗、环境损害、生态效益纳入经济社会发展评价体系，建立体现生态文明要求的目标体系、考核办法和奖惩机制	• 建立生物多样性经济价值评价的理论和方法。 • 开展生物多样性经济价值评估案例研究。 • 制定体现生态文明要求的目标体系、考核办法和奖惩机制	已发布了《生物遗传资源经济价值评价技术导则》，正在制定生态服务功能经济价值评估。1998年完成了全国生物多样性经济价值评估，2010年完成了全国森林生态系统服务功能生物多样性价值评估，并在不同时期开展了典型区域生物多样性价值评估。这为建立生物多样性经济价值评估的理论和方法奠定了基础。2012年11月召开的中国共产党第十八次全国代表大会对建设生态文明作出了部署，提出了"建设美丽中国"的宏大愿景，要求"坚持节约资源和保护环境的基本国策，坚持节约优先、保护优先、自然恢复为主的方针，着力推进绿色发展、循环发展、低碳发展，形成节约资源和保护环境的空间格局、产业结构、生产方式、生活方式"。中国正在制定体现生态文明的目标体系、考核办法和奖惩机制。生物多样性将被纳入这类目标和考核办法中	无	

附表 2　中国实现全球《生物多样性战略计划》(2011—2020) 和 2020 年生物多样性目标的进展评估

全球目标	国家目标	国家行动	所取得的成果	国家指标	总体评估及变化趋势
目标 3：最迟到 2020 年，消除、淘汰或改革危害生物多样性的激励措施（包括补贴），以尽量减少或避免消极影响，制定和执行有助于保护和可持续利用生物多样性的积极鼓励措施，并遵照《公约》和其他相关国际义务，顾及国家社会经济条件	• 加快建立生态补偿机制，加大对重点生态功能区的均衡性转移支付力度，研究设立国家生态补偿专项资金，推行资源型企业可持续发展准备金制度	• 取消了 553 项高耗能、高污染、资源性产品的出口退税。 • 已有 30 个省（自治区、直辖市）建立了矿山环境恢复治理保证金，累计达 612 亿元，用于生态恢复。 • 为林业和生态保护重点工程提供补助。 • 建立森林生态效益补偿基金。 • 初步建立国家重点生态功能区生态补偿机制	(1) 1999 年起实施退耕还林工程，截至 2012 年年底，中央已累计投入 3 247 亿元，2 279 个县 1.2 亿农民直接受益，户均已累计获得 7 000 元政策补助。 (2) 2000 年中国政府在 17 个省启动天然林资源保护工程，国家给予森林管护补助、造林育林补助和社会性支出补助，截至 2010 年年底，一期工程累计投入 1 186 亿元，天然林保护一期工程总投资将有 2 440 亿元。 (3) 退牧还草工程于 2003 年起在 8 省区实施，国家给予草原围栏建设资金补助和饲料粮补助，退牧还草工程 2003—2012 年中央累计投入资金 175.7 亿元，工程惠及 174 个县、450 多万名牧民。 (4) 2004 年建立中央森林生态效益补偿基金，每年达 30 亿元。 (5) 2008 年中央财政设立国家重点生态功能区转移支付资金，2013 年转移支付范围包括 492 个县域和 1 367 个禁止开发区域，转移支付资金达到 423 亿元。 (6) 2006 年以来，"十一五"和"十二五"规划、建立湿地保护财政补助政策，实施了 500 多个湿地保护与恢复项目，全国因此每年新增湿地保护面积 3 000 多 km^2	主要林业重点工程投资	✓ 天然林资源保护工程、野生动植物保护与自然保护区建设工程、湿地保护工程的资金从 2001 年的 97.0 亿元增加到 2011 年的 217.7 亿元，年均增长 13.7%

全球目标	国家目标	国家行动	所取得的成果	国家指标	总体评估及变化趋势
目标 4：最迟到 2020 年，所有级别的政府、商业和利益相关方都已采取措施，实现或执行了可持续生产和消费计划，并将利用自然资源造成的影响控制在安全生态限值范围内	• 到 2015 年，资源节约型、环境友好型社会建设取得重大进展。 • 着力推进绿色发展、循环发展、低碳发展，形成节约资源和保护环境的空间格局、产业结构、生产方式、生活方式	• 将主要污染物排放总量显著减少作为国民经济社会发展的约束性指标，实施污染减排工程。 • 严格执行环境影响评价制度	（1）2006年以来，中国工业废水中化学需氧量、废气中二氧化硫、工业粉尘、工业固体废物的排放量呈下降趋势。近10年，单位GDP污染物排放强度大幅下降了55%以上。2004年以来，单位GDP二氧化碳排放强度下降15.2%。 （2）严格执行环境影响评价制度，2008年以来，采取"区域限批"、"行业限批"等措施，总投资1.1万亿元涉及高污染、高能耗、消耗资源性、低水平重复建设和产能过剩项目332个，这对调整产业结构、优先经济增长发挥了重要作用	污染物排放减少情况	◔ 总体减少，但废水排放在增加
				单位GDP污染物排放量	◔ 近10年来下降55%以上
				单位GDP二氧化碳排放量	◔ 2004年以来下降15.2%
				可持续消费	⋯

附表 2　中国实现全球《生物多样性战略计划》(2011—2020) 和 2020 年生物多样性目标的进展评估

全球目标	国家目标	国家行动	所取得的成果	国家指标	总体评估及变化趋势
战略目标 B：减少生物多样性的直接压力和促进可持续利用					
目标 5：到 2020 年，使所有自然生境，包括森林的丧失速度至少减少一半，并在可行情况下大幅度降低接近零，同时显著减少退化和破碎化程度	● 到 2015 年森林覆盖率提高到 21.66%，森林蓄积量比 2010 年增加 6 亿 m³。 ● 到 2020 年，全国草原退化趋势得到基本遏制，草原生态环境明显改善。 ● 到 2020 年，近海生态环境恶化趋势得到根本扭转，海洋生物多样性下降趋势得到基本遏制。 ● 到 2020 年，水域生态环境逐步得到修复，渔业资源衰退和濒危物种数量增加的趋势得到基本遏制	● 实施林业重点生态工程。 ● 大力开展水土流失综合治理工程。 ● 开展草原生态系统保护与恢复。 ● 实施湿地保护与恢复工程。 ● 开展滨海湿地生态修复与重建	(1) 自启动林业重点工程以来，全国森林资源持续快速增长，共完成造林面积 48.2 万 km²，森林面积较 10 年前增长了 23.0%；森林蓄积量增长了 21.8%，森林覆盖率上升了 3.8 个百分点；促进了野生物种栖息地的恢复和物种数量、种类的增加。2004—2009 年，全国沙化土地年均净减少 1 717 km²，5 年间中度、重度和极重度沙化土地面积共减少 3.59 万 km²，沙化程度减轻，年入黄泥沙减少 3 亿多 t。 (2) 在一些重点地区大力开展水土流失综合治理工程。2009—2012 年全国共治理小流域 1.2 万条，完成水土流失综合防治面积 27.0 万 km²；继续推进水土流失封育保护，全国累计实施封育保护面积 72 万 km²，其中有 45 万 km² 生态环境得到了初步修复。 (3) 草原生态系统保护与恢复成效显著。与非工程区相比，草原植被盖度提高 11%，草丛高度提高 43.1%，鲜草产量提高 50.7%，草原利用状况有较大改善，2012 年全国 268 个牧区县超载率较 2011 年下降 34.5% ~ 36.2%。但是整体看，草原大部分仍处于超载过牧状态，草原退化、沙化、盐渍化尚未得到有效控制。 (4) 近年来，全国每年新增湿地保护面积超过 3 000 多 km²，恢复湿地近 200 km²，自然湿地保护率每年平均增加 1 个多百分点，约一半的自然湿地得到有效保护	活立木总蓄积量	✓ 1988 年为 105.7 亿 m³，2003 年为 136.2 亿 m³，目前为 145.5 亿 m³
				天然林面积	✓ 1988 年为 88.5 万 km²，2003 年为 115.8 万 km²，目前为 119.7 万 km²
				湿地生态系统面积	✓ 2000—2010 年湿地生态系统面积增加
				草地生态系统面积	✗ 2000—2010 年草地生态系统面积减少
				天然草原产草量	✓ 2005—2012 年，年均增长 1.6%
				沙化土地面积	✓ 2005—2010 年，全国沙化土地年均净减少 1 717 km²
				生态退化	…

全球目标	国家目标	国家行动	所取得的成果	国家指标	总体评估及变化趋势
目标 6：到 2020 年，以可持续和合法的方式管理和捕捞所有鱼类、无脊椎动物及水生植物，并采用基于生态系统的方式，避免过度捕捞，对所有枯竭物种制订了恢复计划和措施，使渔业对受威胁鱼群和脆弱生态系统不产生有害影响，物种和生态系统的影响在安全的生态限值范围内	● 到 2020 年，水域生态环境逐步得到修复，渔业资源衰退和濒危物种数目增加的趋势得到基本遏制。 ● 到 2020 年，近海生态环境恶化趋势得到根本扭转，海洋生物多样性下降趋势得到基本遏制	● 实施湿地保护与恢复工程。 ● 开展滨海湿地生态修复与重建。 ● 开展水生生物资源增殖放流。 ● 加强海洋牧场建设	(1) 加强水生生物自然保护区建设。全国已建立各类水生野生动植物及水域生态系统类型自然保护区 200 多处。 (2) 加强水产种质资源保护区建设。划定了 368 个国家级水产种质资源保护区，对 300 多种国家重点保护经济水生动植物和地方珍稀特有水生物种及其栖息繁衍场所提供保护。 (3) 实施大量湿地保护和恢复项目，自然湿地保护率平均每年增加 1 个多百分点。但渔业对水生生物多样性的影响还有待进一步研究	海洋营养指数	✓ 自 1997 年以来持续增加，但仍处于较低水平
				鱼类红色名录指数	✗ 1998 年至 2004 年，中国淡水鱼类的红色名录指数下降
				渔业对生物多样性的影响	⋯

附表2 中国实现全球《生物多样性战略计划》(2011—2020) 和 2020 年生物多样性目标的进展评估

全球目标	国家目标	国家行动	所取得的成果	国家指标	总体评估及变化趋势
目标7：到2020年，农业、水产养殖业及林业用地实现可持续管理，确保生物多样性得到保护	• 到2020年，全国森林保有量达到223万 km² 以上，比2010年增加约22.3万 km²；全国森林蓄积量增加到150亿 m³ 以上，比2010年增加约12亿 m³。 • 到2020年，畜牧业生产方式不断转变，草原可持续发展能力有效增强。 • 到2020年，捕捞能力和捕捞产量与渔业资源可承受能力大体相适应	• 实施测土配方施肥补贴项目。 • 开展生态农业与农村新能源县建设。 • 推动有机农业的发展。 • 开展生态县创建活动。 • 实施天然林资源保护工程。 • 强化草原生态系统保护与恢复	(1) 针对过量施肥、盲目施肥、肥料利用率偏低等问题，启动了测土配方施肥补贴项目，通过项目的实施，初步摸清了所有农业县耕地土壤养分状况，基本掌握了主要作物需肥规律，科学制订了施肥方案，普及了科学施肥技术，对粮食增产、节约成本和控制污染起到重要作用。 (2) 以农业生态高效、农民持续增收、农村环境改善为目标，以秸秆综合利用、农村沼气建设、农村太阳能综合利用、生态循环农业基地建设为重点，开展生态农业与农村新能源县建设，提升农业可持续发展能力。目前中国农村沼气用户4 100多万户，受益人口达1.5亿多人。 (3) 积极推动有机农业的发展，到2012年，中国共有2万 km² 有机农业土地，排名亚洲第一。 (4) 已有15个省（区、市）开展生态省建设，13个省颁布生态省建设规划纲要，1 000多个县（市、区）开展生态县建设。"十一五"以来，命名了38个国家生态县（市、区），建成1 559个国家生态乡镇和238个国家级生态村。 (5) 实施天然林资源保护工程以来，少采伐木材2.2亿 m³，森林面积净增加10万 km²，森林覆盖率增加3.8个百分点，森林蓄积量增加7.25亿 m³。 (6) 实施了退牧还草等工程，与非工程区相比，草原植被盖度提高11%，草丛高度提高43.1%，鲜草产量提高50.7%，草原利用状况有较大改善	活立木总蓄积量	◡ 1988年为 105.7亿 m³，2003年为 136.2亿 m³，目前为 145.5亿 m³
				天然草原产草量	◡ 2005—2012年，年均增长1.6%
				农林渔业对生物多样性的影响	⋯

全球目标	国家目标	国家行动	所取得的成果	国家指标	总体评估及变化趋势
目标 8：到 2020 年，污染，包括营养物过剩造成的污染被控制在不对生态系统功能和生物多样性构成危害的范围内	• 到 2015 年，主要污染物排放总量显著减少，与 2010 年相比化学需氧量、二氧化硫排放分别减少 8%，氨氮、氮氧化物排放分别减少 10%。 • 到 2020 年，单位国内生产总值能源消耗和二氧化碳排放大幅下降，主要污染物排放总量显著减少	• 将主要污染物排放总量显著减少作为国民经济社会发展的约束性指标，实施污染减排工程。 • 严格执行环境影响评价制度。 • 开展废物控制与综合利用	（1）2006 年以来，中国工业废水中化学需氧量、废气中二氧化硫、烟尘、工业粉尘、工业固体废物的排放量呈下降趋势。近 10 年，单位 GDP 污染物排放强度大幅下降了 55% 以上。2004 年以来，单位 GDP 二氧化碳排放强度下降 15.2%。 （2）采取"区域限批"、"行业限批"等措施，2008 年以来，国家层面拒批 332 个，总投资 1.1 万亿元涉及高污染、高能耗、资源消耗性、低水平重复建设和产能过剩项目。 （3）全国烟气脱硫机组装机容量及其占全部火电机组容量的比例、全国城市污水处理率、工业固体废物综合利用率大幅增长。但污染物排放总量仍偏高，近两年工业固体废物综合利用率有所下降	污染物排放减少情况	◯ 总体减少，但废水排放在增加
				单位 GDP 污染物排放量	◯ 近 10 年来下降 55% 以上
				单位 GDP 二氧化碳排放量	◯ 自 2004 年以来下降 15.2%
				烟气脱硫机组装机容量占全部火电机组容量的比例	◯ 从 2007 年的 48% 增加到 2012 年的 92%，年均增长 14%
				废物城市污水处理率	◯ 从 2004 年的 45.6% 提高到 2010 年的 82.3%，年均增长 10.4%
				工业固体综合利用率	◯ 自 2004 年以来年均增长 2.2%

附表2　中国实现全球《生物多样性战略计划》(2011—2020) 和2020年生物多样性目标的进展评估

全球目标	国家目标	国家行动	所取得的成果	国家指标	总体评估及变化趋势
目标9：到2020年，查明外来入侵物种及其入侵路径并确定其优先次序，优先控制或根除，制定措施控制入侵路径，以防止外来入侵物种的引进和加以管理，并防止外来入侵物种种群建立	• 到2020年，全国林业有害生物成灾率控制在4%	• 初步明确了需要优先控制的外来入侵物种。 • 强化监测预警能力建设。 • 开展外来入侵物种清除活动。 • 加强综合防控技术研究。 • 开展宣传和培训	(1) 成立了跨部门的外来入侵物种防治协作组，18个省（自治区、直辖市）成立了外来入侵物种管理办公室或建立了联席会议制度。制订了40种重大外来入侵物种应急防控技术指南，发布了《第二批外来入侵物种名单》，初步确定了需要优先控制的外来入侵物种。 (2) 完善了进出境植物检疫标准体系，初步形成了林业有害生物监测预警网络体系和农业外来入侵物种监测预警网络。 (3) 开展了以豚草、水花生等20余种外来入侵物种为重点的集中灭除，有效控制了外来入侵物种的扩散和蔓延。 (4) 开展了外来入侵物种调查和防治技术示范项目。 (5) 利用广播、电视、报刊、网络等多种媒体，开展了外来入侵物种防治技术与管理知识宣传工作。但外来入侵物种数上升的趋势仍没有得到有效遏制，造成的危害还在加剧	每20年新发现的外来入侵物种种数	✗　新出现的外来入侵物种种数逐步上升的趋势，1950年后的60年间，新出现212种外来入侵物种，占外来入侵物种总种数的53.5%

123

全球目标	国家目标	国家行动	所取得的成果	国家指标	总体评估及变化趋势
目标10: 到2015年，尽可能减少由气候变化或海洋酸化对珊瑚礁和其他脆弱生态系统的多重人为压力，维护它们的完整性和功能	● 到2020年，单位国内生产总值能源消耗和二氧化碳排放大幅下降。 ● 到2020年，基本建成布局合理、功能完善的自然保护区体系，国家级自然保护区功能稳定，主要保护对象得到有效保护	● 调整产业结构，扎实推进污染减排。 ● 实施林业重点工程，保护脆弱生态系统。 ● 加强自然保护区建设，完善就地保护体系	（1）中国政府将主要污染物排放总量显著减少作为经济社会发展的约束性指标，2000—2010年，中国主要污染物年均浓度总体呈下降趋势。近10年，单位GDP污染物排放强度大幅下降了55%以上。2004年以来，单位GDP二氧化碳排放强度下降15.2%。 （2）实施了林业重点工程，森林资源持续快速增长，森林面积较10年前增长了23.0%；森林覆盖上升了3.8个百分点，森林蓄积量增长了21.8%，有效保护了脆弱生态系统。 （3）建立了以自然保护区为主体，风景名胜区、森林公园、自然保护小区、农业野生植物保护点、湿地公园、地质公园、海洋特别保护区、种质资源保护区为补充的保护体系。截至2013年年底，全国共建立各种类型、不同级别的自然保护区2 697个，面积约146.3万 km²，自然保护区面积约占全国陆域面积的14.8%	单位GDP污染物排放量	⊘ 近10年来下降55%以上
				单位GDP二氧化碳排放量	⊘ 自2004年以来下降15.2%
				退耕还林和天然林资源保护工程区森林蓄积量	⊘ 持续增加
				退耕还林和天然林资源保护工程区水土流失	⊘ 总体呈下降趋势
				珊瑚礁的生物多样性	⋯
				气候变化对生物多样性的影响	⋯

附表 2　中国实现全球《生物多样性战略计划》(2011—2020) 和 2020 年生物多样性目标的进展评估

全球目标	国家目标	国家行动	所取得的成果	国家指标	总体评估及变化趋势
战略目标 C: 通过保护生态系统、物种和遗传多样性, 改善生物多样性的现况					
目标 11: 到 2020 年, 至少有 17% 的陆地和内陆水域以及 10% 的沿海和海洋区域, 尤其是对生物多样性具有特殊重要性的区域和生态系统服务的区域, 通过有效公平管理的、生态上有代表性的和连通性好的保护区系统和其他基于区域的有效保护措施得到保护, 并被纳入更广泛的陆地景观和海洋景观	到 2015 年, 陆地自然保护区总面积占陆地国土面积的比例维持在 15% 左右, 使 90% 的国家重点保护物种和典型生态系统类型得到保护。	加强以保护区为主的就地保护	(1) 建立了以自然保护区为主体, 森林公园、自然保护小区、风景名胜区、湿地公园、地质公园、农业野生植物保护点、海洋特别保护区、种质资源保护区为补充的保护体系。截至 2013 年年底, 全国共建立自然保护区 2 697 个, 面积约 146.3 万 km², 自然保护区面积约占全国陆域面积的 14.8%。其中海洋保护区 240 多处。截至 2012 年年底, 建立国家森林公园 2 855 处, 规划总面积 17.4 万 km², 已建立国家级风景名胜区 225 处, 面积约 10.4 万 km²; 建立了自然保护小区 5 万多处, 面积 1.5 万多 km²; 已建湿地公园 468 处; 建立国家级农业野生植物保护点 179 个; 已建水产种质资源保护区 368 个, 面积 15.2 万 km²。但自然保护区的生态代表性、管理有效性还有待提高, 海洋保护区的数量和面积还较低。	自然保护区的数量	☑ 1990 年为 606 个, 2000 年为 1 227 个, 2013 年为 2 697 个, 年均增长约 8.9%
	海洋保护区占管辖海域面积的比例由 2010 年的 1.1% 提升到 2015 年的 3%。	实施了海洋伏季休渔制度和长江、珠江禁渔期制度, 保护了水生生物多样性	(2) 自 1995 年以来, 实施了海洋伏季休渔制度, 实施区域为渤海、黄海、东海及北纬 12 度以北的南海 (含北部湾) 海域, 休渔时间约为三个月; 自 2002 年起在长江、2011 年起在珠江实施禁渔期, 禁渔期为三个月。休渔禁渔制度的实施, 有力地恢复了渔业资源, 保护了水生生物多样性	自然保护区面积占陆地面积的百分比	☑ 1990 年为 4.0%, 2000 年为 9.9%, 2012 年为 14.9%, 年均增长 8.1%
	到 2020 年, 基本建成布局合理、功能完善的自然保护区体系, 国家级自然保护区功能稳定, 主要保护对象得到有效保护			保护区的生态代表性	⋯
				保护区的管理有效性	⋯

全球目标	国家目标	国家行动	所取得的成果	国家指标	总体评估及变化趋势
目标12：到2020年，防止了已知受威胁物种的灭绝，且其保护状况，尤其是其中减少最严重的物种的保护状况改善得到改善和维持	• 到2015年，使80%以上的就地保护能力不足和野外现存种群数量极小的受威胁物种得到有效保护。 • 到2020年，国家级自然保护区功能稳定，主要保护对象得到有效保护。 • 到2020年，使大多数珍稀濒危物种种群得到恢复和增殖，生物物种受威胁的状况进一步缓解	• 加强保护区的建设与管理。 • 合理开展迁地保护。 • 加强受威胁物种的科学研究。 • 促进国际合作。 • 开展公众教育	(1) 有关保护区建设管理情况见目标11的相关说明。 (2) 中国目前已建有各级各类植物园200个、野生植物种质资源保育基地400多处，收集保存了占中国植物区系2/3的2万个物种；还建立了240多个动物园，250处野生动物拯救繁育基地。这些迁地保护设施在保护受威胁物种方面发挥了重要作用。 (3) 有关科学研究和国际合作方面的情况，见目标19的相关说明。 (4) 有关公众教育的情况见目标1的有关说明。	红色名录指数	✗ 1998—2004年，淡水鱼类红色名录指数下降；1996—2008年，兽类红色名录指数下降；1988—2012年，鸟类红色名录指数略有下降
目标13：到2020年，保持了栽培植物、养殖和驯养动物及野生近缘物种以及文化上和社会经济上宝贵的物种的遗传多样性，同时制定并执行了减少遗传侵蚀和保护其遗传多样性的战略	• 到2020年，努力使生物多样性的丧失与流失得到基本控制，基本建成布局合理、功能完善的自然保护区体系，主要保护对象得到有效保护。 • 修订《国家畜禽遗传资源保护名录》，对列入保护名录的珍贵、稀有、濒危的畜禽遗传资源实施重点保护，确保品种不丢失，主要经济性状不降低	• 制定并实施遗传资源保护规划。 • 建设遗传资源就地保护点。 • 建设遗传资源保存库，开展遗传资源的收集、保存和利用研究	(1) 制定并颁布了有关遗传资源保护的战略，包括《全国生物物种资源保护与利用规划纲要》和《全国畜禽遗传资源保护利用"十二五"规划》等。 (2) 建立了国家级水产野生植物种质资源保护区179个和国家级野生植物资源保护区368个，保护了一批珍稀遗传资源。 (3) 建立了作物遗传资源保存体系；建立了150个国家级畜禽遗传资源达42.3万份；建立了150个国家级畜禽保种场、保护区和基因库，有效保护了100个重点畜禽资源；构建了海洋生物资源库，如大型海藻种质资源库和海洋微生物保藏中心。尽管中国在遗传资源丧失和保护方面做了大量工作，但遗传资源丧失的趋势仍没有得到有效遏制	地方品种资源	✗ 据估计，遗传资源丧失和流失的趋势仍没有得到有效遏制

附表2 中国实现全球《生物多样性战略计划》(2011—2020)和2020年生物多样性目标的进展评估

战略目标D：增进生物多样性和生态系统服务给人类带来的惠益

全球目标	国家目标	国家行动	所取得的成果	国家指标	总体评估及变化趋势
目标14：到2020年，提供重要服务（包括与水相关的服务）、以及有助于健康、生计和福祉的生态系统得到恢复和保障，同时顾及了妇女、土著和地方社区以及贫困和弱势群体的需要	• 到2020年，生态系统稳定性增强，人居环境明显改善。 • 到2020年，天然草原基本实现草畜平衡，草原植被明显恢复，草原生产能力显著提高。 • 到2020年，近海生态环境恶化趋势得到根本扭转，海洋生物多样性下降趋势得到基本遏制	• 加强保护区的建设与管理。 • 实施林业重点生态工程。 • 开展草原生态保护与恢复。 • 实施湿地生态恢复工程。 • 开展滨海湿地生态修复与重建	(1) 有关保护区的建设和管理情况，见本表目标11的相关说明。 (2) 有关森林、草原、湿地退化生态系统恢复方面的情况，见本表目标5的相关说明。 (3) 通过保护当地生态系统和服务，依赖于当地生物多样性产品和服务的社区福祉也在改善。天然林保护工程、退耕还林工程、京津风沙源治理工程样本县与样本企业森林蓄积量自1999年起呈持续增加趋势。"天保工程"和退耕还林工程样本县水土流失面积总体呈下降趋势，2011年比2000年增加了40.8%，这在一定程度上得益于"天保工程"和退耕还林工程提供功能数量的增加。"天保工程"样本县贫困人口由1997年的395万人下降至2011年的183万人，退耕还林工程由1998年的830万人下降至2008年的570万人	农村居民家庭人均纯收入 重点生态工程区贫困人口数量 森林蓄积量 重点生态工程区水土流失面积	✓ 持续增加 ✓ 持续下降 ✓ 自1999年起持续增加 ✓ 持续下降
目标15：到2020年，通过养护和恢复行动，生态系统的复原力以及生物多样性对碳储存的贡献得到加强，包括恢复了至少15%退化的生态系统，从而有助于减缓和适应气候变化及防止荒漠化	• 到2020年，与2010年相比新增森林面积5.2万km²，森林蓄积量增加11亿m³，森林碳汇增加4.16亿t。 • 到2020年，累计治理"三化"草原165万km²以上，草原植被明显恢复，草原生产能力显著提高。 • 到2020年，水域生态环境逐步得到修复	• 实施林业重点生态工程。 • 开展草原生态保护与恢复。 • 实施湿地生态恢复工程。 • 开展滨海湿地生态修复与重建	• 有关森林、草原、湿地退化生态系统恢复方面的情况，见本表目标5的相关说明	森林蓄积量 重点生态工程区水土流失面积	✓ 自1999年起持续增加 ✓ 持续下降

全球目标	国家目标	国家行动	所取得的成果	国家指标	总体评估及变化趋势
目标16：到2015年，《生物多样性公约关于获取遗传资源并公正和公平分享其利用所产生惠益的名古屋议定书》已经根据国家立法生效并实施	到2020年，生物遗传资源获取与惠益共享制度得到完善	• 推动遗传资源管理条例的制定。 • 支持遗传资源获取与惠益共享制度的研究	在相关科学研究计划的支持下，中国加强遗传资源获取惠益共享的资料收集和制度研究，目前正在推动遗传资源管理条例的制定和《生物多样性公约关于获取遗传资源并公正和公平分享其利用所产生惠益的名古屋议定书》的批准工作	无	
战略目标E：通过参与性规划、知识管理和能力建设，加强《公约》的执行					
目标17：到2015年，各缔约方已经制定，作为政策工具通过和开始执行了一项有效的并有参与性的最新国家生物多样性战略与行动计划	已发布《战略与行动计划》	• 实施《中国生物多样性保护战略与行动计划》（2011—2030）。 • 各地制定并发布区域生物多样性保护战略与行动计划	《中国生物多样性保护战略与行动计划》（2011—2030），于2010年9月15日由国务院第126次常务会议审议通过，2010年9月17日由环境保护部发布实施。该《战略与行动计划》体现了广泛的代表性和参与性，是中国履行《生物多样性公约》工作协调组和生物物种资源保护部际联席会议各成员单位共同努力的结果，也是国内国际众多组织参与其中精诚合作的典范。中国各省（自治区、直辖市）正在制定区域生物多样性保护战略与行动计划。目前，已有7个省发布实施	政策和规划的实施	⏱
目标18：到2020年，生物多样性保护与可持续利用有关的传统知识、创新和做法以及他们对生物资源的习惯性利用得到尊重，并纳入和反映到公约的执行中，这些应与相一致并符合国际义务，国家立法并由各级层次充分和有效地参与	到2020年，进一步健全国内相关传统知识的文献化编目和产权保护制度	• 设立相关项目，开展传统知识的文献化编目和产权保护制度研究	中国政府尊重各民族长期传承下来的传统知识和做法，设立相关项目，开展了传统知识的文献化编目，并支持传统知识产权保护的研究	无	

附表 2　中国实现全球《生物多样性战略计划》(2011—2020) 和 2020 年生物多样性目标的进展评估

全球目标	国家目标	国家行动	所取得的成果	国家指标	总体评估及变化趋势
目标 19：到 2020 年，已经提高、广泛分享和转让并应用了与生物多样性及其价值、功能、状况和变化趋势以及其丧失可能带来的后果的知识、科学基础和技术	• 到 2020 年，全社会研究开发投入占国内生产总值的比重提高到 2.5% 以上，力争科技进步贡献率达到 60% 以上，本国人发明专利年度授权量和国际科学论文被引用数均进入世界前 5 位。 • 扎实开展环境宣传活动，普及环境保护知识，增强全民环境意识	推动生物多样性保护和持续利用方面的科学研究。 促进生物多样性保护领域的国际合作。 大力开展宣传教育活动	（1）中国政府鼓励并支持有关保护和持续利用生物多样性的研究工作，在国家科技支撑计划、国家基础研究发展规划、国家高技术发展计划、国家自然科学基金、公益性行业科研专项等科技计划中设立有关生物多样性保护与可持续利用的项目。这些研究工作形成了一系列有价值、有影响的科研成果，为中国生物多样性保护提供了科技支撑（详见本书第 3 章 3.3 节的相关说明）。 （2）中国在生物多样性国际合作方面开展了积极探索，取得了可喜成果。中国积极参与生物多样性相关公约的谈判，认真履行相关义务，积极参与国际的对外合作与交流渠道。中国与 50 多个国家建立了以政府间合作为主的多元化合作体系，初步形成了以政府积极开展生物多样性领域的南南合作，与国际众多发展中国家签署了生物多样性相关领域的合作协议。 （3）中国大力开展生物多样性方面的宣传教育活动。通过宣传教育，公众的参与热情高涨，生物多样性保护意识有了明显的提高，生物多样性目标 1 的重要性获得广泛认同（详见本表目标 1 的有关说明）	不同年份公开发表的有关生物多样性的论文 不同年份通过网络检索到中国有关生物多样性的条目	✓ 呈逐年迅速增强趋势 ✓ 详见本表目标 1
目标 20：最迟到 2020 年，依照"资源调集战略"商定的进程，用于有效执行《战略计划》而从各种渠道筹集的财务资源将较目前水平有大幅提高	• 拓宽投入渠道，加大国家和地方资金投入，引导社会信贷、国际资金参与生物多样性保护，形成多元化资金投入机制	• 大幅度增加国内投资。 • 在力所能及的条件下，为一些发展中国家提供帮助	中国已投入巨额资金，用于生物多样性保护，具体情况见本表目标 3 的说明。中国在力所能及的条件下，还为一些发展中国家提供帮助	主要林业重点工程投资	✓ 详见本表目标 3

注：国家指标总体评估结论：✓ 表示状况有改善；≈ 表示状况变化很小或基本没有变化；✗ 表示状况在恶化；⋯ 表示没有足够数据。

附表 3　干旱和半干旱地区生物多样性工作方案执行情况

全球目标、子目标和活动	国家目标	国家采取的行动	所取得的成果	所采用的国家或全球指标	总体评估
见"生物多样性公约"网站	《全国防沙治沙规划》（2011—2020）规定，到2020年全国一半以上可治理的沙化土地得到治理，沙区生态状况明显改善，完成沙化土地治理任务20万km²，其中2011—2015年为10万km²，2016—2020年为10万km²	（1）完善了防沙治沙扶持政策。国家出台了包括集体林权制度改革、森林生态效益补偿、林业贷款贴息、造林补贴、草原生态保护补助奖励等一系列支持沙区生态建设和产业发展的政策措施。（2）推进了防沙治沙重点工程建设。中国继续实施京津风沙源治理、"三北"防护林体系建设、退耕还林、退牧还草、草原保护、小流域综合治理等一系列生态建设重点工程，还相继启动了新疆塔里木盆地防沙治沙、石羊河流域防沙治沙及生态恢复、西藏生态安全屏障保护与建设等区域性防沙治沙工程项目，对沙化重点地区和薄弱环节进行集中治理，推动了全国沙区生态状况的持续好转。（3）提高了防沙治沙支撑保障能力。一是进一步提升科技支撑能力。"沙漠化发生规律及其综合防治模式研究"、"中国北方沙漠化过程及其防治"等科研成果获得国家科技进步奖励。一批防沙治沙科研成果和适用技术得到推广应用。二是制定和完善技术标准。制订颁发了《防沙治沙技术规范》《沙化土地监测技术规程》《京津风沙源治理工程区土壤侵蚀监测技术标准》等一批防沙治沙技术标准。三是强化荒漠化沙化监测和沙尘暴应急处置。完成了第四次全国荒漠化沙化监测、建立了重大沙尘暴灾害应急体系，形成了以遥感监测和地面监测为主、信息员测报为辅的沙尘暴灾害监测体系。（4）强化了防沙治沙部门协作机制。从中央到地方各级政府部门成立了防沙治沙组织协调和领导机构，加强对防沙治沙工作的组织、领导和协调工作。（5）落实了防沙治沙目标责任制。按照《中华人民共和国防沙治沙法》的要求，国务院委托，国家林业局与防沙治沙任务较重的北方12个省级政府和新疆生产建设兵团签订了"十一五"防沙治沙目标责任书。防沙治沙目标责任制在中国防沙治沙史上第一次真正实现了中央政府对省级政府防沙治沙目标责任的履责问责，切实提高了各级政府和有关部门防沙治沙责任意识，进一步推动了全国防沙治沙工作。（6）鼓励发展沙区特色产业。为切实推动沙区特色产业发展，国家林业局制定了《关于进一步加快发展沙产业的意见》。各地结合实际，发展特色优势产业，促进沙区经济发展	监测结果显示，"十一五"期间，全国沙化土地年均净减少1 717 km²，五年间中度、重度和极重度沙化土地面积共减少3.6万km²，沙化程度减轻。局部地区的水土流失得到有效控制，土壤侵蚀模数大幅度下降，年入黄泥沙减少3亿多t。以京津风沙源治理工程为例，据专家评估，工程启动十年来，工程区土壤侵蚀模数（水蚀）平均值下降了68.9%，土壤侵蚀面积减少了39.1%，土壤风蚀总量降低了29%，释尘总量减少了16.2%	荒漠生态系统面积的比例及其变化	有好转

附表4 保护区工作方案执行情况

本报告框架有助于掌握对保护区工作方案 13 项关键目标评估的完成情况以及为落实这些评估结果而采取的具体行动。对评估进展可按 0～4 分进行打分：0 分—没有进展；1 分—计划阶段；2 分—初始进展；3 分—实质性进展；4 分—几乎完成或已经完成。各方可以附上相关评估结果，并选择性地按照三个时间段说明所采取的具体行动（2004 年前；2004—2009 年；2010 年至今）。如果问题不适用，则应填写 "N/A"。保护区工作方案国家联络处应在进行和完成评估后或遵循国家报告的报告周期，填写用户名和密码登录后，将信息上传至《生物多样性公约》网站。

国家：	中华人民共和国
完成调查的人员姓名：	
完成调查的人员电子邮件地址：	
完成调查的日期：	（日期）
请简要介绍参与本次调查信息收集的人员	（姓名和相关组织）
1）是否已成立多利益攸关方顾问委员会以执行保护区工作方案？	是
2）是否已为执行保护区工作方案制订行动计划？	是
3）如果是，请提供战略行动计划的 URL 或附上一份 pdf 文档：	http://www.zhb.gov.cn/gkml/hbb/bwj/201009/t20100921_194841.htm
4）如果是，哪个机构对行动计划的执行负主要责任？	环境保护部
5）如果不是，是否在其他与生物多样性有关的行动计划中包含了保护区工作方案的有关行动？（如果是，请提供 URL 或附件）	

1.1 建立和加强国家及区域保护区系统，并形成全球网络，为实现全球目标作出贡献

1）在对贵保护区网络的代表性、全面性和生态差距进行评估方面已取得哪些进展？	3
2）如果有差距评估报告，请提供其 URL（或附上一份 pdf 文档）：	（URL 或附件）
3）是否在保护区系统方面有具体目标和指标？	是
4）如果是，请提供目标和指标的 URL（或附上一份 pdf 文档）：	http://www.zhb.gov.cn/gkml/hbb/bwj/201009/t20100921_194841.htm

5）为改善保护区网络的生态代表性已采取哪些行动？请核查所有行动，并进行简要描述：

√	行动	2004 年以前	2004—2009 年	2010 年至今
	建立新的保护区	√	√	√
	推动和设置一系列不同类型的保护区[即国际自然保护联盟（IUCN）不同类别保护区]			
	扩大并/或重新划定现有保护区边界	√	√	√
	改变保护区的法律状态和/或治理类型	√	√	√
	为改善网络的代表性和全面性采取的其他行动	√	√	√

中国为改善保护区网络及其生态代表性，在不同时期都制定了全国自然保护区发展规划，明确自然保护区空间布局、建立和管理的目标与要求。2004 年以前，我国共建立各种类型、不同级

别的自然保护区 1 999 个，保护区总面积 144.0 万 km²，约占国土面积的 14.4%，其中国家级自然保护区 226 处。2004—2009 年，中国新建自然保护区 542 处，其中新建国家级自然保护区 93 处；2010 年至 2013 年 6 月底，中国新建自然保护区 128 处，其中新建国家级自然保护区 65 处。自然保护区陆地面积占全国陆地面积的 14.9%。

1.2 将保护区融入广泛的陆地和海洋景观及有关部门，以便维护生态结构和功能			
1) 在评估保护区在陆地和海洋景观上的连通性以及融入有关部门方面已经取得了哪些进展？	3		
2) 如果有，请给出对保护区连通性和部门一体化进行评估的 URL（或附上一份 pdf 文档）：			
3) 为改善保护区的连通性和融入部门，已采取了哪些行动？请核查所有行动，并进行简要描述：			
√	行动	2004 年以前	2004—2009 年	2010 年至今
	改变主要连接区域的法律状态和/或治理	√	√	√
	在主要连接区域建立新的保护区	√	√	√
	为改善连通性而改进自然资源管理	√	√	√
	指定连接走廊和/或缓冲区	√	√	√
	为促进连接而建立市场激励机制			
	改变主要利益攸关方对主要连接区域的认识	√	√	√
	改进主要连接区域内部或周围的法律和政策	√	√	√
	恢复主要连接区域内的退化区域	√		
	改变主要连接区域内的土地使用规划、区划和/或缓冲区		√	√
	扫清实现连接和发挥生态功能上的障碍	√		
	将保护区融入到减贫战略中	√	√	√
	为完善连通性和融入部门而采取的其他行动			

中国为完善保护区网络及其生态代表性，在不同时期都制定了全国自然保护区发展规划，明确自然保护区空间布局和生态廊道建设的要求，并使之被纳入部门发展规划中。例如，中国实施了《全国野生动植物保护及自然保护区建设总体规划》，截至 2006 年，已投入资金 26 亿元，用于自然保护区的建设与管理。中国政府实施了"大熊猫及其栖息地保护工程"，在四川、陕西、甘肃建立了大熊猫保护区网络。截至 2010 年，四川省投入资金 2 亿元，大熊猫自然保护区数量增至 41 个，面积达 2.3 万 km²，使全省 50% 以上的大熊猫栖息地和 60% 以上的野生大熊猫在自然保护区中得到有效保护。在世界自然基金会等国际组织的协助下，实施了多项生态廊道建设、社区发展项目，制定了相应的管理计划，启动了多项扶贫项目，提高了自然保护区之间的连通性及其管理水平。近年来，中国与缅甸、越南、老挝等国合作，开展了大湄公河次区域生物多样性廊道项目，在保护区合作、人员培训、边境防火、亚洲象跨境保护等方面取得了实质性进展。2010 年年底，国务院办公厅印发了《关于做好自然保护区管理有关工作的通知》，明确要求严格限制涉及自然保护区的开发建设活动、加强涉及自然保护区开发建设项目管理、强化监督检查。目前，中国正在编制《全国自然保护区发展规划》，将对自然保护区空间布局和发展目标提出新的要求，并纳入到国民经济和社会发展规划中予以实施。

1.3 建立和加强区域网络和跨国界保护区，与跨国界的毗邻保护区进行合作

1）在识别建立跨境保护区和地区网络的机会和确定保护重点方面已取得哪些进展？	3
2）如果有，请给出对跨境保护区和地区网络相关机会进行评估的URL（或附上一份pdf文档）。	http://politics.people.com.cn/GB/1026/10568760.html

3）为加强地区保护区网络和促进跨境保护区建设，已采取哪些行动？请核查所有行动，并进行简要描述：

√	行动	2004年以前	2004—2009年	2010年至今
	建立跨境保护区	√	√	√
	为地区范围保护走廊的建立作出贡献	√	√	√
	参加地区网络的建设	√	√	√
	制定推动跨境保护区的政策	√	√	√
	建立多国合作机制	√	√	√
	为促进地区网络和跨境区域而采取的其他行动	√	√	√

1994年中国与俄罗斯、蒙古签订了《关于建立中、蒙、俄共同自然保护区的协定》，多年来联合开展了大量调查监测、环境教育、经验交流等活动。2006年以来，中国与俄罗斯设立了专门的政府间跨界自然保护区与生物多样性保护工作组，每年定期召开会议，至今已召开六次工作组会议。双方签订了《中俄黑龙江流域跨界自然保护区网络建设战略》、《中俄关于兴凯湖自然保护区协定》等合作文件，黑龙江三江、洪河、八岔岛等自然保护区与俄罗斯巴斯达克、大赫黑契尔、兴安斯基和博龙斯基等保护区分别签订了协议，积极开展合作。2013年，中俄签订野生东北虎保护协议，加快跨境迁徙通道建设，协议约定双方将各自在边境山区建立东北虎自然保护区，并联合开展野生动物监测与科研、环境教育与宣传、保护区立法与执法、生态旅游规划与管理等方面的深入合作。

2009年，中老双方建立了第一个联合保护区域——中国西双版纳尚勇—老挝南塔南木哈联合保护区域，以更好地保护以亚洲象为代表的跨境迁徙野生动物。2012年年初，中老双方签订了第二个联合保护协议，划定中国勐腊曼庄—老挝丰沙里边境联合保护区域。同年12月又签署了新增联合保护区域及合作协议，保护区域地跨中国西双版纳和老挝北部三省。

近年来，中国与缅甸、越南、老挝等国合作，开展了大湄公河次区域生物多样性廊道等项目，在保护区合作、人员培训、边境防火、亚洲象跨境保护等方面取得了实质性进展。

1.4 大力改进基于现场(site-based)的保护区规划和管理

1）在制订保护区管理计划方面取得了哪些进展？	2
2）具有适当管理计划的保护区占总数的百分比是多少？	
3）管理计划覆盖面积占保护区总面积的百分之多少？	
4）请提供一个体现参与性、基于科学的管理计划的最新实例（URL或pdf附件）。	http://www.doc88.com/p-18967633517.html

5）为改善保护区管理规划已采取了哪些行动？请核查所有行动，并进行简要描述：

√	行动	2004年以前	2004—2009年	2010年至今
	为制订管理计划提供指南和工具			
	在管理规划过程中提供培训和/或技术支持	√	√	√

行动			
为保护区制订管理计划	√	√	√
为加强管理规划而修改法律或政策	√	√	√
改进现有管理计划的科学基础	√	√	√
开展保护区资源普查	√	√	√
为改进管理规划而采取的其他行动	√	√	√

《国家级自然保护区总体规划大纲》(2002年)、《自然保护区总体规划技术规程》(2006年)和《自然保护区生态旅游规划技术规程》(2006年),规定了自然保护区总体规划和生态旅游规划的准则、程序和具体要求。

《国家级自然保护区规范化建设和管理导则(试行)》(2009年)提出了自然保护区建设和管理的要求。

1.5 防止和/或减轻对保护区主要威胁的消极影响

1) 在评估保护区受威胁状态以及减轻、预防和恢复的机会方面取得了哪些进展? 3

2) 如果有,请提供对威胁状态以及减轻、预防和恢复机会进行评估的URL(或附上一份pdf文档)。

3) 为减轻或预防保护区遇到的威胁,或恢复退化区域而采取了哪些行动?请检查所有行动,并进行简要描述:

√	行动	2004年以前	2004—2009年	2010年至今
	改变某保护区的状态和/或治理类型	√	√	√
	为预防和减轻威胁而增加员工数量和/或提高员工技能	√	√	√
	在管理计划中加入应对威胁的措施	√	√	√
	为预防或减少威胁而改进管理做法	√	√	√
	增加减少威胁方面的投资	√	√	√
	为应对气候变化的影响而制订计划	√	√	√
	为减轻或预防威胁而改变市场激励机制			
	改进对威胁的监测和监督	√	√	√
	评价与威胁相关行动的成效	√	√	√
	提高与威胁有关的公众意识和行为	√	√	√
	修改与威胁有关的法律和政策			
	恢复退化区域	√	√	√
	制订和/或执行减轻威胁的战略	√	√	√
	为减轻和预防威胁而采取的其他行动	√	√	√

2004年,原国家环保总局下发了《关于加强自然保护区管理有关问题的通知》,要求"涉及自然保护区的建设项目,在进行环境影响评价时,应编写专门章节,就项目对保护区结构功能、保护对象及价值的影响作出预测,提出保护方案,根据影响大小由开发建设单位落实有关保护、恢复和补偿措施"。2008年环境保护部等7部委联合颁布了《关于加强自然保护区调整管理工作的通知》,强化了自然保护区调整的要求,防止开发建设活动对自然保护区产生不利影响。2011年环境保护部发布了《自然保护区生态环境监察指南》,规范自然保护区的生态环境监察工作。环境保护部会同其他有关部门多次开展了自然保护区专项执法检查,防止不合理的开发建设活动对自然保护区的冲击和破坏。

2.1　促进公平和惠益分享			
1）在评估公平分摊建立保护区的费用和分享其利益方面取得了哪些进展？	2		
2）如果有，请提供对建立保护区的费用进行公平分摊和分享利益进行评估的 URL（或附上一份 pdf 文档）。			
3）为改善利益的公平分享而采取了哪些行动？请检查所有行动，并进行简要描述：			
√	行动	2004 年以前	2004—2009 年	2010 年至今
	建立补偿机制		√	√
	制订并/或实施了获取和分享利益的政策			
	建立公平的利益分享机制			
	使保护区的利益向减贫工作倾斜	√	√	√
	为加强利益的公平分享而采取的其他行动			
4）在评估保护区治理方面取得了哪些进展？	3		
5）已被列为国际自然保护联盟（IUCN）类别的保护区比例？	（%）		
6）如果有，请给出对保护区治理进行评估的 URL（或附上一份 pdf 文档）：	http://www.cnki.com.cn/Article/CJFDTotal-LDGH201006014.htm		
7）为改善和丰富治理类型已经采取了哪些行动？请检查所有行动，并进行简要描述			
√	行动	2004 年以前	2004—2009 年	2010 年至今
	建立具有创新型治理形式的新保护区，例如社区保护区		√	√
	为创造新型治理而对法律或政策进行修改			
	为丰富治理类型而采取的其他行动			

中国实施了天然林资源保护工程，设立了森林生态效益补偿基金，部分自然保护区得到这些资金的支持。中国政府设立了国家重点生态功能区转移支付资金，2012 年转移支付范围包括 466 个县（市、区），转移支付资金达到 371 亿元，这对自然保护区的发展起到了积极作用。2007 年环境保护部发布了《关于开展生态补偿试点工作的指导意见》，要求加快建立自然保护区生态补偿机制。一些地区尝试建立自然保护区生态补偿机制，如 2010 年济宁市颁布了《山东南四湖省级自然保护区湿地生态损失补偿管理办法》。

除自然保护区外，中国还建立了森林公园、风景名胜区、湿地公园、地质公园、自然保护小区、农业野生植物保护点等类型保护区（见第 2 章 2.3.4 节）。这些保护区是中国自然保护区体系的重要补充。

2.2　加强和保证土著与地方社区及相关利益攸关方的参与			
1）土著和地方社区以及其他主要利益攸关方参与主要保护区决策的情况？	2		
2）为改善土著和地方社区的参与情况已经采取了哪些行动？请核查所有行动，并进行简要描述：			
√	行动	2004 年以前	2004—2009 年	2010 年至今
	评估了当地社区参与主要保护区决策的机会和必要性			

	2004年以前	2004—2009年	2010年至今
为改善参与而改进法律、政策和/或做法	√	√	√
制定了重新安置的事先知情同意政策	√	√	√
改进了土著和地方社区的参与机制	√	√	√
增加了土著和地方社区在主要决策过程中的参与程度	√		
为促进参与而采取的其他行动			

中国政府建立了听证制度、公示制度及环境影响评价法中的公众参与制度，加强少数民族和当地社区的能力建设，为当地社区充分有效地参与决策和政策规划创造了条件。

3.1 为保护区提供一个扶持性的政策、体制和社会经济环境

1）在评估建立和管理保护区的政策环境方面取得了哪些进展？　3

2）如果有，请给出政策环境评估的URL（或附上一份pdf文档）：

3）为改善保护区的政策环境采取了哪些行动？请检查所有行动，并进行简要描述：

√	行动	2004年以前	2004—2009年	2010年至今
	为提高管理实效而协调了部门政策或法律	√	√	√
	使保护区价值和生态服务融入国民经济			
	改善决策过程的问责和/或参与	√	√	√
	建立了私营保护区激励机制			
	为支持保护区建立了积极的市场激励机制	√	√	√
	消除妨碍有效管理的消极刺激因素	√	√	√
	加强了建立或管理保护区的法律	√	√	√
	在跨境区域方面与邻国合作	√	√	√
	建立公平的争议解决机制和流程	√	√	√
	为改善政策环境而采取的其他行动	√	√	√

4）在评估保护区对当地和国民经济的贡献方面取得了哪些进展？　3

5）在评估保护区对千年发展目标的贡献方面取得了哪些进展？　3

6）如果有，请给出对保护区为当地和国民经济以及千年发展目标的贡献进行评估的URL（或附上一份pdf文档）：　http://www.cnki.com.cn/Article/CJFDTotal-SAHG201206043.htm

7）为重视保护区的贡献采取了哪些行动？请检查所有行动，并进行简要描述：

√	行动	2004年以前	2004—2009年	2010年至今
	开展交流活动以鼓励决策者认识到保护区的价值	√	√	√
	建立与保护区有关的融资机制（即支付生态系统服务的费用）			

1997年发布了《中国自然保护区发展规划纲要（1996—2010）》，规划目标已实现。2003年国务院批准了《全国湿地保护工程规划（2002—2030）》，"十一五"期间完成总投资30.3亿元，目前正在实施"十二五"规划任务。2000年制定了《全国野生动植物保护和自然保护区建设规划》，

对自然保护区建设给予重点支持。目前正在编制《全国自然保护区发展规划》，将报国务院审批，分期纳入国民经济和社会发展规划中。

3.2 提高保护区规划、建立和管理方面的能力

1）在保护区能力需求评估方面取得了哪些进展？	3
2）如果有，请给出能力需求评估的URL（或附上一份pdf文档）：	http://www.cnki.com.cn/Article/CJFDTotal-BJLY2011S2012.htm

3）为加强保护区的能力已经采取了哪些行动？请核查所有行动，并进行简要描述：

√	行动	2004年以前	2004—2009年	2010年以来
	制订了保护区员工职业发展方案	√	√	√
	对保护区员工进行主要技能培训	√	√	√
	增加保护区员工的数量	√	√	√
	建立重视和分享传统知识的体系	√	√	√
	为提高能力而采取的其他行动	√	√	√

自1998年起，财政部设立了国家级自然保护区能力建设专项资金，截至2012年累计投入资金7.9亿元，专门用于自然保护区的管护能力、科研能力和宣传教育能力建设，对提高保护区管理水平起到非常积极的作用。2008年起，国家还专门设立林业系统国家级自然保护区能力建设补助专项资金。

北京、内蒙古、黑龙江、浙江、江西、福建、山东、湖南、广东、宁夏等省、自治区、直辖市专门设立了自然保护区专项资金。广东省从2000—2009年，投入自然保护区建设资金达3亿多元；福建省采取提高生态补偿标准、加大对省级以上林业自然保护区基础设施建设投资力度等优惠政策，加强自然保护区的建设管理。

环保、林业、农业等自然保护区主管部门多次举办自然保护区管理培训班，针对自然保护区政策法规、规范化管理、规划编制、能力建设项目设计、开发活动监管、管理信息系统建设、资源本底调查等内容进行培训。

在全球环境基金、世界自然基金会等国际组织的帮助下，中国政府先后实施了自然保护区管理项目、湿地生物多样性保护与可持续利用、林业可持续发展等项目，这对于自然保护区能力建设起到了十分重要的作用。

3.3 为保护区发展、应用和转让适当的技术

1）在评估保护区管理对相关适用技术的需求方面取得了哪些进展？	3
2）如果有，请给出技术需求评估的URL（或附上一份pdf文档）：	

3）为改善相关适用技术的获得和使用已经采取了哪些行动？请核查所有行动，并进行简要描述：

√	行动	2004年以前	2004—2009年	2010年以来
	开发并/或使用了适用的栖息地恢复和重建技术	√	√	√
	开发并/或使用了适用的资源测绘、生物普查以及快速评估技术	√	√	√

	开发并/或使用了适用的监测技术	√	√	√
	开发并/或使用了适用的保护和可持续利用技术	√	√	√
	鼓励保护区与各机构间的技术转让与合作	√	√	√
	为改善适用技术的获取和使用而采取的其他行动	√	√	√

科技部设立了"自然保护区建设关键技术研究与示范"等项目，从自然保护区体系构建、功能区划、生境质量与生物资源动态监测、濒危物种保护、干扰生态系统的修复、适应性经营与资源可持续利用六个方面开展研究，为中国自然保护区的建设提供技术支撑。环保、林业、农业等自然保护区主管部门通过培训、会议等多种方式，推广并改进保护区有效管理的各项技术和创新方法，促进经验和技术交流。各自然保护区通过与高等院校、科研院所以及非政府组织的合作，获得并广泛使用了调查、监测、保护和管理等方面的技术，提升了保护区的管理水平。

3.4 确保保护区和国家及区域保护区系统在财政上的可持续性

1）在评估保护区的资金需求方面取得了哪些进展？		3
2）如果有，请给出资金需求评估的URL（或附上一份pdf文档）：		
3）在制订和执行包含多种融资机制组合的可持续融资计划方面取得了哪些进展？		2
4）如果有，请给出可持续融资计划的URL（或附上一份pdf文档）：		

5）为改善贵保护区的可持续融资计划已采取了哪些行动？请核查所有行动，并进行简要描述：

√	行动	2004年以前	2004—2009年	2010年以来
	制订了新的保护区供资机制	√	√	√
	制订了保护区业务计划	√	√	√
	制订了收入共享机制			
	改进了资源分配流程	√	√	√
	提供了金融培训和支持			
	改进了会计和监督	√	√	√
	提高了金融规划能力			
	消除了可持续融资方面的法律障碍			
	明确了机构间财务责任			
	为改善可持续融资能力而采取的其他行动			

见有关2.1和3.2问题的说明

3.5 加强交流、教育和公众意识

1）在开展公众意识和交流活动方面取得了哪些进展？		3
2）如果有，请给出公众意识与交流计划的URL（或附上一份pdf文档）：		http://www.ynly.gov.cn/news/200810/12064.shtml

3）为提高公众意识和加强教育方案已采取了哪些行动？请核查所有行动，并进行简要描述：

√	行动	2004年以前	2004—2009年	2010年至今
	确定与保护区有关的教育、认识和交流方案的核心主题	√	√	√

开展保护区对当地和国民经济以及千年发展目标贡献的意识提高活动	√	√	√
开展保护区对适应和减缓气候变化贡献的意识提高活动	√	√	√
建立并加强与主要目标群体，包括与土著和地方社区之间的交流机制	√	√	√
与教育机构一起开发保护区相关课程	√	√	√
编制公共宣传材料	√	√	√
实施公共宣传方案	√	√	√
为加强传播、教育和意识而采取的其他行动	√	√	√

国家鼓励并要求保护区开展广泛的宣传教育，提高公众对保护区重要性和惠益的认识。在自然保护区内制作宣传碑、宣传木牌，书写宣传标语，印刷自然保护区手册，分发给周边居民和游客。各有关部门和各地通过"世界生物多样性日"、"世界地球日"等节日纪念日，开展了形式多样的宣传活动，介绍自然保护区的价值和重要性。

4.1 制定并使用国家和区域保护区系统的最低标准及最佳做法

1）在开发最佳做法和最低标准方面取得了哪些进展？	3		
2）如果有，请给出保护区最佳做法和最低标准实例的URL（或附上一份pdf文档）。			
3）是否已经建立了一套体系，监督保护区工作方案实施所产生的成果	是		
4）采取了哪些与最佳做法和最低标准有关的行动？请核查所有行动，并进行简要描述：			

√	行动	2004年以前	2004—2009年	2010年至今
	制订了保护区建立与选择的标准和最佳做法	√	√	√
	制订了保护区管理规划的标准和最佳做法	√	√	√
	制订了保护区管理的标准和最佳做法	√	√	√
	制订了保护区治理的标准和最佳做法	√	√	√
	与其他各方和相关组织合作，检验、审查和推广最佳做法和最低标准	√	√	√
	与最佳做法和最低标准有关的其他行动	√	√	√

1999年制定了《国家级自然保护区评审标准》，提出了国家级自然保护区规划、管理等方面必须满足的指标。

2002年发布了《国家级自然保护区总体规划大纲》，指导国家级自然保护区总体规划的编制与实施。

2006年发布了《自然保护区总体规划技术规程》和《自然保护区生态旅游规划技术规程》，规定了自然保护区总体规划以及生态旅游规划的基本准则。

2009年颁布了《国家级自然保护区规范化建设和管理导则（试行）》，进一步规范了国家级自然保护区的建设和管理。

2010年颁布了《自然保护区综合科学考察规程（试行）》，规范了自然保护区综合科学考察活动。

4.2 评价和提高保护区管理的有效性

1)在评估保护区的管理效率方面取得了哪些进展？	3	
2)如果有，请给出评估保护区管理效率的 URL（或附上一份 pdf 文档）：	http://www.zhb.gov.cn/gkml/hbb/bgt/201005/W020100524534788478025.pdf	
3)管理效率已经评估的面积占保护区总面积的百分之几？	62.9%	
4)管理效率已经评估的保护区数量占保护区总数的百分之几？	13.6%	

5)为改进保护区内的管理过程已采取了哪些行动？请核查所有行动，并进行简要描述：

√	行动	2004年以前	2004—2009年	2010年以来
	改进了管理体系和过程	√	√	√
	改进了执法	√	√	√
	改善利益攸关方关系	√	√	√
	改进了游客管理	√	√	√
	改进了自然和文化资源管理	√	√	√
	为提高管理效率已采取的其他行动	√	√	√

关于改进保护区管理方面的规定见1.4节的有关说明。

2008年起，环境保护部等7个部委联合开展了国家级自然保护区管理评估，使保护区管理纳入规范化的轨道。截至2012年，完成了所有300多个国家级自然保护区的评估。环境保护部等部门还多次开展了自然保护区专项执法检查，防止不合理的开发建设活动对自然保护区的冲击和破坏。

4.3 评估和监测保护区的现状与变化趋势

1)在建立一个有效的监测保护区覆盖面积、状态和变化趋势的系统方面取得了哪些进展？	3	
2)如果有，请给出一份最近监测报告的 URL（或附上一份 pdf 文档）：	http://www.shidi.org/sf_A4B06758596347D2A155665A2331390C_151_pyh.html	

3)为改进保护区监测已采取的行动？请核查所有行动，并进行简要描述：

√	行动	2004年以前	2004—2009年	2010年至今
	评估了主要生物多样性的状况和趋势	√	√	√
	监测保护区的覆盖面积	√	√	√
	制订或改进了生物多样性监测方案	√	√	√
	开发了数据库以管理保护区数据	√	√	√
	根据监测和/或研究结果修改了管理计划	√	√	√
	根据监测和/或研究结果改变了管理做法	√	√	√
	开发了地理信息系统（GIS）和/或遥感技术	√	√	√
	其他监测活动	√	√	√

中国政府鼓励和推动自然保护区监测工作。科技部设立了"中国重要生物物种资源监测和保育关键技术与应用示范"重点项目，加强监测网络的顶层设计和各类监测标准的研究。2004年，国家海洋局发布了《海洋自然保护区监测技术规程—总则》，规定了海洋自然保护区监测的内容、

技术要求和方法。环境保护部正在制定《生物物种监测技术指南》。2011年，河南省林业厅发布了《河南省林业自然保护区科研监测方案（试行）》，进一步规范和完善河南省自然保护区的科研监测工作。

2009年起，环境保护部启动了第一期全国自然保护区基础调查，目的是查明全国所有类型、各级别自然保护区的建设管理情况。2011年起，环保部运用环境卫星建立自然保护区的遥感监测体系，及时通过卫星遥感进行监测，并根据这些信息到现场进行核查，实现"天地一体化"监控自然保护区。2012年，环境保护部和中国科学院启动了"全国生态环境十年变化（2000—2010年）遥感调查与评估项目"，该项目设置了"国家级自然保护区生态环境十年变化调查与评估"专题，研究所有300多个国家级自然保护区面临的环境问题和胁迫驱动情况，综合评估中国自然保护区的保护效果。

中国大部分国家级自然保护区拥有一定的监测能力，一些保护区开展了长期的生物多样性监测工作，特别是在长白山、东灵山、神农架、古田山、鼎湖山、西双版纳等多个保护区建立了监测大样地。

4.4 保证使科学知识为建立保护区网络和不断改进其管理作出贡献			
1）为支持保护区的建立和管理而开发了适当的科学与研究计划，在这方面有哪些进展？	3		
2）如果有，请给出一份最新研究报告的URL（或附上一份pdf文档）：			
3）为改进保护区的研究和监测已采取了哪些行动？请核查所有行动，并进行简要描述：			
√ 行动	2004年以前	2004—2009年	2010年至今
确定了主要的研究需求	√	√	√
评估了主要生物多样性的状况和趋势	√	√	√
制订或改进了生物监测方案	√	√	√
对主要的社会经济问题开展保护区研究	√	√	√
推动了保护区研究的普及	√	√	√
根据监测和/或研究结果修改了管理计划	√	√	√
根据监测和/或研究结果改变了管理做法	√	√	√
其他研究与监测活动	√	√	√

经过50多年的全国生物区系调查和100多年的资料积累，中国科学院出版了《中国植物志》、《中国动物志》、《中国孢子植物志》以及大量的地方植物志、动物志，为保护区的建立和不断改进其管理提供了科学依据。

为进一步发展与自然保护区有关的科学知识，北京林业大学创建了自然保护区学院，环境保护部南京环境科学研究所成立了自然保护区研究中心，广东省林业局和华南农业大学共同设立了广东省自然保护区研究中心，在自然保护区学科建设方面取得了较大进展。

在全球环境基金的支持下，中国实施了自然保护区管理项目、湿地生物多样性保护与可持续利用、林业可持续发展等项目，引进了国际先进的自然保护理念和方法。

环保、林业、农业等自然保护区主管部门多次举办培训班和研讨会，推广并改进保护区有效管理的理论、技术和创新方法，提升自然保护区管理水平。

附表 5　全球生物分类倡议能力建设战略和全球植物保护战略的执行情况

缔约方大会有关决定、工作方案和建议活动	全球生物分类倡议能力建设战略	国家执行情况及贡献	进展评估
行动 1：最迟到 2013 年年底，评估国家、次区域及区域层次的生物分类需求和能力，并确定执行《公约》和《生物多样性战略计划》（2011—2020）的优先重点		对亚洲和中国的植物分类能力作了初步评估，2011 年出版了《亚洲植物保护进展（2010）——评估全球植物保护战略的实施进展》，但需求评估有待开展	部分完成
行动 2：截至 2013 年年底，组织多次区域和次区域培训班，向各缔约方及其《生物多样性公约》和全球生物分类倡议国家联络处、教育、科学，及其他相关部门代表介绍生物分类对于执行《公约》和《生物多样性战略计划》（2011—2020）的重要意义以及这一领域的合作需要	中国积极开展生物分类的培训，并介绍生物分类对于生物多样性保护的意义。例如，中国科学院植物研究所植物分类学基础培训班（第一期）于 2011 年 9 月开班。培训班系统介绍了分类学发展史、分类学基础理论、方法论，并重点介绍维管束植物各大类群的基本分类特征，以及常见类群和疑难类群的分类鉴定和鉴定技巧。通过培训，学员掌握了植物分类学的经典方法，并结合现代分类学的新技术、新方法，认识和掌握了植物分类学研究的重要意义。由上海辰山植物园（中国科学院上海辰山植物科学研究中心）主办的"2012 年辰山植物分类学培训班"于 2012 年 10 月开班，主要面向一线工作者，内容包括分类学历史/文献、植物分类学研究方法、种子植物分类、苔藓植物分类、蕨类植物分类、植物标本馆建设管理与配套技术、分子系统发育分析、居群遗传学等。2011 年和 2013 年举行了 3 次菌物多样性与系统进化培训班、中国菌物学会分别在 2010 年、2011 年和 2013 年举行了 3 次菌物、多样性与系统进化培训班，邀请国内外分类专家系统介绍了菌物分类学和种群遗传学基础理论、方法以及新技术的应用，并进行了野外标本采集与分类鉴定。培训班吸引了 50 多个单位的近 130 位学员参加。植物分类建设管理与配套技术，种子植物分类，分子系统发育学等。培训班吸引了 150 余位学员参加		全部完成
行动 3：截至 2014 年之前，组织其他研讨会和培训班，以提高生物分类技能，生物分类知识和信息的质量，并促进生物分类学为执行《公约》作出贡献	正在计划中		

附表5 全球生物分类倡议能力建设战略和全球植物保护战略的执行情况

缔约方大会有关决定、工作方案和建议活动	国家执行情况及贡献	进展评估
行动4：截至2015年之前，在外来入侵物种和生物安全框架内，考虑到已查明的用户需求，制定并继续交流生物分类工具（如野外手册、数字标本集、在线风险分析工具，条形编码等基因和DNA序列检测工具，促进利用这些工具查明有关风险及分析：（1）受威胁物种和（2）外来入侵物种；（3）对农业和水产养殖业有益的物种和特征；（4）可能被非法贩卖的物种；（5）包括微生物在内具有重要社会经济价值的物种	中国已发布了大量生物分类工具。《中国植物志》已提供在线检索服务，其英文修订版（Flora of China）实现了线上全文数据库和印刷版的同步发布。正在编纂中的《泛喜马拉雅植物志》网络版将先于印刷版的信息。自2008年开始，科技部大力推动建设"国家标本资源共享平台"目前网上共享的标本数据达到800万份，文献100万页，野外生态照片400万张，整理了模式标本15 060笔。同时利用信息技术，建立了数字化生物野外调查和信息管理相关技术标本和网络信息系统。2012年9月，《中国生物物种名录》2012版光盘由生物多样性委员会编制完成，并由科学出版社出版发行了《全国植物物种2 000中国节点编制完成。2010年，环境保护部发布了《全国生物物种资源调查技术规定，监测奠定了较好的基础。"生物物种监测技术指南"，并正在制定"生物物种监测技术指南"，农业、质检等部门开发了开展外来入侵生物风险分析工具，一直在开展外来入侵物种的风险分析	大部分完成
行动5：截至2015年之前，以查明并协助监测生物多样性设施进行审查，对人力资源能力和基础设施进行审查，特别是外来入侵物种，研究不足的生物类群，具有重要社会经济价值的受威胁物种。这种审查应当与区域网络建设和国际活动保持协调	中国已评估了现有生物多样性监测的能力和设施，提出了全国生物多样性监测网络建设方案，建立了中国森林生物多样性监测网络，并正在制定"生物物种监测技术指南"。这些标准将涵盖多个生物类群，包括研究相对不足的生物类群	大部分完成
行动6：尽最大可能支持建立国家和专题生物多样性信息设施的现有工作，建立并维护用于印刷和跟踪生物标本特别是模式标本的信息系统和基础设施，并在2016年之前向公众免费和公开提供生物多样性信息	《中国植物志》已提供在线检索服务，其英文修订版（Flora of China）网络版将先于印刷版的同步发布。正在编纂中的《泛喜马拉雅植物志》网络版将先于印刷版的信息。"国家标本资源共享平台"及相关信息的建设，为在线植物志提供了充分的标本、文献、生态图像等信息支撑。这些信息系统都向公众免费开放	已完成
行动7：截至2017年之前，建立充分的人力资源和基础设施，以维持已收藏及今后将收藏的生物标本和活体遗传资源	经过多年的努力，中国已经保存的作物遗传资源达到了3 000多万号。为了妥善保存收集的作物遗传资源，中国政府加强了保存设施建设，扩建和改造了1座国家长期库、1座国家复份库、10座国家中期库、32个国家级种质圃，并新建了7个国家级种质库。这些设施保存的农作物收集品总量已达42.3万份，主要为农作物地方品种及其野生近缘植物资源。中国各有关研究机构和大学建有大量的植物标本馆，并配备专门的人员和设施，开展植物标本的采集、保存和鉴定工作。在标本馆建设中今后应加强中国的动物标本	大部分完成

缔约方大会有关决定、工作方案和建议活动	国家执行情况及贡献	进展评估
行动8：截至2019年之前，提高历史、现有和未来生物多样性收藏记录的质量与数量，并通过生物多样性预测模型供人使用，以提高不同情境下生物多样性预测模型的分辨率和可信度	自2008年开始，科技部大力推动建设"国家标本资源共享平台"，推动植物标本资源的数字化整理、整合和在线共享。目前物标本、岩矿化石标本等各种标本及相关资源数字化共享。目前网上共享的标本数据达到800万份，文献100万页、野外生态照片400万张，整理了模式标本15 060笔，实现了高效率的浏览和统计分析。同时利用信息技术，建立了数字化生物野外调查和信息管理相关的技术体系和网络信息系统。这一平台的建立将为不同情境下生物多样性预测模型提供基础数据	大部分完成
行动9：在生物多样性热点地区、关键生物多样性区域、保护区、社区管理的保护区、可持续生物多样性管理区以及《里山倡议》倡导的社会经济生产景观等国家、区域和次区域优先区域，推动针对所有生物类群的编目	2012年9月，《中国生物物种名录》2012版光盘由物种2 000中国节点编制完成，并由科学出版社出版发行。该名录的建设为生物分类学研究提供了良好的基线数据，并为生物多样性信息的管理和应用提供了核心分类学纲领。在该名录基础上，中国完成了全国生物多样性信息系统，以县域为单元，首次系统采集了全国34 039种野生维管束植物位于2 376个县域的分布数据，并建立了国家生物多样性信息系统，可方便地识别全国植物多样性热点地区、编制热点地区、保护优先区域的维管束植物名录	大部分完成
行动10：2018—2020年，利用与生物分类有关的爱知生物多样性目标指标，评估《全球生物分类倡议能力建设战略》在国家、次区域、区域和全球层次取得的进展，以期在2020年之后继续推动这一战略	正在计划中	
全球植物保护战略（2011—2020）		
目标1：建立一个涵盖所有已知植物的在线植物志	《中国植物志》已提供在线检索服务，其英文修订版（Flora of China）实现了在线全文数据库和印刷版的同步发布。正在编纂中的"泛喜马拉雅植物志"网络版将先于印刷版完成，而目将集成更丰富的信息。"国家标本资源共享平台"及相关信息系统的建设，为在线植物志提供了丰分的标本、文献、生态图像等信息支撑	全部完成
目标2：对所有已知植物物种的保护状况尽可能进行评估，以指导保护行动	即将出版发行的《中国生物多样性红色名录 高等植物卷》，通过建立基础名录，搜集物种信息、小组初评、专家查询四个步骤，按照IUCN红色名录比准对34 450种（含种下单位）进行了评估，结果显示绝灭、野外绝灭和地区绝灭的共52种，极危、濒危、易危的共3 767种。该书的出版将为中国植物保护行动提供重要的基础参考	全部完成
目标3：开发和分享执行本战略所需的信息、研究、相关产出和方法	中国通过植物志编研、自然保护区网络建设、国家标本资源共享平台等各方面的工作推动本目标的实现	大部分完成

附表5　全球生物分类倡议能力建设战略和全球植物保护战略的执行情况

缔约方大会有关决定、工作方案和建议活动	国家执行情况及贡献	进展评估
目标4：通过有效管理和/或恢复，确保每个生态区域或植被类型中至少有15%得以保存	中国已建立包含2 697个自然保护区的保护区网络，良好保护了14.8%的陆地国土面积，有效保护了中国90%的陆地生态系统类型、65%的高等植物群落，涵盖了25%的原始天然林、50%以上的自然湿地和30%的典型荒漠地区。同时通过实施天然林资源保护工程、退耕还林工程、湿地保护工程、沙漠化治理等重点生态工程，促进各种生态系统的恢复	全部完成
目标5：在每个生态区域中至少75%最重要的植物多样性区域得到保护，并建立了保护植物多样性的有效管理措施	在森林、草地、荒漠、湿地等各种生态类型中，中国已建立407个国家级自然保护区对最重要的区域进行了保护	大部分完成
目标6：至少75%的经营土地在各部门得到与保护植物多样性相一致的可持续管理	中国有着精耕细作的传统，轮作、间作等传统耕作方式十分有利于保护生物多样性。2000年以来，中国组织开展了有关目标一致的生态省、市、县创建活动。目前，已有15个省（自治区）、13个省颁布了生态省建设规划纲要，1 000多个县（市、区）开展了生态县建设。"十一五"以来，命名了38个国家级生态县、建成1 559个国家生态乡镇和238个国家级生态村	部分完成
目标7：至少75%的已知受威胁植物种类得到就地保护	据统计，中国85%的国家重点保护野生植物得到了保护。2012年，国家林业局启动了全国极小种群野生植物拯救保护工程，工程将对中国120种极小种群野生植物开展为期5年的拯救保护行动。该行动的实施，将有效改善最濒危的珍稀植物的生存状态	大部分完成
目标8：至少75%的受威胁植物种类得到易地保存，最好是在起源国，而且其中至少20%可用于恢复和繁育项目	中国在云南昆明建立的"西南野生生物种质资源库"，是中国第一个为野生植物、动物和微生物建立的种质资源库，于2008年10月29日正式投入运行，截至2013年4月，已收集和保存10 096种76 864份植物种子。同时以中国科学院北京植物园为核心的植物园网络较好地实施了植物迁地保护。昆明植物园、华南植物园和武汉植物园	大部分完成
目标9：70%的作物遗传多样性包括其野生近缘种和其他具有社会经济价值的植物种类得到保护，相关的地方和土著知识得到尊重和保存	中国国家作物种质资源库建立于1986年，保存农作物及其野生近缘种的种子。到目前为止，保存的种质资源数量已达到42.3万余份。国家作物种质资源及其保存利用将起到非常重要的作用	大部分完成
目标10：采取有效的管理计划以阻止新的生物入侵，保护自然保护区内分布的珍稀濒危植物	中国建立了比较健全的检验检疫管理机制以阻止未自国际贸易的新的生物入侵。植物群落及相关自然保护区网络、保护区内分布的珍稀濒危植物的管理，已纳入各保护区的日常管理	部分完成
目标11：没有野生植物因国际贸易而受到威胁	中国切实履行《濒危野生动植物种国际贸易公约》，对濒危野生植物种类的国际贸易进行监管，没有野生植物种类因为正常的国际贸易而受到威胁	大部分完成

缔约方大会有关决定、工作方案和建议活动	国家执行情况及贡献	进展评估
目标12：基于野生植物的产品的资源得到可持续利用	目前中国来自纯野生植物材料的产品在社会经济中的比重较小。人参、天麻、金银花等大宗植物药材的原料已实现人工栽培，这在保护野生资源的同时实现了资源的可持续利用	部分完成
目标13：与植物资源有关的地方和土著知识、创新和做法得到恰当保存、改进，并支持习惯利用、可持续生计、地方粮食安全和保健事业	中国积极开展地方土著知识的收集、保存和利用。云南省农业科学院生物技术与种质资源研究所出一些具有先进性和实用性的土著知识论文章，鼓励以拥有土著知识的农民为作者发表土著知识文章，已保存记录有关农业生物多样性的土著知识300余条。创建以农民为授课者和资源调查参与式培训模式，共有600余名农民和技术人员直接参加了培训，促进了抗旱野生稻的保护和利用。云南省民族地区特色产业的发展及其土著知识的能力和水平，促进了抗粒野生稻和云南省民族地区特色产业的发展	部分完成
目标14：将植物多样性的重要性和保护植物多样性的必要性纳入传播、教育和大众宣传途径	植物多样性知识已成为全国科学普及的重要内容。在科教领域，除了在中学课程中有生物多样性知识普及外，植物园展示等途径向公众传播。通过电视、广播、报刊等媒体以及全国科普日、植物园展示等途径向公众传播。在教育领域，除了在中学课程中有生物多样性相关内容外，在小学课外科学课中的比重逐渐提高。中国科学院在"标本资源共享平台"建设过程中建立的"中国自然标本馆"、"教学标本子平台"等生物多样性信息系统拥有大量植物科普信息，为社会大众获取植物多样性知识提供了有效的途径	大部分完成
目标15：根据本国需要，具有充足的经过培训并拥有适当设备的从事植物保护的人员，以实现本战略各项目标	环境保护部联合中国科学院、自然保护区等积极举办生物野外调查监测技术培训班。特别是自2009年开始，中国科学院植物研究所连续举办五次"数字化和地标化技术在野外调查工作中的应用与植物摄影技巧"系列培训班，讲授GPS定位、数码摄影信息采集、网络信息管理等内容。通过这些培训，部分自然保护区已开始尝试利用网络信息设备已在植物调查和监测工作中逐步推广，数码相机、GPS轨迹记录器等信息化系统进行植物调查监测数据的管理和科普展示。总体上，在自然保护区管理队伍中得到培训并拥有适当设备的植物保护人员逐渐增加，但人员数量和能力水平还需进一步提高	部分完成
目标16：在国家、区域和国际各级建立或加强植物保护机构、网络和伙伴关系，以推动战略目标的实现	中国成立了由环境保护部牵头、其他相关部委参加的中国履行《生物多样性公约》工作协调组，生物物种资源保护部际联席会议制度和中国生物多样性保护国家委员会，同时由农业部和国家林业局具体负责农业野生植物和森林的保护，体现了国家战略上对生物多样性保护的充分重视。大部分省（自治区、直辖市）人民政府成立了跨部门的协调机制，统筹区域生物多样性保护和管理	大部分完成

参考文献

[1] 安建东, 陈文锋. 中国水果和蔬菜昆虫授粉的经济价值评估. 昆虫学报, 2011, 54(4): 443-450.

[2] 第二次气候变化国家评估报告编写委员会. 第二次气候变化国家评估报告. 北京：科学出版社, 2011.

[3] 樊江文, 钟华平, 员旭疆. 50年来我国草地开垦状况及其生态影响. 中国草地, 2002, 24（5）:69-72.

[4] 高志强, 周启星. 稀土矿露天开采过程的污染及对资源和生态环境的影响. 生态学杂志, 2011, 30(12): , 2011, 20112915-2922.

[5] 国家海洋局. 中国海洋环境状况公报（2000—2012年）.

[6] 国家林业局. 中国林业统计年鉴（1995—2012年）. 北京：中国林业出版社, 2012.

[7] 国家林业局经济发展研究中心, 国家林业局发展规划与资金管理司. 国家林业重点工程社会经济效益监测报告（2003—2012）. 北京：中国林业出版社, 2012.

[8] 国家林业局森林病虫害防治总站. 气候变化对林业生物灾害影响及适应对策研究. 北京：中国林业出版社, 2013.

[9] 国家统计局. 中国统计年鉴（2000—2012年）. 北京：中国统计出版社, 2012.

[10] 国家畜禽遗传资源委员会. 中国畜禽遗传资源·猪志. 北京：中国农业出版社, 2011.

[11] 国家畜禽遗传资源委员会. 中国畜禽遗传资源·牛志. 北京：中国农业出版社, 2011.

[12] 国家畜禽遗传资源委员会. 中国畜禽遗传资源·羊志. 北京：中国农业出版社, 2011.

[13] 国家畜禽遗传资源委员会. 中国畜禽遗传资源·家禽志. 北京：中国农业出版社, 2011.

[14] 国家畜禽遗传资源委员会. 中国畜禽遗传资源·马驴驼志. 北京：中国农业出版社, 2011.

[15] 国家畜禽遗传资源委员会. 中国畜禽遗传资源·特种畜禽志. 北京：中国农业出版社, 2011.

[16] 国家畜禽遗传资源委员会. 中国畜禽遗传资源·蜜蜂志. : 中国农业出版社, 2011.

[17] 环境保护部, 中国科学院. 中国生物多样性红色名录 高等植物卷. 2013.

[18] 环境保护部. 中国环境统计年鉴（1997—2012年）. 北京：中国环境科学出版社, 2012.

[19] 环境保护部. 中国生物多样性保护战略与行动计划. 北京：中国环境科学出版社, 2011.

[20] 环境保护部. 中国环境状况公报（1985—2012年）, 2012.

[21] 景兆鹏, 马友鑫. 云南省西双版纳地区生态系统服务价值的动态评估. 中南林业科技大学学报, 2012, 32(9)：87-93.

[22] 刘瑞玉. 中国海物种多样性研究进展. 生物多样性, 2011, 19 (6): 614-626.

[23] 吕利军, 王嘉学. 滇池水体环境污染研究综述. 水科学与工程技术, 2009, 5：65-68.

[24] 马瑞俊, 蒋志刚. 青海湖流域环境退化对野生陆生脊椎动物的影响. 生态学报, 2006, 26(9)：3061-3066.

[25] 农业部. 中国农业统计资料（1997—2012年）. 北京：中国农业出版社, 2012.

[26] 农业部. 全国草原监测报告（2005—2012年）. 2012.

[27] 农业部.中国农村统计年鉴（1991—2012年）.北京：中国农业出版社，2012.

[28] 农业部草原监理中心.2011年全国草原违法案件统计分析报告.2012, [2012-08-20] http://www.grassland.gov.cn/Grassland-new/Item/3550.aspx.

[29] 农业部渔业局.中国渔业统计年鉴.北京：中国农业出版社，2011.

[30] 欧阳志云，赵同谦，赵景柱，等.海南岛生态系统生态调节功能及其生态经济价值研究.应用生态学报，2004, 15(8)：1395-1402.

[31] 祁如英，祁永婷，郭卫东，等.青海省东部大杜鹃的始绝鸣日期对气候变化的响应.气候变化研究进展，2008, 4(4)：225-229.

[32] 汪松，解焱.中国物种红色名录（第一卷）.北京：高等教育出版社，2004.

[33] 吴春霞，刘玲.加拿大一枝黄花入侵的全球气候背景分析.农业环境与发展，2008,25(5)：95-97.

[34] 吴军，徐海根，陈炼.气候变化对物种影响的研究综述.生态与农村环境学报，2011, 27(4):1-6.

[35] 谢高地，张钇锂，鲁春霞，等.中国自然草地生态系统服务价值.自然资源学报，2001, 16(1)：47-53.

[36] 徐海根，强胜.中国外来入侵生物.北京：科学出版社，2011.

[37] 徐海根，王健民，强胜，等.《生物多样性公约》热点研究：外来物种入侵、生物安全、遗传资源.北京：科学出版社，2004.

[38] 徐海根，吴军，陈洁君.外来物种环境风险评估与控制研究.北京：科学出版社，2011.

[39] 徐海根，曹铭昌，吴军，等.中国生物多样性本底评估报告.北京：科学出版社，2013.

[40] 许存泽.浅析水利工程对鱼类自然资源的影响及对策.云南农业大学学报，2006, (12):31-32.

[41] 张翠英，李瑞英，赵臣道.鲁西南四声杜鹃始、绝鸣期对气候变化的响应.气象科技，2011, 39(1)：114-117.

[42] 赵慧颖，乌力吉，郝文俊.气候变化对呼伦湖湿地及其周边地区生态环境演变的影响.生态学报，2008, 28(3)：1064-1071.

[43] 赵同谦，欧阳志云，郑华，等.中国森林生态系统服务功能及其价值评价.自然资源学报，2004, 19(4)：480-491.

[44] 郑景云，郭全胜，赵会霞.近40年中国植物物候对气候变化的响应研究.中国农业气象，2003, 24(1)：28-32.

[45] 住房和城乡建设部.中国城市建设统计年鉴（2006—2012年）.北京：中国计划出版社，2012.

[46] 住房和城乡建设部.中国风景名胜区事业发展公报（2012年），2012.

[47] Garibaldi L A, Aizen M A, Klein A M, et al. Global growth and stability of agricultural yield decrease with pollinator dependence. PNAS, 2011, 108(14): 5909-5914.

[48] Liu H Z, Gao Xin. Monitoring Fish Biodiversity in the Yangtze River, China. In the Biodiversity Observation Network in the Asia-Pacific Region: Toward Further Development of Monitoring，Ecological Research Monographs, DOI 10.1007/978-4-431-54032-8_12, Springer, Japan, 2012.

[49] Millennium Ecosystem Assessment. Ecosystem and Human Well-being: Biodiversity Synthesis. World Resources Institute, Washington, DC, 2005.

（本书以中文为准文本）

China's Fifth National Report on the Implementation of *the Convention on Biological Diversity*

Coordinating Department: Ministry of Environmental Protection

Participating Departments: National Development and Reform Commission; Ministry of Education; Ministry of Science and Technology; Ministry of Finance; Ministry of Land and Resources; Ministry of Housing and Urban-Rural Development; Ministry of Water Resources; Ministry of Agriculture; Ministry of Commerce; General Administration of Customs; State Administration for Industry and Commerce; General Administration of Quality Supervision, Inspection and Quarantine; State Administration of Press, Publication, Radio, Film and Television; State Forestry Administration; State Intellectual Property Office; China National Tourism Administration; Chinese Academy of Sciences; State Oceanic Administration; State Administration of Traditional Chinese Medicine; the State Council Leading Group Office of Poverty Alleviation and Development.

Project Implementing Institution: Nanjing Institute of Environmental Sciences of the Ministry of Environmental Protection.

List of Acronyms or Abbreviations

AIC Administration for Industry and Commerce
ASEAN Association of Southeast Asian Nations
BSAP Biodiversity Strategy and Action Plan
CAS Chinese Academy of Sciences
CBD Convention on Biological Diversity
CBPF China Biodiversity Partnerships Framework
CCICED China Council for International Cooperation on Environment and Development
CITES Convention on International Trade in Endangered Species of Wild Fauna and Flora
CNCG China National Coordinating Group on Implementation of the CBD
CNTA China National Tourism Administration
COD Chemical Oxygen Demand
COP Conference of the Parties
CSPA China Strategy for Plant Conservation
EIA Environmental Impact Assessment
EU European Union
FFI Fauna and Flora International
FGD flue-gas desulfurization
GAC General Administration of Customs
GAQSIQ General Administration of Quality Supervision, Inspection and Quarantine
GBIF Global Biodiversity Information Facility
GDP gross domestic product
GEF Global Environment Facility
GHG Greenhouse Gas
GMO genetically modified organism
GR genetic resources
GSPC Global Strategy for Plant Conservation
IAS Invasive Alien Species
IPCC Intergovernmental Panel on Climate Change
IUCN International Union for Conservation of Nature
IYB International Year of Biodiversity
MA Millennium Ecosystem Assessments
MEA Multilateral Environmental Agreements
MDGs Millennium Development Goals
MHURD Ministry of Housing, Urban-Rural Development
MLR Ministry of Land and Resources

MEP Ministry of Environmental Protection
MOA Ministry of Agriculture
MOC Ministry of Commerce
MOE Ministry of Education
MOF Ministry of Finance
MOST Ministry of Science and Technology
MPA marine protected areas
MTI Marine Trophic Index
MWR Ministry of Water Resources
NCBC National Committee on Biodiversity Conservation (of China)
NDRC National Development and Reform Commission
NPC National People's Congress (of China)
NPP Net Primary Productivity
PA protected areas
RMB renminbi (Chinese currency)
RLI Red List Index
SAIC State Administration for Industry and Commerce
SATCM State Administration of Traditional Chinese Medicines
SFA State Forestry Administration
SIPO State Intellectual Property Office
SOA State Oceanic Administration
TK traditional knowledge
TNC The Nature Conservancy
TRIPs Agreement on Trade-related Aspects of Intellectual Property Rights
UNCCD United Nations Convention to Combat Desertification
UNDP United Nations Development Programme
UNESCO United Nations Educational, scientific and Cultaral Organization
UNFCCC United Nations Framework Convention on Climate Change
WTO World Trade Organization
WIPO World Intellectual Property Organization
WWF World Wildlife Fund

Executive Summary

China's Fifth National Report on the Implementation of the Convention on Biological Diversity (CBD) was prepared as requested by Article 26 of the Convention and decision X/10 of the Conference of the Parties (COP). The report was prepared by the Ministry of Environmental Protection of China in collaboration with members of China's National Coordinating Group (CNCG) for the CBD Implementation and other relevant institutions. In the process of the report preparation, five national workshops were held with the participation of experts in relevant fields, having discussed about issues related to the report and reviewed initial drafts of the report. Following consultations with members of CNCG, the report was further revised and then the final report has been approved and published by the Ministry of Environmental Protection.

I China's Biodiversity and Its Strategic Importance

Biodiversity refers to the variability among living organisms from all sources, including, inter *alia*, terrestrial, marine, and other aquatic ecosystems and the ecological complexes of which they are part, including diversity within species, between species and of ecosystems. Biodiversity provides conditions for human survival, strategic resources for socio-economic development and important guarantees for ecological and food security. Biodiversity not only provides human beings with many livelihood necessities such as food, clean water, medicine, timber, energy and industrial materials, but also with many ecosystem services, such as carbon sequestration, oxygen release, water regulation, soil conservation, environment purification, nutrient cycling, recreation and tourism.

China is one of the twelve countries in the world with richest biodiversity. Due to its vast land area, China has various and complicated types of ecosystems. Its plant and animal resources are extremely rich. Among others, China's number of higher plant species ranks third in the world; its total number of vertebrate species accounts for 13.7% of the world's total. China's genetic resources are also rich as a place of origin of important crops such as rice and soybeans as well as an important centre of origin and distribution of wild and cultivated fruit trees. Meanwhile, China is also one of the countries facing serious threats to biodiversity. Biodiversity loss can lead to serious consequences, such as worsening health problems, higher food risks, increasing vulnerabilities and fewer development opportunities.Therefore biodiversity conservation is strategically important for China's long-term socio-economic development, well-beings of the present and future generations and the building of an ecological civilization and the implementation of the initiatives such as Beautiful China.

II National Targets for Biodiversity Conservation

The Government of China has laid out its blueprint for building an ecological civilization and Beautiful China. The vision for an ecological civilization is being integrated into economic, political, cultural and social developments, with a view to establish spatial layouts, industrial structures, production and consumption patterns that promote green, cycling and low-carbon development, resource conservation and environmental protection. At the end of 2010, the State Council of China launched *National Plan for Major Function Zones*, according to which the country's land is divided into four major function zones: land for priority development, land for key development, land for limited development and land prohibited for development. 25 key ecological function zones have been included in national-level land zones prohibited for development. Within these zones, large-scale and intensive industrial and urbanization development activities are limited so as to allow for environmental protection and ecological restoration and to enable ecosystems to provide ecological goods. National-level nature reserves, world cultural and natural heritage sites, national-level scenic zones, national forest parks and national geological parks have been also included in national-level land zones prohibited for development, where industrial and urbanization development activities are banned to protect natural, cultural heritages, rare animal and plant genetic resources of China.

In response to the severe situation of biodiversity loss, the Government of China launched on 17 September 2010 *China's National Biodiversity Strategy and Action Plan (2011-2030)* (abbreviated as "NBSAP"). Together with relevant national plans developed with a view to building an ecological civilization, NBSAP has provided a relatively comprehensive set of national targets for biodiversity conservation (Table 1).

III Main Actions

In recent years the Government of China has taken the following main actions to implement the Convention on Biological Diversity:

1 Improving legal and regulatory system and institutional mechanisms

China has basically established a legal and regulatory system for biodiversity conservation, developed and promulgated a series of national, sectoral and local standards for biodiversity conservation. In 2011, China established a National Committee for Biodiversity Conservation (NCBC) to coordinate biodiversity conservation actions at national level. The existing mechanisms such as the Inter-ministerial Joint Conference for Protection of Biological Resources

Table 1   China's National Targets for Biodiversity Conservation

1   Short-term goal: by 2015, the trend of biodiversity decline in key regions will be effectively contained, specifically including:

- Biodiversity status surveys and assessments will be undertaken in 8 to 10 priority areas for biodiversity conservation, and these areas will be effectively monitored;
- *In-situ* conservation will be strengthened and terrestrial protected areas will be maintained at 15% or so of the country's land area, protecting 90% of national key protected species and typical ecosystem types;
- *Ex-situ* conservation will be undertaken on a scientific basis, providing effective protection of 80% of endangered species in areas where *in-situ* conservation is not adequate or whose wild population is very small;
- Forest coverage rate will be increased to 21.66% and forest reserves will be increased by 600 million m^3 over those in 2010;
- A system of monitoring, assessment and early warning of biodiversity, as well as those systems for access to and benefit-sharing of genetic resources and import and export of biological resources will be preliminarily established;
- Main pollutants will be reduced considerably, with COD and SO_2 emission to be reduced by 8%, NO_x and ammonia nitrogen by 10% compared with those in 2010;
- Major progress will be made in building a resource-efficient and environmentally friendly society.

2   Mid-term goal: by 2020, biodiversity decline and loss will be basically controlled, specifically including:

- Biodiversity status surveys and assessments will be completed in all priority areas for biodiversity conservation, with all these areas to be effectively monitored;
- National forest holdings will exceed 2.23 million km^2, an increase of about 223 000 km^2 over that of 2010, and national forest reserves will exceed 15 billion m^3, an increase of 1.2 billion m^3 over that of 2010;
- The cumulative areas of control of degraded, salinized and desertified grasslands will exceed 1.65 million km^2, and grassland degradation trend will be contained, with obvious improvements in grassland ecology and balance between herds and grass supply in natural grasslands achieved;
- The environmental and ecological degradation of the near-shore marine areas will be fundamentally reversed, and the decline of marine biodiversity will be basically contained;
- The aquatic ecosystems will be gradually restored and the depletion of fishery resources and the increase in the number of endangered species will be basically contained;
- A network of nature reserves with reasonable layouts and sound functions will be established, with functions of national-level nature reserves stabilized and main protection targets effectively protected;
- The biodiversity monitoring, assessment and early warning system as well as the system for management of import and export of biological resources, access to genetic resources, benefit-sharing from their use will be improved, and the documentation of associated traditional knowledge and intellectual property rights protection system will be further improved;
- The percentage of total investments from all sources into research and development will be increased to over 2.5% of GDP, with the rate of contributions from science and technology exceeding 60%;
- Energy consumption and CO_2 emissions per unit of GDP will be reduced significantly, and the total amount of main pollutants will be obviously reduced.

3   Long-term goal: by 2030, biodiversity will be effectively protected.

and the National Coordinating Group for Implementation of the Convention on Biological Diversity, are working well. Most provincial governments have reinforced institutions related to biodiversity such as departments of the environment, agriculture, forestry and marine management, and established inter-departmental coordinating mechanisms.

2 Launching and implementing a series of plans for biodiversity conservation

In 2010, the Government of China launched and began implementing *National Plan for Major Function Zones* and *China's National Biodiversity Strategy and Action Plan* (2011-2030). The State Council has also approved a series of plans for promoting actions in biodiversity conservation, such as *National Programme for Conservation and Use of Biological Resources, National Programme of Action for Conservation of Aquatic Biological Resources, National Plan for Water Area Zoning of Important Rivers and Lakes* (2011-2030), *National Plan for Zoning of Marine Areas* (2011-2020), *National Twelfth Five-year Plan for Implementation of Wetland Conservation Projects* (2011-2015), *National Plan for Island Conservation* (2011-2020) *and National Plan for Conservation and Use of Livestock Genetic Resources*. China has been implementing initiatives such as eco-provinces, eco-cities and eco-counties. So far 15 provinces (autonomous regions and province-level municipalities) have begun such initiatives. 13 provinces have launched their programmes for eco-provinces. More than 1,000 counties (cities and districts) have begun eco-county initiatives. As a result, 1,559 eco-towns or communities and 238 eco-villages have been established. Pilot work in building eco-cities with good aquatic ecology has been initiated with the first 46 such cities identified, thus mainstreaming biodiversity into local economic and social development.

3 Strengthening conservation systems

A system of *in-situ* conservation has been established composed primarily of nature reserves and complemented by scenic spots, forest parks, community-based conservation areas, protected sites of wild plants, wetland parks, desert parks, geological parks, special marine protected areas and germplasm conservation farms. By the end of 2013, China has established 2,697 nature reserves, covering an area of about 1.463 million km^2 which accounts for about 14.8% of China's land area. China has also established 2,855 forest parks, covering an area of 174,000 km^2 as well as 225 national-level scenic spots and 737 province-level scenic spots, covering an area of about 194,000 km^2 or 2% of China's land area. In addition, over 50,000 community-based conservation areas have been established, covering over 15,000 km^2; 179 national-level protected sites of agricultural wild plants and 468 wetland parks have been established; 45 national-level special marine protected areas (marine parks) have been established, covering a total area of 66,800 km^2; and 368 national-level conservation areas for aquatic germplasm resources have been set up, covering an area of over 152,000 km^2.

Rescuing and breeding of endangered species have been strengthened. A series of measures have been taken to rescue those endangered wild animals and plants, such as development of breeding techniques, strengthening caring in the wild, habitat restoration and re-introduction to nature. As a result, a group of critically endangered wild animal and plant species have been gradually relieved from risks of extinction. Meanwhile, various and effective measures have been taken to strengthen general protection of other wild animals and plants.

Ex-situ conservation measures have been also undertaken. 200 botanical gardens of various kinds have been established at different levels, collecting and storing 20,000 plants species that account for the two-thirds of China's flora. More than 240 zoos and 250 rescuing and breeding sites for wild animals have been established. A system of protection of livestock genetic resources has been established composed primarily of conservation farms and complemented by protected areas and gene banks, protecting 138 varieties of rare and endangered livestock species. Collection and storage facilities for agricultural genetic resources have been strengthened, with total number of agricultural crops collected coming to 423,000 accessions, an increase of about 30,000 accessions over that in 2007. More than 400 conservation bases for wild plant germplasm resources have been set up. Wild germplasm banks have been established in southwest China to collect and store wild germplasm resources in China.

4 Promoting sustainable use of biological resources

Rules are being implemented for managing the use of key protected wild animals and plants, such as special licensing of hunting, domestication and breeding of key protected wild animals, and licensing of collection of protected wild plants. Other rules have been implemented such as quota system for forest logging, grassland conservation, balancing between grass and herds, grazing ban, licensing of fishery, fishing ban period and areas. Restocking of aquatic species has been increased and marine ranching multiplied. Management and law enforcement of breeding and use of wild animals and plants have been reinforced. Strict technical standards have been developed and specialized labeling systems have been put in place for breeding and use of wild animals and plants. Artificial breeding or cultivation has been undertaken for those endangered species whose population recovery proves relatively difficult. Alternatives have been developed to relieve pressures on their use. Law enforcement has been reinforced to crack down the illegal collection and sales of national key protected wild animals and plants and their products. A number of major cases of smuggling of endangered species have been investigated and punished.

5 Conserving and restoring habitats

A number of key ecological projects continue to be implemented, such as natural forests protection, returning cultivated lands to forests, returning grazing land to grassland, forest belt construction in north, northeast and northwest China as well as in the Yangtze River and coastal areas, control of sandstorms affecting Tianjin and Beijing, comprehensive control of desertification in rocky areas, wetland protection and restoration and integrated control of soil erosion. Since 2001, obvious ecological improvements have been observed in areas where these key projects have been implemented. Forest resources in China have been increasing constantly, with forest areas increased by 23%, forest coverage rate by 3.8% and forest reserves by 21.8% compared with those of a decade ago. A number of wetlands of national and international importance have been rescued and protected, with the protection rate of natural wetlands increasing by over 1% on the average annually. As a result, about a half of natural wetlands has been effectively protected. The area where mangroves and degraded wetlands in the near-shore coastal areas such as tidal flats have been restored has exceeded 2,800 km^2, as a result of an investment of 4.43 billion yuan RMB. The area covered by soil erosion control reached 270,000 km^2 as a result of integrated control measures taken in 12,000 small river basins. The area enclosed for reforestation and conservation has reached 720,000 km^2, with initial ecological recovery occurring in areas of 450,000 km^2. Since 2008, the central government has allocated specialized funds of 19.5 billion yuan RMB for rural environment improvement. These funds supported environmental improvements in 46,000 villages and more than 87 million people in rural areas benefited from these efforts. The implementation of key ecological projects has enhanced recovery of degraded ecosystems and habitats for wild species, thus effectively conserving biodiversity.

6 Developing and implementing incentives favorable for biodiversity conservation

To avoid negative impacts on biodiversity and the environment, the Government of China eliminated in 2007 export subsidies of 553 highly energy-consuming, highly polluting and resource-consuming products, including products from endangered species, leather products, wood products and some disposal wood-made products.

The Government of China has subsidized those rural households involved in key ecological projects. Subsidies were given to those farmers who have returned their cultivated land to forests according to verified areas. By the end of 2012, the central government has cumulatively invested 324.7 billion yuan RMB into this project and 120 million farmers have directly benefited from such investments, with each household being given a subsidy of 7,000 yuan RMB on the average.

For the natural forest protection programme, the Government of China has provided subsidies for forest management, conservation and nurturing and reforestation. The government has also covered pension and other insurances for employees of all forestry enterprises, and subsidized living costs of those laid-off employees and social expenditures of forestry enterprises. During the first phase of the natural forest protection programme the government has invested 118.6 billion yuan RMB. At the end of 2010, the State Council decided to implement a second phase of this programme, which will invest about 244 billion yuan RMB in total from 2011 to 2020.

The Forest Ecological Benefits Compensation Fund was established to subsidize plantation, nurturing, conservation and management of forests for ecological benefits. In 2013, the central government transferred a total of 14.9 billion yuan RMB to various local governments as subsidies for public benefit forests.

Subsidies were also provided to those herdsmen who have returned their grazing land to grassland to cover part of costs for grassland enclosures and forages. During 2003-2012 the central government invested 17.57 billion yuan RMB, benefiting more than 4.5 million herdsmen. In 2011 a mechanism to subsidize and reward grassland ecology conservation was established, and so far subsidies worth 28.6 billion yuan RMB have been provided cumulatively, respectively to areas (820,000 km^2) where grazing bans are implemented and areas (1.737 million km^2) where balancing grass supply with herds is required.

The government has set up a specialized fund to support national key ecological function zones. The funds transferred in 2013 came up to 42.3 billion yuan RMB.

7 Enhancing establishment of biosafety management system

The system of prevention and control of invasive alien species (IAS) has been improved, and a monitoring and early warning system has been established for forest pests and agricultural IASs. Elimination of IASs has been undertaken. Safety assessments, production licensing, commercialization licensing, product labeling and genetically modified product import/export approval have been put in place for agricultural genetically modified organisms (GMOs), and approval of genetic engineering of trees has been undertaken, covering all phases of GMOs from research, development to application.

8 Controlling environmental pollution

Considerably reducing the total amount of major pollutants is one of the binding targets that the Government of China has set for social and economic development and for solving those serious environmental problems. In the past decade, overall the annual average concentration of major pollutants has been going down. The intensity of emission of pollutants per unit of GDP has decreased by over 55%. Since 2004, the density of CO_2 emission per unit of GDP has decreased by 15.2%. The Government of China has been strictly implementing rules of environmental impact assessments (EIAs). Since 2008 the national government has refused to approve 332 projects with total investment of 1.1 trillion yuan RMB, which are projects of high pollution, high energy consumption, high resource consumption, low-level duplicate construction and excessive production capacities.

9 Promoting public participation

China has incorporated relevant biodiversity knowledge into primary and secondary school curriculum, and provided biodiversity-related degree programmes in many universities or colleges. By 2012, more than 556,000 professionals on biodiversity have been trained through such programmes. Relevant government departments and governments of levels have strengthened communication and education in biodiversity. In particular various activities were organized to celebrate the International Year of Biodiversity in 2010, and through various media, reaching out to more than 900 million people. In each of the subsequent years, training activities were organized for journalists and large-scale communication and educational activities organized to promote business engagement with biodiversity conservation. As a result, public awareness of biodiversity conservation has obviously increased, and public participation in biodiversity conservation effectively mobilized.

The conservation actions mentioned above have generated obvious impacts, primarily including:

(1) Constant increase in forest resources, with an increase of 23% in forest areas and of 21.8% in forest growing stock over those of a decade ago.

(2) Comprehensive soil erosion control has been undertaken in 12,000 small river basins, covering an area of 270,000 km^2; and areas of 720,000 km^2 have been enclosed for reforestation and conservation, and among others ecological recovery has started in areas of 450,000 km^2. Since 2006, an additional area of 18,000 km^2 of wetlands has been protected and 1,000 km^2 of wetlands have been restored.

(3) The populations of national key protected animals and plants have been steady and increasing in some cases. The scale of their distribution has been increasing as well and their habitats constantly improving. The number of Giant Pandas (*Ailuropoda melanoleuca*) rose from over 1,000 in the 1980's to 1,590 currently. The number of Crested Ibises (*Nipponia nippon*) has grown from 7 in the 1980's to more than 1,800 at present. The populations of protected plants such as yews, orchids and cycads have been expanding.

(4) By the end of 2013, 2,697 nature reserves have been established, covering a total area of 1.463 million km^2 and accounting for about 14.8% of the country's land area. In addition, a considerable number of scenic spots, forest parks, community-based conservation

areas, protected sites for agricultural wild plants, wetland parks, geological parks, special marine protected areas and germplasm conservation farms have been established. The nature reserves have effectively conserved 90% of terrestrial ecosystem types, 85% of wild animal populations and 65% of higher plant biota, covered 25% of primary forests, more than 50% of natural wetlands, 30% of typical desert areas and nearly 3% of the marine areas under China's jurisdiction.

(5) The total amount of major pollutants has been going down. Since 2000, the intensity of emission of pollutants per unit of GDP has decreased by more than 55%. Since 2004, the intensity of CO_2 emission per unit of GDP has decreased by 15.2%.

In sum, the Government of China has enhanced its efforts in biodiversity conservation and taken various measures, such as improving conservation policies, strengthening establishment of conservation systems, restoring degraded ecosystems, controlling environmental pollution, strengthening science and technology research, promoting public participation and increasing investments. As a result of all these actions, the trend of ecological worsening has been relatively controlled; functions in some ecosystems have recovered and the populations of some key protected species have been increasing. The implementation of updated NBSAP has a good beginning and positive progress is being made. One action has achieved significant progress, 15 actions achieved considerable progress and 14 actions achieved some progress (see Table 2).

In achieving the global 2020 biodiversity targets (20 in total), assessments have shown various degrees of improvements in indicators for targets 1, 3, 4, 5, 7, 8, 10, 11, 14, 15, 17, 19 and 20, except that there are no indicators available to assess progress towards targets 2, 16 and 18. This indicates that achievement of these targets is on track. In particular considerable progress has been made in achieving target 3 (incentive measures), target 5 (habitat loss and degradation reduced), target 8 (environmental pollution controlled), target 11 (protected areas strengthened and managed effectively), target 14 (important ecosystem services restored and ensured) and target 15 (resilience and carbon sequestration of ecosystems reinforced). However, most indicators for target 5 (grassland ecosystem protection among them), target 6 (sustainable fishery), target 9 (control of invasive alien species), target 12 (endangered species protected) and target 13 (protection of genetic resources) have shown worsening trends. This indicates that many more effective policies and measures still need to be taken to achieve these targets, though much has been done so far (see Table 3).

IV Threats to Biodiversity and Main Issues and Priorities for Biodiversity Conservation

Despite various measures taken by the Government of China for biodiversity conservation, the biodiversity decline trend has not been fundamentally contained. The percentage of endangered invertebrates (critically endangered, endangered and vulnerable) is 34.7%. The

Table 2 Assessment of Implementation of NBSAP

Actions	Assessment	Actions	Assessment
1. Develop policies that promote biodiversity conservation and sustainable use	considerable progress	16. Strengthen establishment of conservation farms for livestock genetic resources	considerable progress
2. Improve legal system for biodiversity conservation and sustainable use	considerable progress	17. Develop an *ex-situ* conservation system on a scientific basis	considerable progress
3. Establish and improve biodiversity conservation and management bodies and improve cross-sectoral coordination mechanisms	considerable progress	18. Develop and improve system of storing genetic resources	considerable progress
4. Mainstream biodiversity into regional and sectoral planning processes and plans	considerable progress	19. Strengthen reintroduction of artificially bred species and recovery of wild species	some progress
5. Ensure sustainable use of biodiversity	some progress	20. Strengthen research, development and innovation in use of genetic resources	considerable progress
6. Reduce impacts of environmental pollution on biodiversity	considerable progress	21. Establish a system and mechanism for access to and benefit-sharing of genetic resources and associated TK	considerable progress
7. Undertake baseline surveys of status of biological resources and ecosystems	some progress	22. Establish a system of inspection and examination of import and export of genetic resources	some progress
8. Survey and inventory genetic resources and associated traditional knowledge	considerable progress	23. Upgrade capacities of early warning and monitoring of and emergency response to alien species invasion	some progress
9. Undertake monitoring and early warning of biodiversity	some progress	24. Establish and improve system of biosafety assessment, monitoring and detection of GMOs	considerable progress
10. Enhance and coordinate information systems for genetic resources	significant progress	25. Develop an action plan for addressing climate change impacts on biodiversity	some progress
11. Undertake comprehensive biodiversity assessments	considerable progress	26. Assess impacts of biofuels on biodiversity	some progress
12. Improve and coordinate implementation of protected areas planning across the country	considerable progress	27. Strengthen scientific research in the field of biodiversity	some progress
13. Strengthen protection in priority areas for biodiversity	some progress	28. Strengthen personnel training in the field of biodiversity conservation	some progress
14. Strengthen standardized management of PAs and their management effectiveness	considerable progress	29. Establish mechanisms for broad public participation	some progress
15. Strengthen biodiversity conservation in areas outside PAs	considerable progress	30. Promote establishment of partnerships for biodiversity conservation	some progress

Notes: ● fully achieved; ◔ significant progress; ◐ considerable progress; ◕ some progress; ● no progress

Table 3 Assessment of China's Progress in Achieving 2020 Aichi Targets

Targets	Indicators	Trends	Targets	Indicators	Trends
1. Awareness of biodiversity increased	Items related to China's biodiversity searched through Google or Baidu	✓	10. Pressures on coral reefs and other vulnerable ecosystems reduced	Reductions in pollutants	✓
3. Incentive measures	Ecological compensation and investments into key ecological projects	✓		Forest growing stock	✓
4. Sustainable production and consumption	Reductions in pollutants	✓		Reductions in areas affected by soil erosion	✓
	Indicators for sustainable consumption	•••		Biodiversity of coral reefs	•••
5. Habitat degradation and loss	Forest areas and growing stock	✓		Climate change impacts on biodiversity	•••
	Wetland ecosystem areas	✓	11. Strengthen system of protected areas and management effectiveness	Number and area of protected areas	✓
	Grassland ecosystem areas	✗		Ecological representativeness and management effectiveness of protected areas	•••
	Fresh grass output from natural grasslands	✓	12. Endangered species protected	Red List Index	✗
	Areas of desert ecosystems reduced	✓	13. Genetic resources protected	Number of local varieties	✗
	Ecological degradation	•••	14. Important ecosystem services restored and ensured	Net income per capita of rural households and reduction in number of people living in poverty	✓
6. Sustainable fishery	Marine trophic index	✓		Forest growing stock	✓
	Red List Index of fishes	✗		Reductions in areas affected by soil erosion	✓
	Fishery impacts on biodiversity	•••		Reductions in desertified areas	✓
7. Sustainable agriculture, aqua-culturing and forestry	Forest growing stock	✓	15. Ecosystem resilience and carbon sequestration increased	Forest growing stock	✓
				Reductions in areas affected by soil erosion	✓
				Reductions in desertified areas	✓
	Grass output from natural grasslands	✓	17. NBSAP Implementation	Implementation of policies and programmes	✓
	Agricultural impacts on biodiversity	•••	19. Scientific & technological achievements developed and applied	Academic papers on biodiversity	✓
8. Environmental pollution controlled	Reductions in pollutants	✓		Items related to China's biodiversity searched through Google or Baidu	✓
9. Invasive alien species controlled	Number of new IAS found every two decades*	✓	20. Significant increase in investments	Investments into key ecological projects	✓

Notes: ✓ Increasing; ✗ Decreasing; ••• No adequate data; * IAS negative impacts on biodiversity increasing percentage of endangered vertebrates is 35.9%. The number of endangered plants is 3,767, accounting for 10.9% of the total higher plant species assessed in China. The number of higher plant species that require attention and protection has come up to 10,102, accounting for 29.3% of the total species assessed in China. The loss of genetic resources is also very serious. According to the result from the second national survey on livestock genetic resources, the populations of more than a half of local breeds or varieties have been going down.

1 Threats

Direct pressures that cause biodiversity decline are:

(1) Degradation or loss of habitats. Habitats for wild animals and plants have been destroyed by activities such as reclamation of wetlands and grasslands, coastal development and construction of major transportation and hydropower projects, posing direct threats to reproduction of species and populations.

(2) Excessive exploitation of natural resources. Overgrazing of grasslands led to degradation and desertification of grasslands. High-intensity fishing accelerated depletion of fishery resources. Despite a series of law enforcements, illegal trades in wild animals and plants still occur, and are even very rampant in some regions of China.

(3) Environmental pollution. Water pollution in rivers, lakes and seas directly threats aquatic biodiversity. Use of agricultural chemicals, fertilizers and pesticides has caused increasingly serious environmental pollution. The pollution of coastal and near-shore marine areas is still serious, though overall the marine environmental quality of the areas under China's jurisdiction is not bad. Marine environmental pollution seriously affects marine biodiversity, having caused various marine ecological disasters, such as red tide.

(4) Large-scale cultivation of single species. Only a few agricultural crops are cultivated, with many traditional varieties eliminated or some of them having even disappeared forever.

(5) Invasion of alien species. China is one of the countries in the world most severely affected by the invasion of alien species. There are more than 500 invasive alien species in China, which have caused huge losses to the environment and economy.

(6) Climate change. Climate change has changed the phenology, distribution and migration of species, caused disappearance of some species in their original habitats as well as changed distribution of pests, thus aggravating threats they cause.

2 Main issues

Main issues China faces in biodiversity conservation are: ① inadequate legal and institutional systems; ② low-level awareness of conservation; ③ conflicts between conservation and development and use; ④ inadequate financing or investment; ⑤ inadequate scientific research.

3 Priorities

China has a lot more to do for biodiversity conservation. Next few years are a key period for biodiversity conservation in China. China needs greater determination, more effective measures and more resources to reverse fundamentally the biodiversity loss trend. Future priorities should be given to the following key tasks:

(1) To improve legal and regulatory system for biodiversity and reinforce law enforcement

Existing laws such as *Environmental Protection Law, Wild Animal Protection Law, Wild Plant Protection Regulation* and *Regulation on Nature Reserves* need to be revised or updated. New laws or regulations such as wetland protection regulation, invasive alien species control regulation, regulation on management of genetic resources and regulation on biosafety management of genetically modified trees need to be developed. The ownership of natural resources and their use control system needs to be further defined, with very strict rules to be put in place for source protection, compensation for losses and life-time accountability for causing ecological damage. The payment for ecosystem services should be established as quickly as possible, in particular in those priority areas for biodiversity conservation. Law enforcement needs to be reinforced to further crack down those illegal activities causing damage to biodiversity and strengthen inspection and examination of import and export of biological resources.

(2) To enhance public participation and increase public awareness of conservation

Various forms of communication and education activities should be undertaken. The roles of various civil society organizations and the private sector should be fully played to increase public awareness. Policies and mechanisms for social supervision of biodiversity conservation should be explored and established. Citizen science should be developed and public participation in biodiversity conservation should be enhanced so that an environment should be created for all the public to make joint efforts in conservation and sustainable use of biodiversity.

(3) To implement *National Major Function Zone Plan* and China's updated NBSAP

A system of land spatial development protection should be put in place to improve layouts for spatial development. Biodiversity conservation measures should be proposed for various major function zones. Red lines for ecological conservation should be drawn to ensure ecological security of the country's land. Practical efforts should be made to implement the updated NBSAP. Management and supervision of biodiversity priority areas should be strengthened. Biodiversity should be mainstreamed into national, sectoral and local planning. The environment management of various development and construction activities should be strengthened and responsibilities for developers to restore ecology or ecosystems should be implemented. A system of review, assessment and supervision should be established to promote effective implementation of various plans and programmes.

(4) To further improve networks of *in-situ* conservation and reinforce *in-situ* conservation

Spatial structures of nature reserves and scenic spots should be better-designed for a more effective network of biodiversity conservation. A system of national parks should be established. Major ecological projects should continue, such as natural forest protection, returning cultivation land to forests and grazing land to grasslands, construction of forest belts in north, northeast and northwest China as well as in the Yangtze River Basin, control of origins of sandstorms affecting Beijing and Tianjin, comprehensive control of rockiness in Karst areas, wetland protection and restoration, management of protected areas, and comprehensive control of soil erosion. Major projects for biodiversity conservation should be initiated.

(5) To strengthen institutions and their capacities and improve their management levels

The coordination of the National Committee for Biodiversity Conservation should be strengthened. The National Coordinating Group for Implementation of the Convention on Biological Diversity and the Inter-Ministerial Joint Conference on Conservation of Biological Resources should continue to play their roles. Capacities of relevant departments and bodies involved in biodiversity conservation should be further strengthened, with particular support to be provided to local governments and communities for their efforts in biodiversity conservation so as to improve their management level.

(6) To establish a system of biodiversity survey, monitoring and assessment as well as regularly launch survey and assessment results

Biodiversity surveys should be undertaken on a regular basis. A monitoring and early warning system for biodiversity should be established to capture in time dynamic changes in biodiversity, to launch Biodiversity Red Lists, and to more effectively monitor important species and ecosystems.

Part 1
Current Status of and Threats to China's Biodiversity

1.1 Importance of Biodiversity for Social and Economic Development

Biodiversity refers to the variability among living organisms from all sources, including, *inter alia*, terrestrial, marine and other aquatic ecosystems and the ecological complexes of which they are part, including diversity within species, between species and of ecosystems (the Convention on Biological Diversity). To put it simply, biodiversity refers to all species of plants, animals, micro-organisms and other forms of life on the Earth and their genetic varieties, as well as ecosystems they form in the environment.

Biodiversity provides conditions for human survival and material foundations for sustainable social and economic development as well as guarantees for ecological and food security. Biodiversity is also one of the important sources of literature, arts, scientific and technological innovations. Biodiversity has many values and functions. The primary industries such as agriculture, forestry, husbandry and fishery use biological resources directly, providing essential materials for human livelihood. Most of the second industries in particular pharmaceutical manufacturing, uses biological resources and their products directly as raw materials. More than 50% of medicinal components in the world come from animals and plants. Various and complicated ecosystems not only provide the environment for human survival, but also many ecosystem services. According to relevant estimates in 2000, the values of ecosystem services provided by forests in China, such as goods provided, carbon sequestration and oxygen release, water regulation, soil conservation, environment purification, nutrient cycling, recreation and biodiversity conservation, were about 1.4 trillion yuan RMB/a, equivalent to 14.2% of China's GDP of that year (Zhao et al., 2004). Grassland is a sink of carbons for the earth. The total carbon storage capacity of grassland ecosystems in China is about 44.09 billion tons. Grassland is also a natural water reservoir and energy store. 80% of the water flowing into the Yellow River, 30% of the water flowing into the Yangtze River, and more than 50% of the water flowing into rivers in northeast China come directly from grasslands. The total values of grassland ecosystems in China reach 1.2403 trillion yuan RMB (equivalent to 149.79 billion USD), about 3,100 yuan RMB for per hectare of grassland which far exceeds the value production on grasslands creates (Xie, 2001). Wetlands in China store about 270 million tons of fresh water which accounts for 96% of the total usable fresh water resources in China, so wetlands play an important role in water regulation, hydrological adjustment, water quality purification, underground water supply, flood control and combating droughts. Wetlands provide habitats for 20% of known species in the world and maintain rich biodiversity, therefore they are precious banks of germplasm and genetic resources. Wetlands are also a big sink of carbons whose carbon storage capacity is 35% of the totality of the terrestrial ecosystems. Pollination by insects plays a huge role for fruit and vegetable production in China. The economic value of pollination for fruit and vegetable growth estimated in 2008 was 52.17 billion USD, accounting for 25.5% of the gross output value of 44 varieties of fruits and vegetables (An et al., 2011). In some biodiversity-rich regions, such as Hainan Island, the regulation function of its ecosystems is valued eight times more than the value of goods produced (Ouyang et al., 2004). The value of ecosystem services provided in Xishuangbanna is 11 times more than the total value of GDP of the region (Jing et al., 2012).

Biodiversity attracts increasing attention of the international community due to its important roles in a nation's or a region's social and economic development. It has become another hot topic in the international environmental community, second to the issue of climate change.

1.2 Current Status of China's Biodiversity

China is one of the twelve countries with richest biodiversity in the world, and the country with richest biodiversity in the Northern Hemisphere. China has many types of natural ecosystems, such as forests, shrubs, meadows, grasslands, deserts, tundra, wetlands, marine and coastal ecosystems (The Ministry of Environmental Protection, 2011). According to the statistics from the Project of Remote-sensing Survey and Assessment of National Ecological Changes in the Decade from 2000 to 2010, the ecosystems with areas ranking top four are grassland, forest and agricultural ecosystems and deserts, with the total areas of these four ecosystems accounting for 82.9% of the areas of all ecosystems in China (Table 1-1, Figure 1-1).

China has more than 30,000 higher plant species, ranking third in the world, following Brazil and Colombia. China has over 6,000 vertebrate species, accounting for 13.7% of the world's total (The Ministry of Environmental Protection, 2011). The richness of vascular plants and mammals in China is characterized by highness in south China and lowness in north China, highness in mountains and lowness in plains.

Part 1 Current Status of and Threats to China's Biodiversity

Table 1-1　Distribution of Terrestrial Ecosystems of China and Percentage of Areas of All Ecosystems in 2010

Ecosystem Types	Area/10^3km²	Percentage of Areas of All Ecosystems/%
Grassland ecosystem	283.7	29.9
Forest ecosystem	193.3	20.3
Shrub ecosystem	69.2	7.3
Aquatic and wetland ecosystem	35.7	3.8
Agricultural ecosystem	182.4	19.2
Urban ecosystem	25.6	2.7
Desert ecosystem	127.7	13.4
Others	32.0	3.4

Notes: Data above do not include those from Taiwan Province of China.

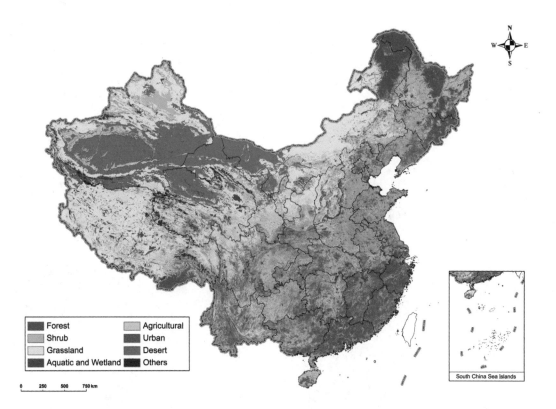

Figure 1-1　Map of Distribution of China's Terrestrial Ecosystems in 2010
（Data of Taiwan Province of China was not included）
Source: Project of Remote-sensing Survey and Assessment of Ecological Changes in
China from 2000 to 2010, courtesy of Ouyang et al.

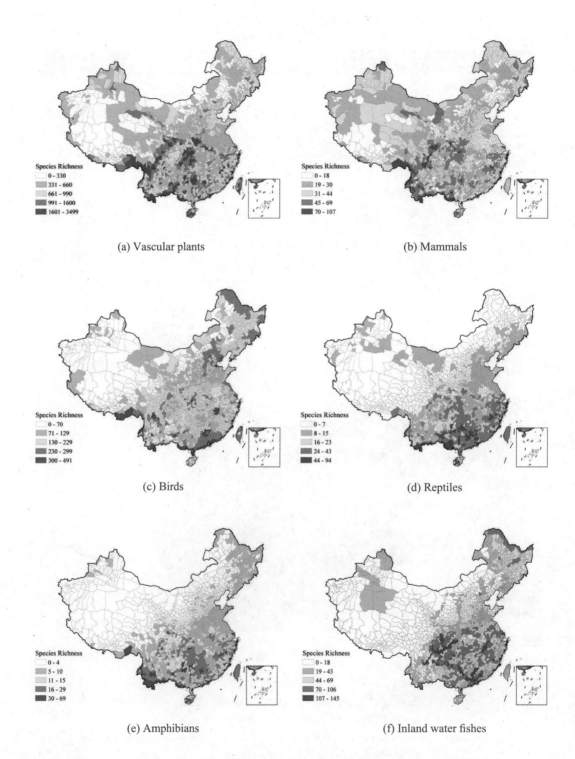

Figure 1-2 Spatial Distribution of China's Wild Vascular Plant and Vertebrate Species
Source: Xu et al. 2013

The main areas rich in vascular plants and mammals are Min Mountain, Qionglai Mountain, Hengduan Mountain, southeastern section of Himalaya Mountain, Qinling Mountain, Daba Mountain, Wuling Mountain, Wuyi Mountain, Xishuangbanna, border areas of southwestern Guangxi, and central and southern parts of Hainan [Figures 1-2 (a), (b)]. Most of the birds in China are migratory, flying to reproduction sites in the spring, and to the wintering sites in the south in the autumn. The distribution of birds is obviously characterized by their migration. The main areas rich in birds are the Bohai Sea rim region, Taiwan Island, coastal areas of Guangdong and Guangxi, Poyang Lake Basin, southeastern part of Tibet, Hengduan Mountain, Gaoligong Mountain in northwestern Yunnan and Xishuangbanna [Figure 1-2 (c)]. Amphibians and reptiles in China are mostly distributed in the region south to Qinling Mountain and Huai River. Regions rich in amphibians and reptiles are mainly Wuyi Mountain, Xishuangbanna, southwest Guangxi, Nanling Mountain and southern and central parts of Hainan [Figure 1-2 (d), (e)]. Areas with richest inland water fishes are the Yangtze River Basin and the Pearl River Basin, followed by the Huai River Basin and Heilongjiang River Basin. Hot areas of inland water fishes are mainly located in main branches in the upper reaches of the Yangtze River and its main branch Jialing River, as well as Wu River, the Pearl River, Min River, Poyang Lake and Dongting Lake [Figure 1-2 (f)] (Xu et al., 2013).

China has rich marine biodiversity. The number of marine species recorded so far has exceeded 28,000, accounting for about 11% of the world's total marine species. Among them, in the prokaryotes there are 9 phylum and 574 genus. In the protista there are 15 phylum and 4,894 genus. In the fungi there are 5 phylum and 371 genus. In the flora, there are 6 phylum and 1,496 genus, and in the fauna there are 24 phylum and 21,398 genus.

China has rich genetic resources, as a place of origin of important crops such as rice and soy beans as well as one of main centers of origin of wild and cultivated fruit trees. According to incomplete statistics, China has 1,339 varieties of cultivated crops, and 1,930 varieties of wild relatives. China's varieties of fruit trees rank top in the world. China is one of the countries with richest varieties of domesticated animals in the world, with 576 breeds of domesticated animals (The Ministry of Environmental Protection, 2011).

The percentage of endangered invertebrates in China (critically endangered, endangered and vulnerable) is 34.7% and the percentage of endangered vertebrates is 35.9% (Wang et al., 2004). The number of endangered plants in China is 3,767, accounting for about 10.9% of the total higher plant species in China. The number of higher plant species that require attention and protection is 10,102, accounting for 29.3% of the total (The Ministry of Environmental Protection and Chinese Academy of Sciences, 2013). China's loss of genetic resources is serious. According to the result of the second national survey on livestock genetic resources, 15 local breeds cannot be found any more, and the populations of more than a half of local breeds have been going down (National Committee on Livestock Genetic Resources, 2011).

1.3 Main Threats to Biodiversity in China

Biodiversity in China is facing various threats. The main pressures come from the rapid population growth and the accelerating pace of industrialization and urbanization, which have caused degradation or loss of habitats for wild species. Other threats include overexploitation of natural resources, environmental pollution, large-scale plantation of single species, invasive alien species and climate change.

(1) Degradation or loss of habitats for wild species

The main factor that endangers wild species is the degradation or loss of habitats (Wei, 2010; the Ministry of Environmental Protection and the Chinese Academy of Sciences, 2013). The land reclamation from wetlands undertaken from the 1950's to the 1990's has drastically shrunk the areas of wetlands. Despite some increases in the area of inland water in recent years, the area of land reclamation from tidal flats is still increasing. The area of land reclamation from the seas from 2008 to 2012 reached 650.6 km^2. As a result of land reclamation from tidal flats, the mangrove of China has decreased by about two-thirds, causing direct damage to habitats and reproduction sites for some important protected species. The total area of land reclamation from grasslands since the 1950's has come up to 193,000 km^2, with 18.2% of the total existing arable land of China coming from grassland reclamation (Fan et al., 2002). Incidents of grassland reclamation are still occurring in recent years. Railway and road construction has fragmented habitats for wild plants and animals, posing direct threat to reproduction of these populations. China's hydropower generation capacity has exceeded 230 million kilowatts, ranking top in the world, however dam building and flood control facilities have fragmented or obstructed rivers and lakes and drastically changed the natural conditions of water courses, causing disastrous impacts on reproduction of fishes (Xu, 2006).

(2) Overexploitation of natural resources

The overexploitation and use of wild biological resources has led to drastic decreases in species and populations as well as depletion or degradation of biological resources. Overgrazing of grasslands is serious. The rate of overgrazing exceeding capacities of key grasslands in China is 28% (Grassland Monitoring and Management Center, Ministry of Agriculture, 2012). Overgrazing for long time has degraded and desertified grasslands. By now 90% of grasslands in China have been degraded or desertified to varying degrees. Marine fishery plays an important role in China's fishery industry, with the marine catch amounting up to 15 million tons (Fishery Bureau, Ministry of Agriculture, 2011). High-intensity fishing has accelerated depletion of marine fishery resources, resulting in increasing the proportion of in catches of small, young and low-value fishes and lowering the nutrition level of fishes. Wild animals and plants have many economic values such as that for pharmaceuticals, food and recreation, and easily become targets of illegal trade. Despite law enforcement actions taken by China, illegal trade is still serious, and even very rampant in some regions.

(3) Environmental pollution

Environmental pollutants can generate various toxicities that can prevent normal growth of biological organisms and their reproductive and survival capabilities. The use of fertilizers, pesticides and herbicides has also caused increasing environmental pollution. Eutrophication caused by water pollution in Dianchi Lake of Kunming since the 1950's has reduced the species richness of higher aquatic plants by 36% and fishes by 25% (Lu et al., 2009). The pollution of China's coastal and near-shore marine areas is serious, though the overall environmental quality of the marine areas under China's jurisdiction is good. The marine environmental pollution has seriously damaged marine biodiversity, causing many marine ecological disasters such as red tide.

(4) Large-scale plantation of single species

With the development and wide application of new varieties, the cultivation of crops uses only a few species, leading to dramatic increases in areas of cultivation of single species. This has led to elimination or even permanent disappearance of many traditional varieties that contain important genetic resources.

(5) Invasive alien species

Invasion of alien species is one of the main causes of biodiversity loss. Due to its vast land area that covers nearly 50 latitudes and 5 climatic zones, as well as its diversified ecosystems, China is more vulnerable to the invasion of alien species, and species from any parts of the world may find suitable habitats in China. China is one of the countries that are most seriously affected by invasive alien species. The number of invasive alien species known so far in China has exceeded 500 (Xu et al., 2011). Invasion of alien species such as *Bursaphelenchus xylophilus, Oracella acuta, Hemiberlesia pitysophila, Hyphantria cunea, Matsucoccus matsumurae, Lissorhoptrus oryzophilus, Liriomyza sativae* and *Achatina fulica*, has caused serious negative impacts on agricultural production, the environment and biodiversity. It is estimated that the total annual cost of invasive alien species to the environment and economy of China is around 119.9 billion yuan (Xu et al., 2004).

(6) Climate change

Climate change shifts the phenology, distribution and migration of species, and causes some species to disappear from their original habitats. The climate in Qinghai Lake area is warming, so 26 bird species such as bean goose have disappeared from the lake area, compared with the situation in the 1950's (Ma et al., 2006). Climate change also expands the scope of distribution of pests and aggravates their harm. For example, climate warming expands the scope of distribution of *Solidago canadensis* (Wu et al., 2008). Climate change modifies the population structure of marine species. The population and its density of cold-water animals in the Yellow Sea of China have decreased as the water temperature rises. The cold water benthic biota diversity in the Yellow Sea has decreased considerably compared with the situation of a half century ago (Liu, 2011).

1.4 Economic and Social Implications of Biodiversity Loss

Biodiversity not only provides human beings with necessary living materials, industrial materials and natural medicinal herbs, but also plays key roles in protecting the environment and maintaining ecological security, in particular in purifying the environment, ensuring water quality and improving soil quality. Biodiversity provides material foundations for human survival and sustainable social and economic development, and therefore very important for human well-being of the present and future generations. According to the Millennium Ecosystem Assessment, biodiversity loss will directly or indirectly cause more health problems, higher food risks, increasing vulnerabilities and fewer development opportunities (Millennium Ecosystem Assessment, 2005).

(1) Direct impacts on human life and property

Biodiversity loss will increase vulnerabilities of ecosystems. Loss of components of biodiversity, in particular decrease in functional diversity and ecosystem diversity at landscape level, will lead to decrease in ecosystem stability. Mangroves and coral reefs are rich sources of biodiversity as well as very good buffers of floods and storms. If mangroves and coral reefs are damaged, floods in coastal areas will increase, seriously affecting mari-culturing and houses of the residents in coastal areas. Excessive deforestation will cause soil erosion which is one of the important causes of landslides.

(2) Impacts on food security

Biodiversity loss will reduce food diversity, forcing human being to depend on only a few main kinds of food, thus break the balance of human food structure and affect human health. For example, loss of pollinating insects will reduce outputs of crops depending on insect pollination (Garibaldi et al., 2011). Wild relatives of crops play an important role in agricultural production. In the 1970's, Academician Yuan Longping, a well-known Chinese expert on rice seed breeding, used wild rice sterile plants found in Hainan to hybrid with cultivated rice and successfully created hybrid rice, making remarkable contribution to food security in China and the world. However if wild rice had disappeared then he would not have made such a huge scientific achievement. Unfortunately, the natural population of wild rice in China is currently rapidly declining, and on the verge of extinction in some sites of distribution. The extinction of wild rice is not just a loss of one species, but will have significant implications for human food security.

(3) Impacts on pharmaceutical sector

China has more than 12,000 kinds of medicinal resources, ranking high in the world. Rare medicinal resources are usually characterized by small areas of distribution, poor capacity of regeneration and long period of growth. Excessive use of wild medicinal resources for long has drastically reduced reserves of many medicinal resources, and even caused extinction of some of them. The development of the pharmaceutical sector will depend on rich medicinal resources, and the sector will lose foundations for development if without these medicinal resources. Despite more and more medicines from lab research and development, a large number of people in the world are still using natural medicines to treat their diseases. The loss of natural medicinal resources will have serious implications.

(4) Impacts on future development

Biodiversity loss will reduce development opportunities of local communities and residents. Biodiversity loss and ecosystem destruction may lead to a decreasing number of tourists in those regions that used to have rich resources for tourism, and local communities and residents will lose opportunities to develop tourism. In some cases, biodiversity loss is irreversible, therefore future generations will lose development opportunities.

To protect biodiversity is a must for ensuring ecological security, and important for maintaining productivity of the natural environment and achieving sustainable development. It is also crucial for changing China's economic development patterns, building an ecological civilization, achieving sustainable development as well as realizing China Dream.

Part 2
National Biodiversity Strategy and Action Plan and Its Implementation

2.1 Development of China's Updated NBSAP

the updated *China's National Biodiversity Strategy and Action Plan* (2011-2030) was approved at the 126th regular meeting of the State Council on 15 September 2010, and promulgated by the Ministry of Environmental Protection on 17 September 2010. This updated NBSAP has identified guiding principles, strategic goals and tasks for biodiversity conservation in China in the next two decades. It has also identified 35 priority regions for biodiversity conservation (Figure 2-1) across China, as well as proposed 10 priority areas, 30 priority actions and 39 priority projects for implementation.

The updating of NBSAP took more than three years. In this process, many thematic studies were undertaken, and many working meetings, consultations and international workshops were held. Consultations were also undertaken with more than 20 central government departments and 31 provincial governments. Therefore the development of this updated NBSAP was a very participatory process by having involved a wide range of stakeholders. It is a result of the joint efforts of the members of China's Coordinating Group for Implementation of the Convention on Biological Diversity and the Inter-ministerial Joint Conference on Conservation of Biological Resources. It is also an example of cooperation between domestic and international institutions or organizations.

China had developed its first NBSAP as early as 1994. Due to its early development, the first NBSAP did not cover some important provisions of the Convention, in particular its third objective (fair and equitable sharing of benefits from use of genetic resources). A few other issues that had emerged after the entry into force of the Convention such as invasive alien species, access to and benefit-sharing from use of genetic resources and associated traditional knowledge and biosafety management of genetic modified organisms (GMOs) were not appropriately addressed in the first NBSAP. In addition the first NBSAP did not really contain national strategies. Therefore the Government of China proposed a new national biodiversity strategy and action plan under the new circumstances, as required by social and economic development in China and the international obligation of biodiversity conservation. The updated NBSAP has identified three major goals for three periods of time as well as 35 priority regions for conservation. Relevant strategies have been added and issues such as invasive alien species, climate change, access to and benefit-sharing from use of genetic resources, traditional knowledge and biosafety management of GMOs have been addressed. The development and implementation of the updated NBSAP will produce positive and far-reaching impacts on biodiversity conservation in China and even in the world.

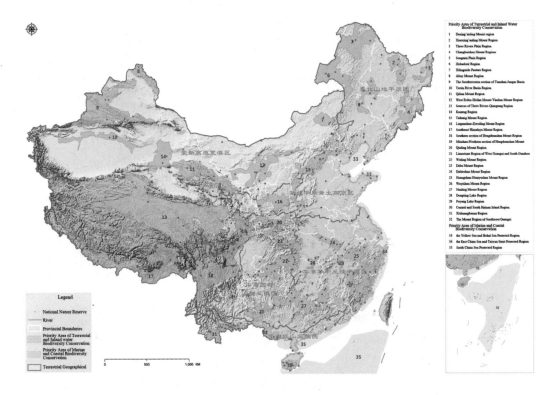

Figure 2-1　35 Priority Regions for Biodiversity Conservation

2.2　National Targets for Biodiversity Conservation

The tenth meeting of the Conference of the Parties held in Japan in October 2010 adopted the *Strategic Plan for Biodiversity* (2011-2020), which identified 2020 Biodiversity Targets (also called "Aichi Targets", hereafter "2020 Targets"). These targets provided roadmaps and time tables for biodiversity conservation in the world as well as a flexible framework for setting national targets. The 2020 Targets consist of 5 strategic goals and 20 specific targets (Table 2-1).

At the end of 2010 the State Council issued *National Plan for Major Function Zones*, which divides the country's land into four major function zones, such as zones for priority development, zones for key development, zones for limited development and zones prohibited for development. 25 key ecological function zones have been included in those for limited development. Large-scale and high-intensity industrial and urbanization development activities will be limited within these zones to allow for the conservation and restoration of the environment and to provide ecological goods and ecosystem services. Meanwhile, national-level nature reserves, world natural and cultural heritages, national scenic spots, national forest parks and geological parks have been included in those zones prohibited for development, where industrial and urbanization development activities will be prohibited to protect China's natural and cultural resources and genetic resources of rare animals and plants.

The Eighteenth National Congress of the Chinese Communist Party held in November 2012 laid out a blueprint for building an ecological civilization, and adopted a grand vision of "Building Beautiful China". The meeting also proposed that priority would be given to building an ecological civilization by integrating it with various aspects and processes of economic, political, cultural and social developments. The meeting required that in doing so the principles below should be followed:

(1) Continuing to implement the fundamental national policy of environmental protection and resource conservation;

(2) Giving priorities to conservation, protection and natural restoration;

(3) Promoting green, cycling and low-carbon development;

(4) Creating resource-efficient and environmentally friendly spatial layouts, industrial structures, production and consumption patterns as well as lifestyles to reverse the environmental worsening trends at source.

The updated NBSAP proposed the following fundamental principles: ① conservation as a priority; ② ensuring sustainable use; ③ public participation; ④ benefits for all. The plan has identified the short-term (2015), mid-term (2020) and long-term (2030) goals (Table 2-1). Though the updated NBSAP was launched before the tenth meeting of the Conference of the Parties held in 2010 adopted the *Strategic Plan for Biodiversity* 2011-2020, its development had fully considered goals and tasks contained in the draft Strategic Plan for the period after 2010.

The vision and strategic goals that the Government of China had proposed for building an ecological civilization and Beautiful China, together with the updated NBSAP, have formed a relatively comprehensive set of national targets for biodiversity conservation in China (Table 2-1). However there are no specific national targets in line with Aichi Targets 7, 9, 10, 13, 16 and 19, and effective measures and means to achieve these targets are lacking. Therefore, the Government of China should give more attention to addressing issues such as agricultural and forest sustainable development, prevention and control of invasive alien species, protection of genetic resources and benefit-sharing from their use, addressing climate change impacts on coral reefs and other vulnerable ecosystems. Meanwhile, China needs to further increase its investment in research, development and application of science and technologies for biodiversity conservation.

Table 2-1 2020 Global Biodiversity Targets and China's National Targets

Strategic Plan for Biodiversity 2011-2020	National Goals and Targets	Sources of Information
The vision of this Strategic Plan: a world of "Living in harmony with nature" where "By 2050, biodiversity is valued, conserved, restored and wisely used, maintaining ecosystem services, sustaining a healthy planet and delivering benefits essential for all people." The mission of the Strategic Plan: "take effective and urgent action to halt the loss of biodiversity in order to ensure that by 2020 ecosystems are resilient and continue to provide essential services, thereby securing the planet's variety of life, and contributing to human well-being, and poverty eradication. To ensure this, pressures on biodiversity are reduced, ecosystems are restored, biological resources are sustainably used and benefits arising out of utilization of genetic resources are shared in a fair and equitable manner; adequate financial resources are provided, capacities are enhanced, biodiversity issues and values mainstreamed, appropriate policies are effectively implemented, and decision-making is based on sound science and the precautionary approach."	**Long-term Goal:** by 2030, biodiversity will be effectively protected. **2020 Goal:** by 2020, biodiversity loss will be basically controlled. **2015 Goal:** by 2015, biodiversity decline in key regions will be effectively contained	*Updated NBSAP* [c]
2020 Global Biodiversity Targets (Aichi Targets)		
Target 1: By 2020, at the latest, people are aware of the values of biodiversity and the steps they can take to conserve and use it sustainably	● Practical efforts will be made in environmental education and communication, popularizing environmental knowledge and increasing public environmental awareness. ● By 2030, biodiversity conservation will become voluntary action of the public	*National Programme for Environmental Education and Communication (2011-2015)* [d] *Updated NBSAP* [c]
Target 2: By 2020, at the latest, biodiversity values have been integrated into national and local development and poverty reduction strategies and planning processes and are being incorporated into national accounting, as appropriate, and reporting systems	● Resource consumption, environmental damage and ecological benefits will be incorporated into the system of assessing social and economic development, and a system of goals and targets, as well as related assessment methods and reward/penalty mechanisms that meet requirements for building an ecological civilization will be established	*Report of Eighteenth National Congress of the Chinese Communist Party (CPC)* [a]

Strategic Plan for Biodiversity 2011-2020	National Goals and Targets	Sources of Information
Target 3: By 2020, at the latest, incentives, including subsidies, harmful to biodiversity are eliminated, phased out or reformed in order to minimize or avoid negative impacts, and positive incentives for the conservation and sustainable use of biodiversity are developed and applied, consistent and in harmony with the Convention and other relevant international obligations, taking into account national socio economic conditions	• Establishment of mechanisms for ecological compensation and increasing fiscal transfers to key ecological function zones will be accelerated; and studies will be undertaken on the establishment of national specialized funds for ecological compensation and the system of reserves for sustainable development of resource-consumption enterprises will be promoted	*National 12th Five-Year Plan for Social and Economic Development of the People's Republic of China* [b]
Target 4: By 2020, at the latest, Governments, business and stakeholders at all levels have taken steps to achieve or have implemented plans for sustainable production and consumption and have kept the impacts of use of natural resources well within safe ecological limits	• By 2015 considerable progress will be made in building a resource-efficient and environmentally friendly society. • Efforts will be made to promote spatial layouts, industrial structure, production and consumption patterns and lifestyles that promote green, recycling and low-carbon development, natural resources conservation and the environmental protection	*National 12th Five-Year Plan for Social and Economic Development* [b] *Report of Eighteenth CPC National Congress* [a]
Target 5: By 2020, the rate of loss of all natural habitats, including forests, is at least halved and where feasible brought close to zero, and degradation and fragmentation is significantly reduced	• By 2015, forest coverage rate will be increased to 21.66% and forest reserves will be increased by 600 million m^3 over that in 2010. • By 2020, grassland degradation trend will be basically contained and grassland ecological environment will be obviously improved. • By 2020, the environmental and ecological worsening trends in coastal and near-shore areas will be fundamentally reversed and marine biodiversity decline trend will be basically contained. • By 2020, aquatic environment and ecology will be gradually restored and decline of fishery resources and increase in endangered species will be basically contained	*National 12th Five-year Plan for Social and Economic Development* [b] *National Master Plan for Conservation and Use of Grasslands* [c] *National Twelfth Five-year Plan for Marine Development* [c] *National Programme of Action for Conservation of Aquatic Species* [c]

Strategic Plan for Biodiversity 2011-2020	National Goals and Targets	Sources of Information
Target 6: By 2020 all fish and invertebrate stocks and aquatic plants are managed and harvested sustainably, legally and applying ecosystem based approaches, so that overfishing is avoided, recovery plans and measures are in place for all depleted species, fisheries have no significant adverse impacts on threatened species and vulnerable ecosystems and the impacts of fisheries on stocks, species and ecosystems are within safe ecological limits	• By 2020, aquatic environment and ecology will be gradually restored and decline of fishery resources and increase in endangered species will be basically contained. • By 2020, the environmental and ecological worsening trends in coastal and near-shore areas will be fundamentally reversed and marine biodiversity decline trend will be basically contained	*National Programme of Action for Conservation of Aquatic Species* [c] *National Twelfth Five-year Plan for Marine Development* [c]
Target 7: By 2020 areas under agriculture, aquaculture and forestry are managed sustainably, ensuring conservation of biodiversity	• By 2020, national forest holdings will exceed 2.33 million km², an increase 223,000 km² over that of 2010; and national forest reserves will be increased to 15 billion m³, an increase of about 1.2 billion m³ over that of 2010. • By 2020, husbandry production pattern will be changed and grassland sustainability will be effectively enhanced. • By 2020, fishing capacities and outputs will be generally consistent with carrying capacities of fishery resources	*National Programme for Conservation and Use of Forestland 2010-2020* [c] *National Master Plan for Conservation and Use of Grasslands* [c] *National Programme of Action for Conservation of Aquatic Biological Resources* [c]
Target 8: By 2020, pollution, including from excess nutrients, has been brought to levels that are not detrimental to ecosystem function and biodiversity	• By 2015, the total amount of emission of main pollutants will be significantly reduced, with COD and SO_2 reduced by 8%, and ammonia and NO_x reduced by 10% compared with the levels of 2010. • By 2020, energy consumption and CO_2 emission per unit of GDP will be reduced significantly, with the total amount of main pollutants considerably reduced	*National Twelfth Five-year Plan for Social and Economic Development* [b] *Report of Eighteenth National Congress of CPC* [a]
Target 9: By 2020, invasive alien species and pathways are identified and prioritized, priority species are controlled or eradicated, and measures are in place to manage pathways to prevent their introduction and establishment	• By 2020, forest pest disaster rate will be controlled at 4%	*National Plan for Forest Pest Control 2011-2020* [d]

Strategic Plan for Biodiversity 2011-2020	National Goals and Targets	Sources of Information
Target 10: By 2015, the multiple anthropogenic pressures on coral reefs, and other vulnerable ecosystems impacted by climate change or ocean acidification are minimized, so as to maintain their integrity and functioning	• By 2020, energy consumption and CO_2 emission per unit of GDP will be reduced significantly. • By 2020, a system of nature reserves with reasonable layouts and comprehensive functions will be established, with functions of national-level nature reserves stable, and main targets of protection effectively protected	*Report of Eighteenth National Congress of CPC* [a] *Updated NBSAP* [c]
Target 11: By 2020, at least 17 per cent of terrestrial and inland water areas, and 10 per cent of coastal and marine areas, especially areas of particular importance for biodiversity and ecosystem services, are conserved through effectively and equitably managed, ecologically representative and well connected systems of protected areas and other effective area-based conservation measures, and integrated into the wider landscapes and seascapes	• By 2015, the total area of terrestrial nature reserves will be maintained at 15% or so of the country's land area, protecting 90% of national key protected species and typical ecosystem types. • The percentage of the area of marine protected areas out of the marine areas under China's jurisdiction will be increased from 1.1% in 2010 to 3% in 2015. • By 2020, the total area of marine protected areas will exceed 5% of the marine areas under China's jurisdiction, with the area of coastal marine protected areas exceeding 11%. • By 2020, a system of nature reserves with reasonable layouts and comprehensive functions will be established, with functions of national-level nature reserves stable, and main targets of protection effectively protected	*Updated NBSAP* [c] *National Twelfth Five-year Plan for Marine Development* [c] *National Marine Zoning Plan 2011-2020* [c] *Updated NBSAP* [c]
Target 12: By 2020 the extinction of known threatened species has been prevented and their conservation status, particularly of those most in decline, has been improved and sustained	• By 2015, more than 80% of endangered species whose wild populations are very small and for which *in-situ* conservation capacities are inadequate will be effectively protected. • By 2020 functions of national-level nature reserves will be maintained stable, and main targets of protection effectively protected. • By 2020, the majority of rare and endangered species and populations will be restored and reproduced, relieving the situation of species endangerment	*Updated NBSAP* [c] *National Programme for Conservation and Use of Biological Resources* [c]

Strategic Plan for Biodiversity 2011-2020	National Goals and Targets	Sources of Information
Target 13: By 2020, the genetic diversity of cultivated plants and farmed and domesticated animals and of wild relatives, including other socio-economically as well as culturally valuable species, is maintained, and strategies have been developed and implemented for minimizing genetic erosion and safeguarding their genetic diversity	● By 2020, biodiversity loss will be basically contained, and a system of nature reserves with reasonable layouts and comprehensive functions will be established, with main targets of protection effectively protected. ● *National List of Protection of Livestock Genetic Resources* will be revised so as to accord key protection to rare and endangered livestock genetic resources in the list and ensure that protected varieties will not be lost and their economic values will not be decreased	*Updated NBSAP* [c] *National Twelfth Five-year Plan for Conservation and Use of Livestock Genetic Resources*
Target 14: By 2020, ecosystems that provide essential services, including services related to water, and contribute to health, livelihoods and well-being, are restored and safeguarded, taking into account the needs of women, indigenous and local communities, and the poor and vulnerable	● By 2020, the stability of ecosystems will be strengthened, and the human environment will be considerably improved. ● By 2020, grass-herd balance will be achieved in natural grasslands, grassland habitats will be obviously restored and grassland productivity will be significantly enhanced. ● By 2020, the environmental degradation of the coastal and near-shore marine areas will be reversed, and decline of marine biodiversity will be basically contained	*Report of Eighteenth National Congress of CPC* [a] *National Master Plan for Conservation and Use of Grasslands* [c] *National Twelfth Five-year Plan for Marine Development* [c]
Target 15: By 2020, ecosystem resilience and the contribution of biodiversity to carbon stocks has been enhanced, through conservation and restoration, including restoration of at least 15 per cent of degraded ecosystems, thereby contributing to climate change mitigation and adaptation and to combating desertification	● By 2020, forest areas will be increased by 52,000 km^2 over that in 2010, and forest reserves net increased by 1.1 billion m^3 over that in 2010, and forest carbon sinks by 416 million tons. ● By 2020, the total areas of control of degraded grasslands will exceed 1.65 million km^2, with grassland habitats obviously restored and grassland productivity significantly enhanced. ● By 2020, the aquatic environment and ecology will be gradually restored	*Plan for Second-Phase of Project on Natural Forest Resources Protection* [c] *National Master Plan for Conservation and Use of Grasslands* [c] *National Programme of Action for Conservation of Aquatic Biological Resources* [c]

Strategic Plan for Biodiversity 2011-2020	National Goals and Targets	Sources of Information
Target 16: By 2015, the Nagoya Protocol on Access to Genetic Resources and the Fair and Equitable Sharing of Benefits Arising from their Utilization is in force and operational, consistent with national legislation	● By 2020, the system of access to genetic resources and benefit-sharing from their use will be improved.	*Updated NBSAP* [c]
Target 17: By 2015, each party has developed, adopted as a policy instrument, and has commenced implementing an effective, participatory and updated national biodiversity strategy and action plan	Updated NBSAP has been promulgated	
Target 18: By 2020, the traditional knowledge, innovations and practices of indigenous and local communities relevant for the conservation and sustainable use of biodiversity, and their customary use of biological resources, are respected, subject to national legislation and relevant international obligations, and fully integrated and reflected in the implementation of the Convention with the full and effective participation of indigenous and local communities, at all relevant levels	● By 2020, documentation of relevant traditional knowledge within China and the intellectual rights protection system will be further improved	*National Programme for Conservation and Use of Biological Resources* [c]
Target 19: By 2020, knowledge, the science base and technologies relating to biodiversity, its values, functioning, status and trends, and the consequences of its loss, are improved, widely shared and transferred, and applied	● By 2020, the percentage of investment in research and development activities will exceed 2.5% of national GDP, with contributions from science and technology to GDP reaching 60%, and the number of annual patent grants to the Chinese individuals and groups and of citations of academic papers by international journals ranking top five in the world. ● Environmental education will be undertaken to popularize environmental knowledge and increase public environmental awareness	*National Mid and Long-term Plan for Science and Technology Development 2006-2020* [c] *National Programme of Action for Environmental Education 2011—2015* [d]

Strategic Plan for Biodiversity 2011-2020	National Goals and Targets	Sources of Information
Target 20: By 2020, at the latest, the mobilization of financial resources for effectively implementing the Strategic Plan for Biodiversity 2011-2020 from all sources, and in accordance with the consolidated and agreed process in the Strategy for Resource Mobilization, should increase substantially from the current levels. This target will be subject to changes contingent to resource needs assessments to be developed and reported by Parties	● Channels of investment will be broadened and investments from local and central governments will be increased and financing from the banking sector, international donors and the civil society will be attracted to biodiversity conservation, with diverse financing mechanisms established	*Updated NBSAP* [c]

Notes: (a) docs approved by National Congress of Chinese Communist Party; (b) docs approved by National People's Congress; (c) docs approved or issued by the State Council; (d) docs promulgated by six relevant government departments including the Ministry of Environmental Protection.

2.3 Main Actions to Implement the Convention on Biological Diversity

2.3.1 Laws and regulations

The Government of China has promulgated in recent years a number of new laws and regulations such as *Island Conservation Law* and *Regulation on Protection of New Plant Varieties*, in addition to more than 50 existing laws and regulations related to biodiversity. China has developed and launched a series of national, sectoral and local standards for biodiversity conservation. All this has further improved the legal and regulatory system for biodiversity conservation and use.

2.3.2 Cross-sectoral working mechanisms

To implement the Convention on Biological Diversity (CBD), the State Council has approved the establishment of China's Coordinating Group for Implementation of the Convention on Biological Diversity, headed by the Ministry of Environmental Protection and composed of 24 departments. CBD Implementation Office was established in the Ministry of Environmental Protection. To strengthen conservation and management of biological resources, the State Council has also approved the establishment of an Inter-ministerial Joint Conference on Conservation of Biological Resources, headed by the Ministry of Environmental Protection and composed of 17 ministries and commissions. An office for this joint conference was established in the Ministry of Environmental Protection. To organize celebration activities for "the International Year of Biodiversity in 2010", the Government of China established National Committee for the International Year of Biodiversity (2010), headed by one of the Vice Premiers of the State Council responsible for the environment and composed of 25 departments. In 2011, the State Council decided to change this Committee into National Committee on Biodiversity Conservation. This Committee consists of 25 departments and the Secretariat of this Committee was established in the Ministry of Environmental Protection. These three implementation coordination bodies headed by MEP and participated by relevant departments are mutually supportive while exercising their unique, important roles in enhancing biodiversity conservation in China.

Most provincial and municipal governments have also strengthened their biodiversity-related departments or institutions such as those responsible for agriculture, environment, forestry and marine management, as well as established inter-departmental coordinating mechanisms.

Case 2.1 Sichuan Provincial Biodiversity Strategy and Action Plan (2011-2020)

Sichuan Province is located in the upper reaches of the Yangtze River, and one of the world's 25 biodiversity hotspots. Sichuan's formulation and implementation of its Biodiversity Strategy and Action Plan(BSAP) is significant for implementing the Convention on Biological Diversity in China. With the support of the Ministry of Environmental Protection, United Nations Development Programme (UNDP) and The Nature Conservancy, Sichuan Environment Department and Forest Department, together with other relevant departments, initiated in 2007 the development of Sichuan Provincial BSAP. It took more than two years to complete this BSAP as a result of joint efforts of relevant departments, institutions and experts involved, following many studies, surveys and consultations. This BSAP was approved at the 89th session of Sichuan Province Government in December 2011.

This BSAP for the first time identified 13 priority areas for biodiversity conservation in the province, 9 priority areas for actions (such as developing regulations and policies, establishing baseline information, protecting wild species and their habitats, monitoring and strengthening research on conservation etc.) as well as 46 priority actions.

At present, relevant departments of Sichuan Province are effectively promoting the implementation of this BSAP, and making great efforts in establishment of biodiversity information system, biodiversity recovery following the earthquake in Wenchuan, and rescuing rare, endangered wild species.

> **Case 2.2 Biodiversity Conservation in Yunnan Province: From Northwest Yunnan to Whole Province**
>
> Yunnan is one of the provinces with richest biodiversity in China. Northwest Yunnan is located in the transition zone between Qinghai-Tibet Plateau and Yunnan-Guizhou Plateau, and has been therefore identified one of the priority areas for biodiversity. It covers 18 counties (cities or districts) in Yunnan Province. To strengthen biodiversity conservation in northwest Yunnan, in February 2008, the People's Government of Yunnan Province held a meeting in Lijiang on biodiversity conservation in northwest Yunnan. The meeting developed a number of recommendations and adopted a declaration. Subsequently, the Government of Yunnan Province developed a programme and an action plan for biodiversity conservation in northwest Yunnan. The province also established a mechanism of joint conference of relevant departments for biodiversity conservation. In May 2010 this joint conference as well as a forum on Yunnan Actions for the International Year of Biodiversity (IYB) was held in Tengchong, where a programme of action for 2010 IYB was declared. Yunnan has also established a Biodiversity Fund and an Institute of Biodiversity. In April 2012, Yunnan held another joint conference in Xishuangbanna where an agreement was reached for biodiversity conservation in Yunnan Province and ten measures were proposed to enhance biodiversity conservation. In 2013, Yunnan Province approved and issued its *Biodiversity Strategy* and *Action Plan* (2012-2030). This document provides strategic guidance for biodiversity conservation in Yunnan in the next two decades. From the *Lijiang Declaration, Tengchong Programme of Action, Xishuangbanna Agreement* and *Yunnan Provincial Biodiversity Strategy* and *Action Plan*, Yunnan has been putting more and more efforts into its biodiversity conservation.
>
> During the eleventh five-year plan period, various departments of Yunnan had invested cumulatively a total of nearly 7 billion yuan RMB into biodiversity conservation and sustainable use. Since the twelfth five-year plan period began, Yunnan has been increasing its investment into biodiversity conservation. Yunnan Biodiversity Fund has received a donation of 32.3 million yuan RMB. In March 2013, the Government of Yunnan Province has established a specialized fund for biodiversity conservation totaling 50 million yuan RMB, with allocations from the province's budget.

2.3.3 Survey and monitoring

(1) Inventorying national forest resources

National forest resources consensus is undertaken at province level every five years. Sampling techniques are used to select 415,000 fixed sites on the ground and 2.84 million sample sites using remote-sensing. Consensus is undertaken to get the current status of forest resources including dynamic changes in all provinces (autonomous regions and province-level municipalities) with surveys undertaken at the same time following the same procedures and requirements. The consensuses have generated rich results, intensive information and reliable data, and therefore are considered the most authoritative data that reflect the status of forest resources at national and province level. So far China has undertaken seven consensuses on forest resources. During 2009-2013, China conducted the eighth consensus and its result will be launched shortly.

(2) Biodiversity survey

From 2006 to 2008, China undertook a survey on coastal and near-shore marine species, which helped get the baseline data on marine biological resources. As a result China's Marine Species and Atlas was published, which includes more than 28,000 marine species and pictures of more than 18,000 species.

Surveys were also undertaken on plant, animal and microorganism diversity in key regions, including Pan-Himalaya region, Qinghai-Tibet Plateau, Xinjiang, Luoxiao Mountain Range, hilly areas of south China, areas in southwest China inhabited by ethnic minorities, ecologically sensitive areas in plains and hills along coastal areas in southeast China, tropical islands and coastal areas, the Yangtze River Basin, arid areas in northwest China, the Loess Plateau, Major and Minor Xing'anling Mountains in northeast China, and grasslands in northeast China. From 2011 to 2012, systematic wild surveys were undertaken on biological resources in 26 counties (cities and districts) of Yunnan, Guangxi and Guizhou Provinces. These surveys resulted in discoveries of 19 new (or probably new) plant taxa, three new records of distribution in China and 49 new records of distribution at provincial level.

During the eleventh five-year plan period, a second national survey on livestock genetic resources was undertaken, and *Annals of China's Livestock Genetic Resources* was published. The survey showed that 15 local varieties of livestock genetic resources were no longer found, and the populations of more than a half of local varieties took a downward trend.

Since 2009, China has organized a second national survey on all wetlands with area above 0.08 km^2, by using "3S" technologies and following the criteria set by the Ramsar Convention.

In addition China had undertaken a remote-sensing survey and assessment of ecological changes in the decade from 2000 to 2010 as well as a survey on agricultural biological resources. China is now undertaking a number of surveys, including a second survey on key wild animals, a second survey on national key protected wild plants, the fourth survey on habitats for Giant Pandas and a pilot consensus on Chinese medicinal resources.

(3) Biodiversity monitoring

A proposal has been developed for establishing a national network of biodiversity monitoring based on stratified sampling. Technical guidelines have been developed for monitoring plants, mammals, birds, amphibians and reptiles, fishes, soil animals, butterflies and large fungi. Training activities were also organized on monitoring techniques.

China Forest Biodiversity Monitoring Network (CForBio) (http://www.cfbiodiv.org) was established in 2004, covering different types of forest vegetation at different latitudes, including coniferous and broad-leaved mixed forests, deciduous forests, evergreen deciduous and broad-leaved mixed forests, ever-green broad-leaved forests and tropical rainforests. By 2012, CForBio has covered 12 major monitoring sites, with each site covering an area ranging from 0.09 to 0.25 km^2.

Since 2011, China has been undertaking pilot monitoring of birds and amphibians. More than 200 monitoring sites have been established in different regions and ecosystems, with more than 450 line transects and more than 430 point transects established.

Since 2004, China has established 18 marine ecological monitoring zones in a number of ecologically vulnerable and sensitive coastal and near-shore areas, and been undertaking systematic biodiversity monitoring, assessment and conservation in these zones. The area being monitored has reached 52,000 km^2, including typical marine ecosystems such as bays, estuaries, coastal wetlands, coral reefs, mangroves and sea grass beds.

Since 2005, China has established an ecological monitoring system at sources of three major rivers in Qinghai Province. This system includes 5 ecological monitoring systems, 14 ecological monitoring stations, 486 on-the-ground monitoring points, 3 soil conservation monitoring communities, 2 mobile monitoring stations of hydrological resources, 4 monitoring team of hydrological resonrces and 2 automatic meteorological stations.

China has been monitoring 15 protected sites of wild plants, and has obtained relevant data for five consecutive years.

(4) Biodiversity assessments

During 2007-2012, China completed its national biodiversity assessment. This assessment was undertaken at county level, and as a result for the first time data have been collected of county-level distribution of 34,039 vascular plants and 3,865 wild vertebrates. Based on that a national biodiversity information system has been established, with almost all the information concerning the status, spatial distribution and main threats of terrestrial biodiversity covered. In addition, through this assessment national biodiversity hotspots and major gaps in conservation have been identified, having preliminarily resolved the difficult situation that China did not know its own biodiversity status for long. China has developed a set of indicators for assessing the health, functions and values of ecosystems in order to evaluate scientifically and accurately the status of the ecosystems.

2.3.4 *In-situ* Conservation

A conservation system has been established, primarily composed of nature reserves and complemented by scenic spots, forest parks, community-based conservation areas, protected sites of wild plants, wetland parks, geological parks, special marine protected areas and germplasm conservation areas. By the end of 2013, a total 2,697 nature reserves of various categories have been established at different levels, covering about an area of 1.463 million km^2 and accounting for about 14.8% of the country's land area. Among them, there are 407 national-level nature reserves, covering an area of about 940,000 km^2 which accounts for 64.3% of the total area of nature reserves and 9.8% of the country's land area. Since 2008 the number of marine protected areas in particular national-level MPAs has increased substantially. By the end of 2012, more than 240 marine protected areas of various types have been established at different levels, with the total area (MPAs) covering 87,000 km^2, accounting for nearly 3% of the marine areas under China's jurisdiction.

By the end of 2012, China has established 2,855 forest parks, covering a total area of 174,000 km^2. Among them there are 764 national-level forest parks and 1,315 province-level forest parks. 225 national-level scenic spots have been established, covering an area of 104,000 km^2, and 737 province-level scenic spots established, covering an area of about 90,000 km^2, areas with both combined accounting for 2% of China's land area. More than 50,000 community-based conservation areas have been established, covering an area of over 15,000 km^2. 179 protected sites of various wild plants have been established. 468 wetland parks have been established. From 2007 to 2012, 368 national-level aquatic germplasm conservation areas have been established, covering an area of more than 152,000 km^2.

Nature reserves have become key zones among China's major ecological function zones, and constitute main parts of "zones banned for development". They have effectively protected 90% of terrestrial ecosystem types, 85% of wild animal populations and 65% of higher plant biota in China. They have also covered 25% of primitive and natural forests, more than 50% of natural wetlands and 30% of typical desert regions, thus playing a crucial role in maintaining ecological security and promoting sustainable social and economic development of China.

Case 2.3 Compulsory Conservation Model of Wuyi Mountain

Wuyi Mountain has the most complete, most typical and the biggest area of subtropical forest ecosystem compared with those located at the same latitude of the world. It was included on the World's Heritages List in December 1999. Through innovative approaches described below, a new sustainable development model has been established in Wuyi Mountain, which is "using ecological industry development in 10% of its area in exchange for biodiversity conservation in 90% of its area".

(1) Strengthening regulatory development and promoting regulation of tourism industry. A few regulations have been promulgated, such as *Rules on Management of Scenic Spots in Wuyi* and *Provisional Rules on Management of Jiuqu Stream in Wuyi Mountain*. The implementation of these regulations has significantly promoted ecological conservation and sustainable tourism development in Wuyi Mountain.

(2) Setting up a joint mechanism for conservation and establishing a complete system of conservation. A joint conservation committee was established composed of governments of all levels and local communities. A joint 200-km-long protection line was established around protected areas, and 272 forest guards were hired and nearly 10% of residents in local communities were directly involved in conservation and management. Specialized forest fire fighting team and police patrolling points were established. As a result of joint efforts of all stakeholders involved, no forest fires, no illegal crimes and no major forest pest happened in the past 25 years. Wuyi has become a national example in this regard.

(3) Adopting compulsory conservation measures in "ecologically significant areas" for harmony between conservation and development. In recent years, 10% of the total area of protected areas has been set aside as "production areas", in which local people can grow bamboo, tea and bee and develop ecologically friendly industries consuming less natural resources. Meanwhile, measures have been taken to ensure that forests and biodiversity in 90% of the area will be effect-

ively protected. This model has been praised by UNESCO(United Nations Educational, Scientific, and Cultural Organization) as "a successful example of addressing conflicts between development and conservation in China's protected areas". The forest coverage rate in the protected area has increasing from 92.1% initially to 96.3% currently, with win-win for both development and conservation.

(4) Setting up a research platform to demonstrate achievements in biodiversity conservation. In 2010, one-decade-long survey results were digitalized, and on that basis a biodiversity research and information platform was set up, which is relatively complete in China. The platform uses and integrates database, GIS, virtual animation, audio and visual techniques to demonstrate a complete and vivid picture of the distribution of rare animals and plants in Wuyi Mountain. The platform plays an important role in supporting biodiversity research and education.

Case 2.4 Huangshan Mountain Model-Conservation through Enclosing and Alternating Opening Scenic Spots

Huangshan Mountain was listed as one of the world's natural and cultural heritages by UNESCO in December 1990. It was also selected as one of the candidates for the world geological park in February 2004. It is one of the first "5A" national tourism areas in China. Huangshan Mountain Tourism Administration used enclosing and alternating opening of scenic spots as entry point to explore a new model of sustainable development by combining the proper use and conservation of the world heritage.

(1) Strengthening regulations concerning management of scenic spots. Anhui Province People's Congress and relevant provincial government departments adopted a regulation on management of Huangshan Mountain scenic spot, a master plan for Huangshan scenic spot for 2007-2025, and an environmental plan for Huangshan Mountain scenic spot. These regulations and plans standardize the development of the scenic spot.

(2) Introducing the practice of alternating opening scenic spots. Protection measures have been taken in main scenic spots such as Tiandu Mountain, Lianhua Mountain and Shixin Mountain, such as enclosing and alternating opening these spots for 2 to 4 years.

(3) Introducing a new model of providing services to tourists. Since 2007, Huangshan Mountain Tourism Administration has moved its offices and staff housing facilities out of the mountain to minimize impacts on the scenic spots. For energy use in the core scenic spots electricity is mainly used, with liquefied natural gas as supplementary.

(4) Establishing an effective mechanism for investment into conservation. Specialized funds established for heritage protection, with contribution from 10% of entrance ticket sales. During 2007-2011, the cumulative investment into heritage protection exceeded 600 million yuan RMB.

(5) Strengthening international cooperation and exchange. In 2008, World Tourism Organization set up a monitoring station in Huangshan Mountain of sustainable tourism development of world heritages. Since 2009, Huangshan Mountain has joined a number of tourism and conservation organizations, such as IUCN, Global Sustainable Tourism Council (GSTC), World Tourism and Travel Council (WTTC) and Pacific-Asia Tourism Association (PATA). In 2010, Huangshan Mountain was awarded by WTTC the prize for management of global tourism destinations. At the end of 2011, as the only candidate from Asia, Huangshan Mountain was included as one of the first experimental zones for sustainable tourism development in global tourism destinations, and developed standards for sustainable tourism in global tourism destinations, together with experts from relevant international organizations.

Case 2.5 Model of Coordination between Nature Conservation and Economic Development in Dujiangyan, Sichuan Province

The City of Dujiangyan is in mountain areas located at the western side of the Sichuan Basin, (its northern latitude is 30°45′-31°22′ and east longitude is 107°25′-107°47′) covering an area of 1,208 km^2. Dujiangyan has various types of topography, with big differences in altitudes, heavy cloudiness and moisture, little sunshine and short frosty period. The forest coverage rate in 2003 was 50.1% and increased to 58.9% in 2012.

The City of Dujiangyan attaches high importance to biodiversity conservation. In 1992, the city established a Protected Area (PA) in Longxi-Hongkou covering an area up to 310 km^2, with neighboring towns and communes identified as buffer zones, covering an area of 117 km^2. In 1993, this protected area was upgraded to a province-level PA, and again to a national-level PA in 1997. In 2003, the city developed a municipal biodiversity strategy and action plan, with the support of UNDP, UN Foundation, FFI and the Biodiversity Working Group of China Council for International Cooperation on Environment and Development (CCICED). This BSAP provided guidance for biodiversity conservation and promoting coordinated social and economic development at local level. In 2006, land, covering an area of 195 km^2 from Qingchengshan and Zhaogongshan within the city, was included as part of the international natural heritage for Giant Pandas in Sichuan. So far, land of 622 km^2 within the city has been included as strictly protected areas, with the target set in local BSAP having been achieved.

The city Forest Department and Longxi-Hongkou PA Administration have been strongly supporting rural economy development in mountain areas by helping local farmers establish a number of rural economic cooperation organizations, such as cooperatives for medicinal materials, cropping and cultivation in forests and wild edible vegetables in mountains. In addition they also helped with multiple businesses in forestlands so as to reduce damage to forest resources. The Cooperative for Medicinal Materials currently has membership of more than 2,300 households, covering an area of more than 100 km^2 and output value exceeding 600 million yuan RMB.

Dujiangyan City develops tourism based on its natural resources, with income from tourism having reached 7.74 billion yuan RMB. Hongkou Commune in the buffer zone of Longxi-Hongkou PA started organizing tourism activities in the buffer zone since 2000, and upgraded it to a "4A" scenic spot in 2011. In 2013, a total of 723,000 tourists were received and income of 86.04 million yuan RMB generated from tourism.

Local farmers living in the buffer zones organize tours to rural areas using the good environment in the PA and the buffer zones. So far, 192 operators have begun their business in such tours. These tour operators and their employees have changed from their traditional way of living by cultivation, harvesting and logging to organization of rural area tours. The income per capita in 2012 has reached 10,542 yuan RMB. Meanwhile, higher income level brings about change in awareness and as a result many farmers have voluntarily participated in nature conservation activities.

2.3.5 *Ex-situ* conservation (including genetic resources)

(1) Botanical gardens and zoos

Botanical gardens are main bases for implementing *ex-situ* conservation of plant species. In accordance with incomplete statistics, so far China has established 200 botanical gardens of various kinds at different levels, having collected and stored over 20,000 plant species, two-thirds of China's flora. China has also established over 400 conservation and breeding sites for wild plant germplasms, as well as conservation centers for cycads and orchid germplasms, which have collected more than 240 varieties of cycads and 500 varieties of orchids respectively. According to incomplete statistics, China has established more than 240 zoos (including animal demonstration areas) and 250 rescuing and reproduction sites for endangered wild animals.

(2) Genetic resources of crops

By December 2012, the total of 423,000 accessions of agricultural crops have been collected in China, an increase of about 30,000 accessions over those in 2007. China has reinforced the construction of storage facilities to ensure that collected genetic resources are well stored. On one hand, China has expanded or renovated a few existing facilities, including 1 national long-term storage bank, 1 national duplicates bank, 10 national mid-term storage banks, 32 national germplasm nurseries (including 2 plantlets libraries). On the other hand, China has built 7 new national germplasm nurseries, as well as core germplasm banks for genetic resources of important crops. So far, China has built core germplasm banks for rice, wheat, maize, soybeans, cotton, barley and millet, and mini core germplasm banks for rice, wheat, maize and soybeans, with a large number of important functional genes discovered.

(3) Forage germplasm

A system of conservation and use of forage germplasm has been preliminarily established, with the amount, varieties, distribution and use of forage germplasm identified to varying extent. China has established 2 mid-term banks, 8 to 10 short-term storage banks and 5 germplasm nurseries, which store more than 240,000 accessions of germplasm, with 18,783 accessions verified. All this provides a basic condition for *ex-situ* storage of forage germplasm and their genetic diversity.

(4) Livestock genetic resources

A system of conservation of livestock genetic resources has taken shape, composed primarily of conservation farms and complemented by protected areas and gene banks. The system provides key protection to 138 rare and endangered varieties of livestock. As a result of implementation of the good breed selection project, China has established and expanded more than 120 key conservation farms, protected areas and gene banks. As a result of the implementation of the project on protection of livestock germplasm resources, more than 100 local varieties have been effectively protected every year, and molecular-level assessments of genetic resources have been undertaken. By the end of August 2012, China has established 150 national-level gene banks, conservation areas and farms.

(5) Forest germplasm

Analysis and assessment has been undertaken of the genetic diversity and variation of nearly 100 key tree species such as cedar, pine, poplar, arborvitae, spruce, birch, Mongolian oak, liriodendron, fagus, alder, amine tree, plum flower, bloom, clove, peony and bamboos. As a result important information concerning genetic variations and distribution of these tree species has been obtained, and strategies have been developed for genetic improvements and storing forest germplasm. 31 provinces, 295 cities (regions) and 1,569 counties (cities) have established tree seedling management stations or bodies to manage forest germplasm, thus forming a relatively complete system of forest germplasm management. A number of *ex-situ* storage banks particularly for forest germplasm have been established, which store more than 2,000 species, over 120 of which are key tree species. Currently a national programme is under development for collection, storage and use of forest germplasm, which will provide guidance for forest germplasm conservation.

(6) Germplasm of wild biological resources

By the end of 2012, China Southwestern Germplasm Bank of Wild Species has collected and stored 76,864 accessions of seed materials of 10,096 plant species, 9,123 accessions of non-seed *in-vitro* reproductive materials of 844 plant species and 45,980 accessions of active plant materials of 437 species. The bank has also collected and stored 13,805 accessions of animal germplasms of 354 animal species, mainly those of rare and endemic wild vertebrate species, as well as 330 accessions of 319 species of large fungi, 8,235 accessions of 815 species of micro-organism germplasms and 12,155 pieces of DNA materials of 1,311 species.

(7) Marine genetic resources

China has established a germplasm bank of marine species. The big seaweed germplasm bank located in China Ocean University has collected and stored nearly 500 stems of seaweed germplasm of 60 species. China National Storage Center of Marine Micro-organism Strains located in the Third Institute of Oceanography under the State Ocean Administration has stored more than 14,000 strains of bacteria. Genetic grouping or sequencing has been undertaken. In July 2010 Chinese scientist announced that they had completed a whole genome sequence map of oysters, which is the world's first such map of shellfish. Meanwhile Chinese scientists also announced completion of a whole genome sequence map of gunther, which is also the world's first such map of flounder fishes.

2.3.6 Key ecological projects

Key ecological projects continued to be implemented, such as natural forest resources protection, returning cultivated land to forest, construction of forest belts in north, northeast and northwest China as well as in the Yangtze River Basin, control of origins of sandstorms affecting Beijing and Tianjin and control of desertification in rocky Karst areas. Since 2001, ecological conditions of key project regions have improved obviously. Forest resources across the country have increased constantly, with reforestation area increased by 482,000 km^2 and forest coverage area increased by 23.0% over that of a decade ago. The current forest coverage rate has reached 20.4%, 3.8% increase over that of a decade ago. The forest reserves have reached 13.72 billion m^3, 21.8% increase over that of a decade ago. These projects have also enhanced restoration of habitats of wild species and the rise in the population and number of species.

The project of returning grazing land to natural grasslands has been implemented. Since the beginning of the project in 2003, by 2012, grassland fences covering an area of 606,000 km^2 have been established, among them, 262,000 km^2 for grazing ban fences, 317,000 km^2 for fences for temporarily stopping grazing, 27,000 km^2 for alternating grazing and 153,000 km^2 for reseeding seriously degraded grasslands. The average vegetation rate in the project implemented areas is 64%, 12% higher than that of the non-project areas. The fresh grass output per mu in the project areas has reached 212 kg, 70% or so higher than that from the non-project areas. Vegetation structure is gradually stabilizing with biodiversity being improved and good-quality grass percentage obviously going up.

Great efforts were put into soil conservation projects in some key regions. During the period from 2009 to 2012, such control projects were implemented in a total of 12,000 small river basins, covering an area of 270,000 km^2. Enclosing for soil conservation continued, with total enclosed area having reached 720,000 km^2. Among them, ecological conditions in areas of 450,000 km^2 have been restored to some extent. This impact of enclosing is increasingly obvious in sources of three major rivers in Qinghai, river basins in Xinjiang and Tibet, while the ecological functions of these areas are effectively protected.

205 wetland protection and restoration projects have been implemented since 2006. A number of wetlands of international and national importance have been rescued, with rate of protection of natural wetlands increased by 1% annually. More than a half of natural wetlands have been effectively protected. The capacities for wetland protection and management have been obviously strengthened and livelihood in wetland project areas has further improved.

Ecological restoration and re-construction in coastal and marine areas has been undertaken.

Work is on-going to restore and reconstruct coastal reed wetlands, mangroves, coral reefs, sea grass beds and Suaeda wetlands. Since 2010, a total investment of nearly 3.875 billion yuan RMB (from marine expenditures of the central government) has been made to restore mangroves and tidal flats and other important wetlands, with areas restored exceeding 2,800 km^2.

China has been promoting creation of eco-provinces, cities and counties. By now 15 provinces have started such initiatives, among which 13 provinces have developed programmes for building eco-provinces. More than 1,000 counties (cities, districts) are implementing eco-county programmes. 38 counties (cities, districts) have been awarded national eco-counties (cities, districts), with 1,559 national eco-towns and 238 national eco-villages established. Since 2008, the central government has arranged a specialized fund of 19.5 billion yuan RMB for rural environment improvement. By following the policy of "using rewards instead of compensation to promote rural environment improvement", support was provided to 46,000 villages for such efforts and more than 87 million farmers benefited.

2.3.7 Environmental pollution control

(1) Enhancing pollution emission abatement

The Government of China has adopted "significant reduction in the total emission amount of major pollutants" as one of binding targets for social and economic development, with a view to addressing outstanding environmental problems. During the period from 2000 to 2010, the annual average concentration of major pollutants has been going down overall. In particular since 2006, 10% reduction in the emissions of two major pollutants SO_2 and COD has been set as binding targets for social and economic development to enhance efforts in the total amount control of major pollutants. Since 2006, the emissions of COD from industrial waste water, SO_2 from effluent gases, soot, industrial dust and industrial solid wastes have been decreasing constantly. In the last decade, the intensity of pollutant emission per unit of GDP has significantly dropped by more than 55%. Since 2004, energy consumption per unit of GDP has decreased by 19.6%, and the intensity of CO_2 emission per unit of GDP by 15.2%. However, overall the intensity of pollutant emission and energy consumption per unit of GDP in China is still high, and the amount of waste water discharge is still increasing.

(2) Strictly implementing strategic environment assessments for plans or programmes and environmental impact assessments for projects

EIAs for projects have been strictly implemented. Measures such as "limited approvals for certain regions and certain sectors" have been implemented. Since 2008, governments have refused to approve 332 projects with total investment of 1.1 trillion yuan RMB, mostly of high pollution, high energy consumption, resource consumption, low-level duplication and excessive production capacities. This has played an important role in industrial restructuring and optimizing economic growth.

(3) Enhancing pollution prevention and control

China has promulgated and been implementing a *Plan for Pollution Prevention and Control in Major River Basins* (2011-2015) and *Twelfth Five-year Plan for Air Pollution Prevention and Control in Key Regions*. China has also revised its air quality standards. Active measures have been taken to implement *National Plan for Prevention and Control of Underground Water Pollution*, and a programme for underground water pollution prevention and control in north China has been developed. Pollution prevention and control in rivers and lakes has been further enhanced, with the percentage of seven major rivers with water quality better than that of Grade III increased from 41% in 2005 to 64% in 2012. The percentage of the rivers with water quality lower than Grade V decreased from 27% in 2005 to 12.3% in 2012. Great efforts have been made to dispose of chromium accumulated over years. A total of 6.7 million of chromium stored across the country for several decades or even half a century have been disposed of.

2.3.8 Prevention and control of invasive alien species

(1) Improving management system and mechanisms

A national programme has been developed for invasive alien species (IASs) emergency preparedness and response. 18 provinces (autonomous regions and province-level municipalities) have established IASs Management Office or a mechanism of joint conference (of relevant departments). 27 provinces (autonomous regions, province-level municipalities) have announced emergency responses for managing IASs. So far China has initiated 12 second-level and above emergency responses. In 2011, Hunan Province promulgated a provincial regulation on alien species. China has developed and issued 57 technical guidelines for monitoring, assessing, preventing and controlling invasive alien species as well as the first and second lists of invasive alien species for monitoring and tracking. China has also developed and issued 352 sectoral standards, 104 national standards and 2 international standards for plant quarantine.

(2) Strengthening capacity building for monitoring and early warning

A network of monitoring and early warning of forest pests and agricultural invasive alien species has been preliminarily established. Surveys have been undertaken of invasive alien species so as to know their distribution and damages they have caused. A system of IASs risk assessment has been established, with more than 1,500 alien species assessed. Mail inspection for tracking IASs has been undertaken across the country.

Case 2.6 Innovative Management Model of Liaohe Protected Area

For some time, due to overexploitation, water pollution and biodiversity loss of Liaohe River have become serious problems and therefore the river was on the national list of rivers for priority control. In 2010, Liaoning Province Government set up a protected area in Liaohe River and established a management body for this PA. Liaoning Province People's Congress promulgated a regulation on this PA, which mandated its management body to coordinate the pollution prevention and control, natural resources conservation and ecological conservation in this protected area, undertaking supervision powers and law enforcement responsibilities that are usually exercised by departments of the environment, water resources conservation, land management, transportation, forestry, agriculture and fishery.

Liaohe PA covers the main part of the river basin and 14 counties (districts) in Liaoning Province, with a total area of 1,869.2 km^2 and total length of 538 km. Liaohe PA management body has undertaken the following measures, with engineering, management and conservation measures combined:

(1) Controlling pollution at sources. 134 municipal wastewater treatment plants have been established or renovated. 121 plants treating wastewater from towns and villages and 34 waste disposal plants have been established.

(2) Implementing ecological projects. 167-km-long Watercourses have been treated comprehensively through measures like river dredging, bank protection with tree plantation and restoring aquatic plants. 16 water accumulation projects for ecological conservation have been implemented, and 53.3 km^2 of new wetlands added.

(3) Enclosing main flood control areas. Cultivated land in main flood control areas have been reclaimed to the river, with 386.7 km^2 of land reclaimed and 22.7 km^2 of river bank areas reforested.

(4) Implementing strict accountability system. Management is exercised by enforcing relevant laws. 123 sites of sand collection in the watercourses have been closed and more than 2,000 m^3 of wastes have been cleaned and disposed of in non-hazardous ways.

After more than three years of efforts, obvious results have been achieved in the treatment and conservation of Liaohe River: (a) Water quality has been improving constantly year by year. By the end of 2009, one year ahead of planned time, Liaohe River has no sections with water quality lower than Grade Ⅴ. 80% of the time periods in 2012 the river had the water quality of Grade Ⅳ, and the water quality of some sections reached Grade Ⅲ at some periods of time. (b) Ecosystems are improving as well. The vegetation coverage in Liaohe PA has increased from 13.7% to 63%, with biodiversity obviously increasing. The seal population at the estuary of Liaohe River is increasing, river saury began to swim back and the number of bluff fish and white bait reproduced obviously growing.

Kaiyuan Gujiazi Wetland before treatment (2010)

Kaiyuan Gujiazi Wetland after treatment (2011)

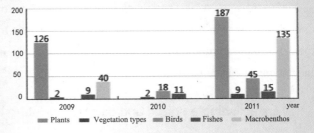

Increase in species richness of Liaohe Protected Area

(3) Eliminating IASs

Elimination actions have been taken in more than 600 counties or cities of 22 provinces (autonomous regions, province-level municipalities), focusing on 20 key invasive alien species such as *Ambrosia artemisiifolia* and *Alternanthera philoxeroides*. Over 42.72 million persons-time have been mobilized for these actions and the total area with IASs eliminated or controlled has reached over 57,300 km^2-time. The elimination rate in key regions has exceeded 75%, having effectively prevented the expansion of IASs. In February 2010, China issued a revised technical programme for preventing and controlling *Bursaphelenchus xylophilus*, with a view to further strengthen prevention and control of *Bursaphelenchus xylophilus*. As a result the area affected by *Bursaphelenchus xylophilus* has decreased from 846.7 km^2 of the peak period to 453.3 km^2 in 2011.

(4) Education and training

Various media such as radio, television, journals and internet are used for education and training on techniques for preventing and controlling IASs. More than 200,000 copies of relevant publications such as *100 Questions on Agricultural Invasive Alien Species* have been published and disseminated. National training workshops have been organized on emergency responses to invasive alien species.

2.3.9 Biosafety management of GMOs

The Government of China attaches high importance to biosafety management of genetically modified organisms (GMOs) and has undertaken the following actions:

(1) Establishing strict and rule-based management system

The State Council has promulgated *Regulation on Management of Agricultural Genetically Modified Organisms*, which provides rules for risk assessment, production licensing, business licensing, product labeling and import/export approvals. China has also issued other relevant regulations and rules to manage the whole processes from research, development to application of GMOs. The State Forestry Administration has issued rules for approval of genetic engineering of trees which regulates relevant activities in this regard. The State Council has established a mechanism of joint conference of relevant ministries and departments, which is responsible for studying and coordinating major policies and legal issues concerning biosafety management of agricultural GMOs.

(2) Establishing a sound and complete system of assessment

Safety assessments are undertaken by Biosafety Committee of Agricultural GMOs composed of experts from different fields. The members of this committee are recommended by relevant ministries and commissions and appointed by the Ministry of Agriculture. Currently there are 64 members sitting on this committee. The science-based, case-by-case, precedent-based, and step-by-step principles are followed to undertake assessments of agricultural GMOs at different levels and in different phases.

(3) Strengthening technical support capacity building

The Government of China pays high attention to technical capacity building for risk assessment and detection of GMOs. So far, 39 institutions undertaking GMOs risk assessment and detection have been approved by the Ministry of Agriculture. 82 technical standards for GMOs safety have been developed and regular detection of GMOs is being undertaken. Some results have been highly recognized by the international science community, and provide strong technical support for safety supervision of GMOs in China.

2.3.10 Incentive measures

(1) Eliminating subsidies unfavorable to biodiversity

To avoid negative impacts on biodiversity and the environment, China eliminated in 2007 export tax rebates of 553 products of high energy consumption, pollution and resources consumption, including endangered animals and plants and their products, leather products, some wood products and disposal wooden products.

(2) Establishing guarantee funds for ecological restoration and environmental improvement of mining sector

In 2006, the Ministry of Finance, together with the Ministry of Land Resources and the State Environmental Protection Administration developed guidance for establishing a responsibility system for ecological restoration and environmental improvement of the mining sector. The guidance requires the mining sector to provide guarantee funds out of their mining product sales incomes for ecological restoration and environmental improvement. So far, 30 provinces (autonomous regions, province-level municipalities) have established such funds for ecological and environmental restoration in the mining areas. By the end of 2012, 80% of the mines have paid their guarantee funds, totaling 61.2 billion yuan RMB and accounting for 62% of the total funds that should be paid.

(3) Subsidizing households that return cultivated land to forests

Since 1999, the central government has been subsidizing those households that have returned their cultivated land to forests according to the actual areas returned and verified. These households also have the ownership of forests that grow on returned land, with contract period for owning and using returned land being as long as 70 years, while enjoying preferential tax incentives for benefits from use of returned land. In 2007, the State Council issued a notice on improving the policy of returning cultivated land to forests, with a view to increase the subsidies to related households. According to this notice, households living in the Yangtze River Basin and South China can be subsidized in cash by 1,575 yuan RMB per hectare of land annually, while households living in the Yellow River Basin and North China can get a cash subsidy of 1,050 yuan RMB per hectare of land. Farmers that return land to forests with ecological functions can be compensated for eight years, while those that return land to forests with economic functions can be compensated for five years. From 2008 to 2011, the central government provided specialized grants totaling 46.2 billion yuan RMB. By the end of 2012, the central government has invested cumulatively 324.7 billion yuan RMB, and 124 million farmers in 2,279 counties directly benefited from this investment, with per household being subsidized 7,000 yuan RMB on the average.

(4) Subsidizing the projects on natural forest protection

Natural forest resources protection projects were initiated in 17 provinces in 2000. The central government subsidized forest management and conservation as well as seedling cultivation and reforestation. The central government also provided subsidies by covering pension insurances for forest enterprise employees and social expenditures of forest enterprises, and providing basic life guarantees for laid-off forest workers. The total investment for the first phase of this project went up to 118.6 billion yuan RMB. At the end of 2010, the State Council decided to implement a second phase of this project from 2011 to 2020, with 11 more counties (cities, districts) to be included in the project. The subsidy provided for reforestation will be 4,500 yuan RMB per hectare, and those for enclosing mountains for forest conservation and aerial seeding will be 1,050 yuan RMB per hectare and 1,800 yuan RMB per hectare respectively. Education subsidy is 30,000 yuan RMB per person per year. Sanitation subsidy for forest areas in the upper reaches of the Yangtze River, the upper and middle reaches of the Yellow River and Inner Mongolia is 15,000 yuan RMB per year and 10,000 yuan RMB per year respectively. For state-owned forests, the central government provides 75 yuan RMB per hectare annually as forest conservation fee. For those collectively-owned forests that also belong to national-level public benefits forests, during 2011-2012, the central government provided 150 yuan RMB per hectare annually as part of the funds for ecological compensation. Since 2013, this rate has been increased to 225 yuan RMB per year. For local benefits forests the compensation funds are provided mainly from local government budgets, while the central government also provides 45 yuan RMB per hectare per year as forest conservation fee. The total investment of the second phase of this project will be around 224 billion yuan RMB.

(5) Subsidizing projects of returning grazing land to grasslands

Since 2003, such projects have been implemented in 8 provinces such as Inner Mongolia, Sichuan, Qinghai and Xinjiang. The central government has been subsidizing the construction of fences and the provision of forages. In 2011, the central government raised the subsidy standards and percentages. 300 yuan RMB per hectare is provided to fence building in Qinghai-Tibet Plateau while 240 yuan RMB per hectare to other regions. A subsidy of 300 yuan RMB per hectare is provided to reseeding grass; 2,400 yuan RMB per hectare to artificial forage farming and 3,000 yuan RMB per household for building feeding stables and rings. The central government invested cumulatively a total of 17.57 billion yuan RMB in this project during the period 2003-2012, with projects having benefited 174 counties, more than 900,000 farm households and more than 4.5 million farmers and herdsmen.

(6) Establishing subsidies and incentives for ecological conservation of grasslands

Since 2011, eight provinces or autonomous regions with most of the grasslands in China have established incentive measures for grassland ecology conservation, with a subsidy of 90 yuan RMB per hectare annually for grasslands where grazing is banned; 22.5 yuan RMB per hectare annually for grasslands where balance of herds and grass supply is implemented. Subsidies are also provided to herdsmen for their production, with 150 yuan RMB per hectare annually for grass seed and 500 yuan RMB per household annually for production materials. Herdsmen are also trained to promote their shift to new jobs. The subsidies for grassland ecology conservation increased from 13.6 billion yuan RMB in 2011 to 15 billion yuan RMB in 2012, with cumulative total investment reaching 28.3 billion yuan RMB. By the end of 2012, the areas covered by subsidies for grassland grazing bans have reached 820,000 km^2, and the areas where rewards are given for keeping the herd-and-grass balance have reached 1,737,000 km^2.

(7) Subsidizing wetland conservation

In 2010, the Ministry of Finance together with the State Forestry Administration initiated subsidies for wetland conservation, which covered 27 wetlands of international importance, 43 natural wetland nature reserves and 86 national wetland parks. Some local governments also increased support to wetland conservation from government budgets, and gradually included important wetlands as part of ecological compensation.

(8) Establishing funds for compensating forest ecological benefits

In 2004, China established national funds for compensation of forest ecological benefits, which subsidize plantation, nurturing, conservation and management of national-level public benefits forests, with funding allocated from the central government budgets. Among them, a subsidy of 75 yuan RMB per hectare is provided annually for state-owned national-level public benefits forests, and 225 yuan RMB per hectare annually for national-level public benefits forests owned collectively and privately. Currently the areas that have received such subsidies have reached 924,000 km². In 2013 the central government provided a total of 14.9 billion yuan RMB for compensation for ecological benefits of forests. Local governments also compensated for local public benefits forests.

(9) Establishing national mechanisms of ecological compensation for national key ecological function zones

Since the central government budget established in 2008 an item of fiscal transfers for national key ecological function zones, the scope of transfers has been constantly expanding. In 2013, funds were transferred to 492 counties and 1,367 land zones prohibited for development, with the total of funds transferred reaching 42.3 billion yuan RMB. In 2013, scenic spots in Yunnan, Guizhou, Sichuan and Xinjiang were included in the pilot work on ecological compensation.

Case 2.7 Pilot Ecological Compensation in Xin'an River Basin

Xin'an River originates from Huangshan City, Anhui Province, and flows into Qiandao Lake scenic spot in Zhejiang Province, and eventually into East China Sea, via Fuchun River and Qiantang River, with the river basin covering an area of 11,674 km². Xin'an River has the biggest water flow to Qiandao Lake while the lake is an important drinking water source for Zhejiang Province. To protect the water environment of Xin'an River, Huangshan City and other places in Anhui Province have sacrificed their own development for years by delaying their industrialization and urbanization processes.

In March and September 2011, the Ministry of Finance and the Ministry of Environmental Protection issued a notice on initiating pilot work in ecological compensation for the water environment of Xin'an River as well as a programme of implementation in this regard. Funds of 300 million yuan RMB were allocated in 2011 for compensation and specially used for water pollution control and water quality improvement in the upper reaches of Xin'an River. Among them 200 million yuan was allocated from the central government budget and 100 million yuan provided by Zhejiang Province. Anhui and Zhejiang Provinces were the first in China among those that have established a mechanism of compensation for the water environment in a trans-province river basin, to protect water resources for Xin'an River and Qiandao Lake.

2.3.11 Scientific research

The Government of China encourages and supports scientific research in conservation and sustainable use of biodiversity. Projects related to conservation and sustainable use of biodiversity have been included in a number of science and research plans or programmes, such as National Plan for Support to Science and Technology, National Plan for Development of Key Fundamental Research, National High-tech Development Plan, National Natural Sciences Fund and Specialized Funds for Research in Public Benefit Sectors. For example, in the Eleventh Five-year Plan for Support to Science and Technology, there were key projects such as "Monitoring of Important Biological Resources in China and Demonstration and Application of Key Conservation Techniques", "Techniques for Restoration of Typical Vulnerable Ecosystems and Demonstration" and "Comprehensive Monitoring and Assessment of China's Important Terrestrial Ecosystems in Support of Decision Making". National Plan for Development of Key Fundamental Research also included projects such as "Research on Evolution and Protection of Biodiversity in the Himalaya Region", and "Theory and Methods of Pest Control by Agricultural Biodiversity and Protection of Germplasm Resources". In building information platforms for nature-related science and technology, support has been provided to survey and collection of animals, plants, micro-organisms and germplasms as well as associated information system development and sharing of relevant information and specimens. All these research projects have generated a number of valuable and influential research results, thus providing scientific and technical support to biodiversity conservation in China.

2.3.12 Public participation

To popularize biodiversity knowledge among primary and middle school students, curriculum requirements for biology for middle and high schools published in 2011 have incorporated biodiversity-related knowledge. Through classroom teaching in primary and middle schools, biodiversity knowledge of primary and middle school has been upgraded overall. 1,908 universities in China have established majors or programmes on biology. Nearly 50 universities have undergraduate programmes on ecology, 38 universities have master-degree programmes on ecology and 22 universities have PhD degree programmes on ecology. These degree programmes have trained professionals for biodiversity conservation and research, with the total number of professionals having exceeded 556,000 by the end of 2012.

Various departments have organized various communication and education activities through media such as television, internet, newspapers, journals and radios as well as training workshops, lectures and dissemination of training materials. In particular, while organizing China Biodiversity Communication Trip to celebrate the International Year of Biodiversity in 2010, 40 large-scale communication and education activities of various kinds were organized at national level, with more than 370,000 communication and educational materials of various kinds disseminated and audience influenced by various media reaching 804 million person-time. At local levels, 191 large-scale communication and educational activities were organized, with more than 350,000 materials disseminated and 25 movies with biodiversity themes produced. About 20,000 institutions including nature reserves, zoos, botanical gardens, parks, institutions undertaking environmental education and research as well as televisions, newspapers and internet were involved in a series of educational activities open for primary and middle school and university students, reaching out to 100 million person-time. Through communication and education, the public has shown higher enthusiasm for participation and their awareness of biodiversity conservation has obviously increased, with importance of biodiversity more widely recognized.

2.3.13 International cooperation and exchanges

China has been undertaking various ways of cooperation such as bilateral and multilateral cooperation and South-South cooperation and has achieved satisfactory results and promoted the implementation of the CBD and its Protocols.

(1) Multilateral cooperation moving forward stably

China is a Party to the Convention on Biological Diversity, the Convention on International Trade in Endangered Species of Wild Fauna and Flora, the Convention on Wetlands of International Importance especially as Waterfowl Habitat and the United Nations Framework Convention on Climate Change. China has been actively participating in the negotiations of these Conventions and seriously implementing relevant obligations as well as participating in the building of relevant multilateral systems. China has been contributing what it can to international developments of the CBD and its implementation processes including provision of funding. China has implemented a number of biodiversity projects in collaboration with the Global Environment Facility, the World Bank, UNDP and UNEP. Through these projects China has introduced useful concepts and technologies of and attracted funds for biodiversity conservation and management, which have promoted biodiversity conservation in China. Since the 32nd session of the Global Environmental Facility (GEF) Council held in November 2007 approved China Biodiversity Partnerships Framework (CBPF), 9 projects have been implemented one after another.

(2) Breakthroughs made in bilateral cooperation

China has established broad channels of cooperation and exchanges with over 50 countries such as Germany, USA, Russia, UK, Norway, Canada and Australia. A system of various forms of cooperation has been established with cooperation mainly between governments. The

Case 2.8 Eu-China Biodiversity Programme Achieved Rich Results

First, the programme contributed significantly to strategic planning. The programme provided important support to updating China's NBSAP (2011-2030). The programme had significant influence on mainstreaming biodiversity at local level. Four-fifths of the provinces (autonomous regions or province-level municipalities) that have been involved in this programme have incorporated biodiversity into their own development plans for the twelfth five-year plan period.

Second, the programme strengthened coordination mechanisms and promoted the processes of implementing the CBD. At national level the programme strengthened China National Coordinating Group on Implementation of the CBD and promoted the synergies in the CBD implementation among relevant departments such as the National Development and Reform Commission, the Ministry of Land Resources, the Ministry of Agriculture, the Ministry of Water Resources, the General Administration on Quality Supervision, Inspection and Quarantine, the General Administration on Chinese Traditional Medicine and the State Forestry Administration. At local level the programme promoted incorporation of biodiversity into relevant regional planning for social and economic development. For example, *Hainan Province Master Plan for Land Use*, which is consistent with requirements for biodiversity conservation, was approved by the State Council in 2010.

Third, the programme supported high-level decision making. The programme supported relevant strategic research under China Council for International Cooperation on Environment and Development (CCICED) and provided policy recommendations related to ecosystem services and biodiversity, which have been presented to high-level policy makers. The programme for the first time proposed that biodiversity should be included in the strategic environmental assessment for national key industry development. This plays a positive role in promoting transition in patterns of economic development and building an environmentally friendly and resource-efficient society.

Fourthly, the programme strengthened capacities. The project provided technical and financial support to assist governments of different levels in developing policies, regulations, plans, standards and guidelines related to biodiversity. So far 46 of these have been approved by the State Council, relevant local governments, People's Congresses or government departments, while 27 are waiting for approval. All these have extensively improved capacities to implement the CBD at different levels.

Fifthly, the programme supported communication and education and increasing the environmental awareness. The programme helped establish public platforms for communication and education, which communicate and popularize concepts and knowledge of biodiversity for the governments, the private sector, local communities, schools and the media, by using internet, movies, publications, newsletters and various forms of participatory training activities. They have increased to great extent the public knowledge and awareness of biodiversity. They have in particular supported activities in China celebrating the International Year of Biodiversity, which have generated significant social impacts.

Finally, the programme supported team building and personnel training for implementation. A big number of policy makers, managers, experts and media professionals from various levels, being from government departments to research institutions, have been deeply involved in the implementation and management of the programme. As a result they have learned about new concepts, broadened their vision, strengthened their capacities and upgraded their levels, thus becoming influential core team members with stronger capacities for the implementation of the CBD.

Case 2.9 Incorporating Biodiversity into Land Use Planning and Land Reclamation

With the support of EU-China Biodiversity Programme, the Ministry of Land Resources initiated in October 2008 a project on biodiversity conservation in land use planning and land reclamation, and Guizhou and Hainan Provinces were identified as pilot sites for the project. The project supported the development of guidelines for incorporating biodiversity into land use planning and land reclamation as well as proposed ecologically friendly engineering techniques for land reclamation that are suitable for China's circumstances. This was a useful attempt to integrate biodiversity into land use. In 2010, Hainan Province Department of Land Resources and the Environment issued *Technical Guidelines for Incorporating Biodiversity into Land Use Planning*, which require that concepts of biodiversity conservation should be integrated in master planning for land use. Meanwhile Hainan Province approved two local master plans for land use that had incorporated biodiversity considerations, one for Lingshui County and another for Ledong County, both of which are resided by people of Li ethnicity. Guizhou Province Department of Land Resources also issued a notice on strengthening biodiversity and ecological conservation while comprehensively treating land. In September 2010, Guizhou Province approved two local land reclamation plans that incorporated biodiversity consideration, one for Guanling County and another for Libo County. The issuance of these plans and documents has improved the structure and layout of land use, thus promoting biodiversity conservation.

Case 2.10 Biodiversity Conservation and Sustainable Management of Grasslands in Hulunbei'er

Hulunbei'er Grassland, located in Inner Mongolia, is facing threats of large-scale land degradation and desertification, due to overgrazing, deforestation and irrational mining as well as arid climate. With the support of EU-China Biodiversity Programme and relevant international organizations and research institutions, Hulunbei'er City implemented a project on biodiversity conservation and sustainable management of grasslands and achieved obvious results.

(1) Established a committee on biodiversity conservation headed by vice mayor. Meanwhile counties and districts under the city's administration have also set up their biodiversity committees, thus forming an operational mechanism for biodiversity at the city level and incorporating biodiversity into relevant government work.

(2) Issued the first local regulation on biodiversity, which clearly requires that biodiversity will be incorporated into the city's twelfth five-year plan for social and economic development, providing a policy basis for grassland biodiversity conservation.

(3) Developed the city's biodiversity strategy and action plan. The city has gradually undertaken grassland biodiversity conservation activities according to BSAP.

(4) Signed a MOU of cooperation with Mongolia and established a transboundary cooperation platform with a province of Mongolia. Many transboundary conservation actions have been taken which play a very important role in protecting the habitat, reproduction and migration of

Part 2 National Biodiversity Strategy and Action Plan and Its Implementation

> Gazelles endemic to the Mongolian Plateau.
> (5) Developed a brochure of guidance for grazing in Hulunbei'er Grassland and demonstrated models for best practices for grazing in degraded grasslands, which are of great significance for grassland biodiversity conservation.
> (6) Established demonstration sites of restoration of degraded grasslands and summarized 8 models from relevant work. All these results and models have been applied for the desertification control of an area of 667.7 km^2 of the city in 2009, providing technical support for desertification control at larger scale.
> (7) Developed a technical handbook for monitoring of grassland biodiversity and increased biodiversity monitoring capacities of the ecological monitoring stations of the city and Dalai Lake and Hui River national-level nature reserves.
> (8) Disseminated biodiversity knowledge to the public through TV, radio, newspapers, government bulletins and large-scale activities and increased the public awareness of biodiversity.

six-year-long Eu-China Biodiversity Programme (ECBP) was successfully completed in 2011, which played an important role in promoting biodiversity conservation and sustainable use in China. The project was considered an indication of breakthrough in bilateral cooperation.

(3) New progress made in South-South cooperation

In recent years, the Government of China has been actively undertaking South-South cooperation in the field of biodiversity, by signing agreements of cooperation with many developing countries in areas related to biodiversity. China has organized a number of capacity development workshops for developing countries from the sub-regions such as South and Southeast Asia. China has also established a Center for China-ASEAN Environmental Cooperation, the first platform China has established for South-South environmental cooperation and regional environmental cooperation.

2.4 Overall Assessment of Progress in Implementing NBSAP

Since the updated NBSAP (2011-2030) was launched in 2010, China had a good beginning of implementation of this new strategy and action plan and is moving towards the right direction. One action has achieved major progress, i.e. Action 10: "promoting and coordinating the establishment of information system on genetic resources". 15 actions have achieved considerable progress and 14 actions have achieved some progress (details in Annex Ⅰ). Main progress is achieved in the following three areas.

2.4.1 China has established a system of biodiversity conservation and management with its own characteristics

(1) The legal and regulatory system for biodiversity conservation and sustainable use is being increasingly improved.

(2) Coordination mechanisms for biodiversity conservation are almost in place and management capacities of governments have been further upgraded.

(3) A network of nature reserves has been established, with various categories, relatively reasonable layouts and relatively sound functions. A large number of scenic spots, forest parks, community-based conservation areas, protected sites of wild plants, wetland parks, geological parks, special marine protected areas, and germplasm conservation areas have been also established. Nature reserves have protected 90% of China's terrestrial ecosystem types, 85% of wild animal populations and 65% of higher plant biota as well as covered 25% of primitive natural forests, over 50% of natural wetlands and 30% of typical desert areas.

(4) The public enthusiasm for and capacities of participation in conservation have been considerably enhanced.

(5) The innovation capacities of colleges, universities and research institutes have been considerably upgraded.

(6) New progress made in international cooperation and exchanges.

2.4.2 The ecological degradation trend is slowing down and ecosystems in some regions are being restored

(1) The forest resources have been constantly increasing, with forest areas increased by 23% over that of a decade ago and forest reserves by 21.8% over that of a decade ago.

(2) The areas where soil erosion has been prevented and controlled have reached 270,000 km^2, covering 12,000 small river basins. The areas where enclosure for conservation was implemented have reached 720,000 km^2, and among them, ecological recovery is occurring in areas of 450,000 km^2.

(3) The number and populations of some national key protected wild animals and plants are stable, and some even going up, with their scope of distribution expanding and quality of habitats constantly improving. The number of giant pandas increased from over 1,000 in the 1980's to 1,590 currently. The number of crested ibises grew from 7 in the 1980's to over 1,800 now. The populations of protected plants such as yews, orchids and cycads are constantly increasing.

(4) The annual emissions of main pollutants are going down overall. Since 2000, the density of pollutant discharging per unit of GDP has fallen by more than 55%. Since 2004, energy consumption per unit of GDP has decreased by 19.6% and the density of CO_2 emission

per unit of GDP down by 15.2%.

2.4.3 Comprehensive social and economic development at local level while biodiversity is conserved

Well-beings of local communities are improved while ecosystems are conserved and restored. The net income of rural households in 2011 increased by 40.8% over that in 2000, with the number of people living in poverty significantly going down.

Part 3
Sectoral and Cross-sectoral Integration of Biodiversity

Conservation and use of biodiversity involves many departments and sectors. This part covers biodiversity integration into relevant sector planning and main actions and measures taken by relevant member departments of China's National Committee on Biodiversity Conservation.

3.1 Development and Reform Commission

National Development and Reform Commission (NDRC) has taken into full consideration important roles of ecological conservation in social and economic sustainable development. Following the principles specified in China's updated NBSAP (2011-2030), NDRC has developed relevant policies and regulations for returning arable land to forests, returning grazing land to grasslands, control of desertification in rocky areas, control of sources of sandstorms and protecting key ecological zones.

(1) Building strategic layout for land use and development based on national ecological security

As the first national land spatial plan, *National Plan of Major Function Zones* divides the whole country into three major areas, i.e. urban areas, agricultural production areas and ecological function areas, and identifies four types of regions according to their functions: regions for priority development, regions for key development, regions for limited development and regions prohibited for development. The Plan outlines a strategic layout of ecological security (Table 3-1) composed primarily of ecological barriers in Qinghai-Tibet Plateau, the Loess Plateau and Sichuan-Yunnan Provinces, forest belts in northeast China, sand control belts in north China and hilly areas of south China. The Plan also contains a list of national major ecological function zones and a list of banned development zones, as well as relevant assessment and zoning maps.

Table 3-1 China's Strategic Ecological Security Layout ("Two Barriers and Three Belts")

Regions	Key Priorities for Ecological Conservation
Qinghai-Tibet Plateau	To protect various, unique ecosystems to allow them to play roles in regulating water for big rivers and climate
Loess Plateau and Yunnan-Sichuan	To strengthen control of soil erosion and protection of natural habitats to ensure ecological security in the Yangtze River Basin and the middle and lower reaches of the Yellow River
Northeast China Forest Belt	To protect forest resources and biodiversity to allow northeast China plains to play the role of ecological security barriers
North China sand control belt	To strengthen construction of forest belts, grassland conservation, sand fixing and prevention of sandstorms; and to enclose for protecting those areas where desertified land cannot be controlled for the time being; and to allow forest belts to play the role of ecological security barriers
Hilly areas of south China	To strengthen habitat restoration and control soil erosion, so that south China and southwest China can play the role of ecological security barriers

(2) Promoting establishment of mechanisms for ecological compensation

In 2010, NDRC together with relevant ministries and commissions drafted a regulation on ecological compensation. This draft regulation provides principles, areas, targets, approaches and criteria for ecological compensation. Currently, NDRC is studying and drafting a set of "recommendations for establishing and improving mechanisms for ecological compensation".

(3) Integrating ecological conservation into local planning for economic development and transition

To help local governments create a favorable policy environment for economic transition and ecological conservation, NDRC and State Forestry Administration (SFA) and other relevant ministries jointly issued in 2010 *Plan for Economic Transition and Ecological Conservation in Major and Minor Xing'anling Forest Areas* (2010-2020). Currently NDRC is also developing other important plans such as *Plan for Development of Border Areas in Heilongjiang Province and Eastern Inner Mongolia* and *Plan for Economic Transition and Ecological Conservation in Changbai Mountain Forest Areas*.

(4) Enhancing ecological conservation in key regions

NDRC together with other relevant ministries and departments has developed the *Twelfth Five-year Plan for West China Development*. This plan has identified strategies and action plans for implementing major ecological projects in West China during 2011-2015. NDRC has also developed and issued other important plans such as *Plan for Building Tibet's Ecological Security Barrier* (2008-2013) and *Plan for Second-Phase Implementation of the Project on Control of Sources of Sandstorms Affecting Beijing and Tianjin* (2013-2020). NDRC arranged funds from the central government budget to support key ecological projects such as natural forest protection, control of desertification in rocky areas and construction of forest belts in north, northwest and northeast China.

(5) Integrating biodiversity conservation into responses to energy conservation and emission reduction, climate change

NDRC has developed consecutively a number of plans or programmes in this regard, such as *Comprehensive Programme of Work for Energy Conservation and Pollution Abatement for* 2011-2015, *Plan for Energy Conservation and Pollution Abatement for* 2011-2015 and *Programme for GHG Emission Control for* 2011-2015. All these plans or programmes are intended to fully synergize efforts in climate change adaptation and ecological and biodiversity conservation, through implementing measures such as adjusting industrial structures, improving energy efficiency, developing low-carbon energy resources, optimizing energy structures, increasing carbon sinks and strengthening ecological conservation.

3.2 Education

The education sector attaches high importance to education, teaching and training in the field of biodiversity, through strengthening basic education, science popularization activities, professional education and training.

(1) Disseminating biodiversity knowledge in basic education and science popularization activities

To allow primary and secondary school students to understand the importance of biodiversity, biodiversity-related knowledge has been incorporated into middle and secondary school curriculums and textbooks such as those published in 2011 on biology for middle and secondary schools. One chapter on biodiversity was included in the school textbook on biology for middle schools published in 2011, where content and requirements related to biodiversity are clearly specified. The secondary school curriculum criteria require students to summarize the evolutionary theory and the formation of biodiversity as well as the importance of biodiversity and measures that could be taken to protect it. Students are encouraged to get involved in biodiversity conservation activities after school. Various places and schools organized various activities to widely disseminate knowledge of biodiversity sciences and related laws and regulations, such as creation of green communities, green schools and green families, and organizing knowledge contests, lectures, eco summer camps, writing contests and various celebration activities.

(2) Providing undergraduate programmes and professional training on biodiversity

In accordance with the list of programmes and majors for universities published in 2012, majors or programmes related to biodiversity include biological sciences, marine sciences, nature conservation and environmental ecology, forestry sciences and grass sciences. Currently, nearly 2,000 universities or colleges (about 80% of the total number of universities in China) in China have programmes or majors related to biodiversity. Among them there are about 800 universities and about 1,200 colleges. A total of 298 universities have majors or programmes on biology; 52 universities have majors in ecology and most of comprehensive universities and teachers' colleges have courses on life sciences and ecology. The number of university graduates from biodiversity-related programmes or majors has increased over years. The number in 2008 was about 106,000, and rose to 117,000 in 2012, an increase by 10% over that of 2008. By the end of 2012, the total number of professionals that have received university education in biodiversity-related sciences has reached 557,000. Meanwhile, Chinese universities invest significantly in specimen museums, and by now there are about 180 specimen museums of various kinds in Chinese universities, providing platforms for research and education in the field of biodiversity.

(3) Strengthening disciplines related to biodiversity

In China's current Catalogue of Degree Programmes and Disciplines for Professional Training, there are many Class A subjects related to biodiversity. For 5 Class A subjects, i.e. biology, ecology, biological engineering, forestry and grass sciences, at present Chinese universities have 172 PhD programmes and 225 master programmes. From 2008 to 2012, Chinese universities with these degree programmes (on these five subjects) have granted PhD degrees to 17,110 candidates and master degrees to 63,634 candidates. China has encouraged those universities with such degree programmes to set up Class B subjects related to biodiversity as required by needs for their programme development and broader needs for social and economic developments, with a view to strengthen research, development and professional training in the field of biodiversity.

(4) Nurturing and attracting more talents

In recent years, China has been implementing a series of plans to attract talents and chief scientists to better support and advance biodiversity research. In particular since One Thousand Talent Plan was implemented in 2008, Chinese universities have attracted 118 biodiversity experts and 78 "Yangtze River" scholars. The Ministry of Education established platforms and created conditions for talents to upgrade their innovation capacities and competitiveness through a plan for senior, innovative talents. In 2013, the Ministry of Education supported 18 innovation teams and more than 100 new-century talents in the fields related to biodiversity.

3.3 Science and Technology

The National Programme for Mid and Long-term Development of Science and Technology (2006-2020) has included in its key areas for the environment themes such as "restoration of ecosystem functions in ecologically vulnerable regions" and "monitoring of global environmental change and responses". In its key areas for agriculture one theme is "ecological security of agriculture and forestry and modern forestry". In its fundamental research programmes, there is one research programme on mechanisms of human activities impacting the Earth's systems. All these themes and research programmes have provided directions for biodiversity research.

The Ministry of Science and Technology (MOST) has included projects on conservation and sustainable use of biodiversity in the National Plan for Support to Science and Technology, National Programme for Key Fundamental Research and Development, National Plan for High-tech Development and other plans using specialized funds. In the Eleventh and Twelfth Five-year Plans for Support to Science and Technology, there are 32 key projects on ecological and biodiversity conservation, such as "building ecological security barriers in southwest China and demonstration (Phase I)","research on and demonstration of techniques for biodiversity conservation and breeding of endangered species", and"research on and demonstration of techniques for monitoring and conservation of important species". A total of 1.37 billion yuan RMB was invested into these projects. The National Programme for Key Fundamental Research and Development also included projects such as "research on evolution and protection of biodiversity in Himalaya Region", and "theory and methods of pest control and protection of germplasm resources in support of agricultural biodiversity". Some projects in the National High-tech Development Plan ("863" Plan) have also included technology development for biodiversity conservation and sustainable use of biological resources. The international cooperation plan of the Ministry of Science and Technology has included a few key projects such as "key techniques for ecosystem management in Dongting Lake Basin and their demonstration applications". In building platforms of information resources for nature-related science and technology, support has been provided to survey and collection of animals, plants, micro-organisms and germplasms as well as associated information system development and sharing of relevant information and specimens. All these research projects have generated a number of valuable and influential research results, thus providing scientific and technical support to biodiversity conservation in China.

From August to December 2010 and from September to November 2013, the Ministry of

Science and Technology organized a series of site visits for scientists and experts in the field of ecological conservation. Senior experts were invited to visit more than 20 typical areas of ecological degradation, such as rocky area in Bijie, Guizhou Province, and degraded grassland in areas of origin of three major rivers in Qinghai Province and deserts in southern Xinjiang. Discussions were held following the site visits. From the discussions, a number of technical approaches for ecosystem conservation were summarized and recommended for promotion and application.

Since the eleventh five-year plan period, the Ministry of Science and Technology supported, through specialized funds for fundamental research, scientific surveys in the field of biodiversity and climate change undertaken by those research institutions affiliated with the Ministry of Education, the Ministry of Agriculture and the Chinese Academy of Sciences (CAS). By the end of 2013, the Ministry of Science and Technology has supported nearly 70 projects and invested nearly 600 million yuan RMB into such surveys. The regions surveyed include Pan-Himalaya region, Qinghai-Tibet Plateau, Hainan Island and Xisha Islands, Luoxiao Mountain Range, hilly areas of south China, ecologically sensitive hilly areas and plains in coastal areas of southeast China, tropical islands and coastal areas, arid areas in north and northwest China, and temperate conifer forest areas in northeast China. The surveys covered lakes, islands, seas and oceans, forests, species and germplasm in special habitats as well as invasive alien species. As a result of these surveys, Fauna of China, Flora of China, China's vegetation map including vegetation composition and distribution, and a list of marine species were compiled and developed. In addition, China Forest Biodiversity Monitoring Network and China Ecosystem Research Network have been further improved, and standards and specifications have been developed for monitoring plants, animals and micro-organisms.

By using rich digitalized information available, CAS and other institutions have established National Specimens Information Infrastructure (NSII), Asian Biodiversity Conservation and Development Network (ABCDNet) and China node for the Global Biodiversity Information Facility (GBIF). CAS has completed assessments of the status of endangerment of more than 30,000 higher plant species in China and based on that, issued China Plant Red List as well as identified hot spots for plant conservation based on the distribution of protected and endemic plants in China. CAS has also initiated certification of carbon budget and relevant projects in response to climate change.

3.4 Land and Resources

Land and resources sector gives strategic importance to biodiversity while developing and implementing its national land planning, land use planning and land remediation planning.

(1) Integration of biodiversity into national land planning

The Ministry of Land and Resources (MLR) has comprehensively undertaken national land planning, with a view to coordinate between land development, use, conservation and remediation and to improve spatial layouts of land development. The *National Land Planning Strategy* requires that forests, grasslands, lakes, wetlands and coastal and marine ecosystems should be strictly protected, and the protection of functions of various national-level nature reserves and important ecological function zones should be strengthened. The strategy also calls for establishment of a system of *in-situ* conservation for important biological resources primarily composed of nature reserves and complemented by various forest parks, germplasm conservation areas, hunting ban areas, deforestation ban areas and protected sites for original habitats for some species.

(2) Integration of biodiversity into land use planning at various levels

The National Master Plan for Land Use (2006-2020) issued by the State Council stressed the guiding principle of coordinating land use for production, livelihood and ecology conservation by giving priority to nature and ecology conservation. Chapter V of this plan clearly requires that land essential for ecology should be protected and greater efforts will be put into ecological and environmental improvements with land ecology improved based on local conditions. In 2010, MLR issued rules for developing master plans for land use at municipal, county and town levels. The rules require local governments to control land use for urban and rural development and identify core areas of nature reserves, forest parks, geological parks, natural habitats for wild animals and plants included in the provincial and above protected lists, and core areas of protected water source areas as zones banned for development. Those construction or development activities inappropriate to major functions of these areas should be strictly prohibited. In *Hainan Province's Master Plan for Land Use*, which was approved by the State Council, biodiversity has been incorporated into the land use master plan. Meanwhile, Hainan Province Government approved biodiversity-based land use master plans for Lingshui County and Ledong County, both of which are autonomous counties for Li ethnic people.

(3) Strengthening biodiversity conservation while planning land remediation

The *Regulation on Land Reclamation* issued by the State Council in 2011 provides that those having the obligation to reclaim land should follow standards for land reclamation and relevant environmental standards to protect soil quality and ecology and avoid soil and underground water contamination. *The National Plan for Land Remediation* (2011-2015) issued by the State Council has identified ecological and biodiversity conservation as one of important targets. Specifically the target is to promote self-recovery capacities of degraded land ecosystems primarily through natural recovery as well as conservation and comprehensive environmental and ecological improvements. To improve the ecology of mining areas, MLR launched an action called 'Re-greening Mines' nationwide in June 2012, which was intended to address geological environmental problems caused by mining activities in important nature reserves and scenic spots, through engineering and biological measures. The action will help restore damaged ecosystems. Guizhou Province incorporated biodiversity conservation concepts into land remediation. In September 2010, Guizhou approved biodiversity-based land remediation and reclamation plans for Guanling Autonomous County and Libo County.

(4) Establishing mechanisms of multilateral cooperation

MLR, together with the Ministry of Commerce, the German Ministry of the Environment, EU and UNDP, implemented a few projects such as "biodiversity conservation in land use planning and land remediation", and "China-Germany project on low-carbon land use". MLR also undertook technical exchanges with Belgium, Germany and EU for biodiversity conservation in land use planning, land remediation and reclamation. These collaborative activities have helped the Chinese personnel involved improve their biodiversity knowledge and broaden platforms for China and other countries to collaborate on research on land use and biodiversity conservation.

(5) Strengthening communication on biodiversity

During the World Expo 2010 in Shanghai, MLR organized thematic lectures and a series of communication and educational activities on biodiversity conservation in land use planning and land remediation to improve public awareness of biodiversity conservation in land management. In celebration of the Earth Day (22 April) and National Land Day (25 June) and other dates, MLR popularized biodiversity concepts and knowledge among land managers and researchers, university students and farmers, providing technical support for biodiversity conservation in land remediation.

3.5 Housing, Urban and Rural Development

(1) Improving relevant regulations and policies, sectoral standards and development plans

The Ministry of Housing and Urban-Rural Development (MHURD) has incorporated into its Twelfth Five-year Plan "strengthening biodiversity conservation and adaptation to climate change in planned urban zones and scenic spots". In 2010, MHURD issued *Recommendations Concerning Further Strengthening Management of Zoos*, and in 2013, MHURD developed a *National*

Programme for Development of Zoos, identifying targets and measures for species and population expansion. MHURD has also planned to implement sector-wide management of those species or populations identified as national levels Ⅰ & Ⅱ protected species, endangered wild flora and fauna and listed in annexes Ⅰ and Ⅱ of the CITES. MHURD developed guidelines for designing urban botanical gardens and zoos in cities and technical guidelines for zoo management, all of which are intended to put urban biodiversity conservation into legal and standardized tracks.

(2) Incorporating biodiversity conservation into evaluation of national garden cities and eco-cities

Biodiversity has been incorporated as an important element in the national standards for garden cities issued by MHURD in 2010 and the methods and standards for classifying and evaluating eco-cities issued in 2012. Indicators such as comprehensive species index, local woody plant index, rate of naturalization of water body coasts and urban ecological conservation have been included for evaluation. Through creation of garden or eco-cities, most cities in China have completed consensuses on species, developed plans for protecting biodiversity in cities and implementation measures as well as effectively protected landscapes, hydrology, vegetation and species, which lead to increasingly rich biodiversity in cities.

(3) Using botanical gardens, zoos and wetland parks as bases for species conservation and research

In accordance with incomplete statistics, within the HURD system there are 200 botanical gardens that conserve *ex-situ* 20,000 plant species. In 2012, SFA, MHURD and CAS jointly issued recommendations on strengthening *ex-situ* conservation of plant germplasm in botanical gardens. More than 240 zoos including animal demonstration zones have been established, with 26 pedigree depositaries and 37 species lineages established. Considerable achievements made in *ex-situ* conservation of rare, endangered species, in particular giant pandas and south China tigers. The number of giant pandas raised in captivity has reached 332 by the end of 2011, with the main work having shifted from rapid growth in the number of individuals to upgrading the quality of the whole population. The number of south China tigers raised in captivity has increased from 6 to more than 100 currently, bringing back hope of recovery of its wild population. In response to human impacts on urban ecosystems, MHURD issued *Provisional Rules for Management of Urban Wetland Parks* and *Guidelines for Designing Urban Wetland Parks*. So far, 49 urban wetland parks have been established across the country, with the capacities increased and enhanced for conserving and managing wetlands and wetland species in cities.

(4) Strengthening the system of monitoring and management of species in scenic spots

Scenic spots are important areas for biodiversity conservation. Since 2002, MHURD has established a system of information for monitoring and managing national-level scenic spots, using remote-sensing and GIS technologies to monitor natural resources conservation and implementation of relevant plans in scenic spots. Systems of dynamic monitoring and verification using remote-sensing technologies and inspectors for urban and rural planning have been established. So far, monitoring has been undertaken in 208 national scenic spots. In 2012 and 2013, MHURD organized inspections of law enforcement of conservation and management of national-level scenic spots. In 2013, pilot work on ecological compensation was undertaken in some scenic spots in Yunnan, Guizhou, Sichuan and Xinjiang. All these measures have played an important role in protecting species and the ecological environment and maintaining biodiversity within scenic spots.

(5) Establishing collaboration with international institutions and organizations

In recent years, MHURD has established close collaboration with a number of international organizations and institutions such as UNESCO, IUCN and US Department of the Interior. In recent years, due to the fact that more and more habitats for rare species and typical ecosystems are included in the world's heritage sites, MHURD has strengthened conservation and research of ecosystems and biodiversity in scenic spots and the world's heritage sites. In 2012, MHURD together with IUCN undertook pilot work on China Green List, and worked with UNESCO in monitoring and conserving biodiversity in the world's natural heritage sites such as Sichuan giant panda habitat, Libo Zhangjiang and Wuyi Mountain. MHURD is helping implement an action plan for biodiversity in one of the world's heritage sites in Dujiangyan-Qingcheng Mountain, Sichuan Province. Since China joined the World Convention on Natural and Cultural Heritages, by the end of 2013, 21 sites in China have been successfully named as the world's heritage sites, attracting the attention worldwide.

3.6 Water Resources Management

In recent years, the water resources management sector has actively undertaken biodiversity conservation as required by the updated NBSAP (2011-2030) and in accordance with the sectoral situation.

(1) Improved water-related laws, policies and programmes in support of biodiversity

Water resources allocation and protection, ensuring water use for ecology and soil conservation related to biodiversity have been legalized and regulated, through improving regulations/rules associated with the *Water Law* and revising the *Water and Soil Conservation Law*, as well as issuing the State Council decision on accelerating reforms in water resources management and recommendations for implementing the strictest rules for water resources management. Requirements for biodiversity conservation have been upgraded to legal and policy requirements, which clearly identify responsibilities and duties of water resources management, project implementers and the public in biodiversity conservation. The water resources sector has taken biodiversity into account while developing sectoral development strategies, plans or programmes. For example, *National Plan for Water Resources* (2010-2030), *National Programme for Protection of Water Resources*, *National Programme for Protection and Use of Underground Water, National Flood Control Plan* (2011-2015) and the revised integrated plans for seven river basins have covered mainly water use quantity and quality for ecology, soil erosion prevention and control, soil conservation and ecological compensation. In 2011, the State Council approved *National Water Zoning Plan of Important Rivers and Lakes*. This plan identified functions and water quality targets of 4,493 important water function zones. This plan provides an important basis for use and protection of water resources, preventing and controlling water pollution and improving the water environment.

(2) Ecological worsening trend in many important protected areas contained through ensuring water use for ecology

In recent years, the water resources management sector has optimized the use of water resources in the Yellow River, Talimu River, Shiyang River and Hei River. As a result, no water flow shortage has happened in the Yellow River for 14 consecutive years; water resumed in Taitema Lake in the lower reaches of Talimu River, which had been dry for more than 20 years; Dongjuyanhai in the lower reaches of Hei River did not run dry for 9 years; and the water quantity flowing downstream from the section of Minqincai County of Shiyang River increased gradually. For nine years water from the Yangtze River was transferred to Tai Lake so as to improve the lake water quality by increasing water flow and volume to reduce pollution and avoid drought. Water is also supplied to lakes and wetlands such as Zhalong Wetland, Nansi Lake, Baiyangdian and Hengshui Lake to improve local ecological conditions and maintain water security for the ecologically vulnerable areas. Optimized water management was attempted in the Three Gorges Reservoir and other reservoirs in the river basin to meet ecological needs.

(3) Local ecology significantly improved by increased efforts in soil erosion control

Soil erosion control accelerated in recent years. Big soil erosion control projects have been implemented in key regions such as the upper and middle reaches of the Yangtze River and the Yellow River, the upper reach of the Pearl River, the reservoir area and upper reaches of Danjiang River, water source areas for Beijing, origins of sandstorms affecting Beijing and Tianjin, rocky areas of Shanxi, Shaanxi and Inner Mongolia, rocky Karst areas and black soil areas in northeast China. From 2009 to 2012, areas with soil erosion controlled have reached 270,000 km^2, covering 12,000 small river basins. Enclosures for soil erosion control continue to be implemented, with areas with enclosures implemented having reached 720,000 km^2, among which ecological conditions have started to improve in areas of 450,000 km^2, in particular sources of three major rivers in Qinghai, inland river basin of Xinjiang and Tibet. Impacts from enclosures for soil erosion control are increasingly obvious with ecological functions of enclosed areas effectively protected.

(4) Aquatic environmental and ecological quality in some regions significantly improved as a result of aquatic ecosystem conservation and restoration

14 cities such as Wuxi, Wuhan, Guilin and Harbin have started aquatic ecosystem conservation and restoration through taking measures such as connecting water bodies, diverting polluted water, dredging rivers and lakes, treatment and restoration of coastal lines and protection of water sources. These measures have achieved good results and improved the aquatic environment and ecology. Wuhan has improved water quality of 16 lakes in downtown areas through connecting rivers and lakes. Guilin has ensured water supply for Li River in dry season through optimizing water supply for effective biodiversity conservation. The rate of water functions in Wuxi having met relevant standards increased from 11.8% in 2005 to 46.8%. Lishui City reduced areas affected by soil erosion by 567 km^2 through control measures in small river basins like Oujiang River, and the water quality of main branches that has met relevant standards has increased up to 98.7%. Laizhou City reduced areas eroded by marine waters from 261 km^2 to 228.5 km^2, through cracking down overexploitation of underground water.

3.7 Agriculture

Agricultural biodiversity is an important component of biodiversity. The Government of China always attaches great importance to conservation and sustainable use of agricultural biodiversity and has been developing and implementing relevant regulations, plans and programmes.

(1) Strengthening leadership and organization

The Ministry of Agriculture (MOA) has strengthened the following bodies or institutions as well as their capacities for conservation and management of agricultural biodiversity: Leading Group on Protection of Agricultural Wild Plants, Expert Review Committee on Wild Plant Protection, National Committee on Livestock Genetic Resources, MOA Office for Aquatic Wild Flora and Fauna Conservation, Scientific Committee on Endangered Aquatic Wild Flora and Fauna, Fishery Resources Management Committees of the Yangtze River Basin, the Yellow River Basin and the Pearl River Basin, National Coordinating Group on Prevention and Control of Invasive Alien Species (IASs), Office of Management of IASs, and MOA Research Center for IASs Prevention and Control. MOA has also established National Committee on Agricultural Crops Germplasm to coordinate the management of agricultural crops germplasm including agricultural wild plants, to propose development strategies and policies in this regard and to guide the development of mid and long-term plans in this regard.

(2) Improving regulations and standards

On the basis of existing laws and regulations, the Ministry of Agriculture has issued a series of regulations to improve the legal system for protection of livestock genetic resources, such as *Rules for Approval of Import and Export of Livestock Genetic Resources* and *Collaborative Research with Foreign Entities, Rules for Management of Livestock Genetic Resources Conservation Farms and Gene Banks, Rules for Determination of New Livestock Varieties and Identification of Livestock Genetic Resources, Rules for Licensing of Livestock Genetic Material Production* and *Provisional Technical Requirements for Importing Livestock Genetic Resources*. In 2010, NDRC and MOF issued a notice on collecting fees for grassland vegetation recovery, requiring that those using grasslands for mineral prospecting, mining and project construction should pay fees for grassland vegetation recovery. In 2012, China's Supreme Court issued an interpretation of several legal issues concerning the trial of criminal cases against grassland resources, which set criteria for conviction and sentencing of such crimes. The Ministry of Agriculture developed four sectoral standards for protection of wild plants such as technical specifications for monitoring and early warning of protected sites of wild flora and technical guidelines for *ex-situ* storage of wild flora. MOA also issued *Provisional Rules for Management of Protected Areas of Aquatic Germplasm Resources, Rules on Management of Restocking of Aquatic Species, Regulations on Impact Assessment of Construction Projects on National-Level Protected Areas for Aquatic Species*, and *a Notice on Management of Marine Summer Fishing Moratorium*. MOA launched a programme for emergency response to invasion of alien species and major pests, the first list of IASs under national management as well as 17 sectoral standards for preventing and controlling IASs, such as technical specifications for controlling mile-aminute weed and apple snails, as well as over 40 technical guidelines for emergency responses to IASs. All these have further improved the legal system for conservation of agricultural biodiversity.

(3) Integrating biodiversity into relevant plans and programmes

The *National Programme for Modern Agriculture Development* (2011-2015) issued by the State Council set a number of goals for strengthening conservation of agricultural resources and ecology, grassland, aquatic resources, aquatic ecological restoration, livestock genetic resources and agricultural wild plant resources. For grassland biodiversity, MOA issued *National Master Plan for Grassland Conservation and Use* and *National Twelfth Five-year Plan for Husbandry Development* as well as a joint notice (together with MOF) on implementing rewards or subsidies for grassland ecological conservation and rules for assessing effectiveness of such mechanisms. A number of grassland conservation projects were initiated such as returning grazing land to grassland, grassland conservation to control sandstorms affecting Beijing and Tianjin, permanent settlement for nomadic herdsmen and grassland fire prevention. The *Twelfth Five-year Plan for Husbandry Development* proposed to continue implementing livestock breeding projects, support infrastructure construction for conservation farms, protected areas and gene banks for livestock genetic resources, and to improve the national system of conservation of livestock genetic resources. The *National Twelfth Five-year Plan for Fishery Development* issued by MOA contains goals and measures for developing environmentally friendly fishery, improving aquatic environment, strengthening fishery moratoriums, restocking of rare species, strengthening protected areas of aquatic species and protecting aquatic biodiversity. The *National Twelfth Five-year Plan for Agricultural Science and Technology Development* required further strengthening collection, protection and identification of agricultural germplasm resources and agricultural genetic resources as well as development and improvement of breeding materials. The *National Twelfth Five-year Plan for Plantation Industry* also proposed to improve the national system for storing and using germplasm resources. The issuance and implementation of all these plans and programmes have strongly promoted the conservation and sustainable use of agricultural biodiversity.

(4) Strengthening protection of agricultural wild plant resources

One measure is to survey and monitor wild plant resources. China has obtained information concerning the distribution of 172 wild plant species across the country, and based on that, established a national information system of wild plant resources. Another measure is to promote establishment and monitoring of protected sites for wild plants. 42 endangered wild species are protected and 15 protected sites for wild rice, soybeans and wheat in Guangxi, Hainan, Yunnan,

Henan, Jilin, Heilongjiang, Xinjiang and Ningxia are being monitored. The third measure is to strengthen *ex-situ* conservation of wild plant resources. Wild major crops have been collected and rescued, and a group of new germplasm nurseries have been established, with more than 30,000 accessions of major crops wild relatives having been collected and stored. The system of *ex-situ* conservation of agricultural crops germplasm composed of germplasm nurseries, long-term banks, copy banks, mid-term banks and plantlets libraries has been further improved, providing reliable support for conservation and use of wild plant biodiversity.

(5) Strengthening protection and restoration of grassland ecosystems

MOA has undertaken a second remote-sensing rapid survey of grasslands. MOA has also improved bodies for grassland supervision and management. By the end of 2012, there are 844 such organizations at the county and above levels, including one at national level, 23 at provincial level, 126 at municipal level and 694 at county level. During 2003-2009 more than 90,000 illegal cases of various kinds of damaging grasslands have been penalized. During 2010-2012 more than 20,000 such cases have been convicted annually. China has established a network of grassland conservation with reasonable layout, relatively complete types, wide distribution and coverage and high representativeness. To protect grassland ecosystems, China has been implementing many projects such as returning grazing land to grassland, control of origin of sandstorms affecting Beijing and Tianjin, ecological conservation of sources of three major rivers in Qinghai, grassland conservation in rocky areas of southwest China and prevention and reduction of grassland disasters. According to the monitoring results by MOA in 2012, grassland conservation projects have achieved significant results. Compared with areas without such projects implemented, vegetation coverage of grasslands in project areas increased by 11%, height of grass increased by 43.1%, fresh grass output increased by 50.7%. The situation of grassland use is considerably improving, with the rate of overcapacity use of grasslands in 268 counties in 2012 down by 34.5%-36.2% compared with the situation in 2011. However, overall most of grasslands are being used beyond their capacities, and degradation, desertification and salinization of grasslands have not been effectively controlled. Challenges for grassland ecological conservation remain to be huge.

(6) Preventing and controlling invasive alien species

One measure is to survey and monitor invasive alien species. In recent years, MOA has undertaken consensus on invasive alien species across the country through combination of point and non-point surveys, focusing on 22 agricultural invasive species such as *Ageratina adenophora, Mikania micrantha, Parthenium hysterophorus, Amaranthus palmeri* and *Xanthium spinosum*. Another measure is to eliminate agricultural IASs. In the past five years, MOA has organized more than 10 activities eliminating IASs like *Pomacea canaliculata,*

Mikania micrantha, Ageratina adenophora, Alternanthera philoxeroides and *solanum rostratum* in Yunnan, Hunan, Hubei, Sichuan, Guizhou, Jiangxi and Jilin Provinces, and so on, having effectively controlled the expansion of major IASs. The third measure is to undertake researches on technologies for IASs prevention and control. High-efficiency spraying technologies such as low-volume sprayers and static electricity ultralow volume sprayers have been developed, and ecological regulation technologies using grass, bushes and crops as alternatives to invasive species have been developed. Natural enemies to *Brontispa longissima* namely *Tetrastichus brontispae* and *Asecodes hispinarum* have been selected with control effects exceeding 85%. Four natural enemy factories were established in Hainan with daily production scale of parasitoids reaching 2 million, and areas controlled reaching 1,000 km^2. Breakthroughs were made in developing technologies for using IASs for other purposes.

(7) Protecting aquatic resources in accordance with relevant laws

One measure is to establish protected areas of aquatic species. China has established more than 200 protected areas for various kinds of aquatic wild plants and animals and aquatic ecosystems, forming a network of conservation of aquatic species with reasonable layout and various types. The second measure is strengthening establishment of protected areas for aquatic germplasm. By the end of 2013, China has identified 428 national-level protected areas for aquatic germplasm, which will play an important role in protecting key aquatic plants and animals of economic value and endemic, rare local aquatic species as well as their habitats and reproduction sites. The third measure is increasing restocking of aquatic species. Restocking is undertaken in suitable water areas, and the varieties, number and scope of restocking gradually increased. As a result, the population of important fish species is recovering to some extent. For example, in Bo Sea and parts of northern Yellow Sea, some species that used to disappear such as Chinese shrimp, jellyfish and blue crab are now coming back in the fall fishing season. The fourth measure is strengthening marine ranching. Various places in China undertake marine ranching using artificial fish reefs as carriers for bottom sowing proliferation and seagrass plantation and complemented by restocking. In connection with ship capacity reduction, abandoned fishing boats are used to reduce costs of artificial fish reefs or nests. The cumulative artificial fish reefs of various kinds installed across the country reach nearly 2 million m^3, providing habitats and reproduction sites for marine species in important regions.

(8) Promoting biosafety assessment and management of agricultural GMOs

Please see details contained in section 2.3.9 of Part 2 in this book.

Case 3.1 Protection of Crop Wild Relatives

From 2007 to 2013, with the support of the Global Environmental Facility (GEF), the Ministry of Agriculture and UNDP jointly implemented a project on conservation and sustainable use of crop wild relatives. Eight sites in eight provinces with different social and economic conditions were selected, for protection of wild rice, wild soybeans and wild relatives of wheat. The strategies for the project were: first, to regulate human behaviour through policy guidance and legislation to minimize damage to crop wild relatives and their habitats; second, to develop alternative livelihood for local farmers and herdsmen to reduce their dependence on crop wild relatives and their habitats; third, to provide financial support to guide local farmers and herdsmen to develop biodiversity-friendly household economy; and fourth, to increase their awareness and encourage local farmers and herdsmen to participate in activities of protecting crop wild relatives.

The result of project implementation indicates that compared with the baseline data of 2008, the status index of resources in all project sites has increased to some extent, with the overall trend going up and the expected results achieved. The threat reduction index of all project sites is above 80%, indicating that threats to crop wild relatives are being reduced and protection activities have good sustainability (see figures below). More importantly, income per capita of residents in all project sites has gradually increased, with the growth rate higher than that of neighboring villages with similar conditions, so local sustainable economic development has been promoted while protecting crop wild relatives.

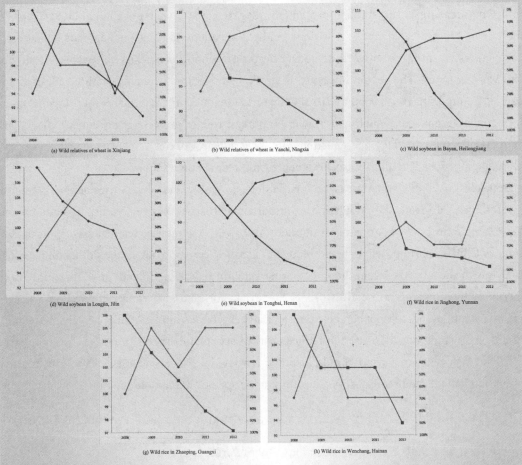

Note: blue curves reflect status of change in resources, red curves reflect changes in threat reduction index. (a) wild relatives of wheat in Xinjiang; (b) wild relatives of wheat in Yanchi, Ningxia; (c) wild soybean in Bayan, Heilongjiang; (d) wild soybean in Longjin, Jilin; (e) wild soybean in Tongbai, Henan; (f) wild rice in Jinghong, Yunnan; (g) wild rice in Zhaoping, Guangxi; (h) wild rice in Wenchang, Hainan.

3.8　Commerce

(1) Enhancing synergies of CBD with multilateral trade system

China has been actively promoting discussions in the WTO Council on Intellectual Property Rights about the relationship between the Trade-related Intellectual Property Rights Agreement (TRIPs) and the Convention on Biological Diversity. Together with other developing country members, China proposed that the TRIPs should be revised to be consistent with the CBD. Article 19 in the Doha Declaration required the WTO Council on Intellectual Property Rights to review the relationship between the TRIPs and the CBD and examine the protection of traditional knowledge and folklore. Based on that, developing country members proposed that the TRIPs should be consistent with the CBD by following the three principles of national sovereignty, prior informed consent and benefit-sharing. Developing country members also proposed that it would be mandatory to disclose the origin of genetic resources in patent applications for those inventions relying on genetic resources to ensure that the right to know of those providers of genetic resources and their right to benefit from use of genetic resources. China, Brazil, India, EU, Switzerland and other members that supported the mandatory disclosure of the origin of genetic resources submitted in May 2008 a joint proposal (W52), which requested for enhancing the multilateral system of registration of geographical indicators, expanding the scope of protection of geographical indicators and disclosure of origin of genetic resources in patent applications. In April 2011, China, Brazil, India and other members that had submitted the W52 proposal submitted to WTO a draft decision concerning supporting the mutual supportiveness of the TRIPs and the CBD (TN/C/W/59), which suggested that in Article 29 of the TRIPs one more provision be added "requiring the disclosure of origin of genetic resources and/or associated traditional knowledge", making the disclosure of origin of genetic resources and associated traditional knowledge an essential element in patent applications.

(2) Incorporating biodiversity into policies, regulations and guidance issued by the Ministry of Commerce

Since March 2009, the Ministry of Commerce (MOC) together with other ministries and commissions has issued a number of guidelines or guidance concerning biodiversity conservation, requiring Chinese companies or enterprises to protect biodiversity in their domestic and international operations, and encouraging them to take into account biodiversity in the international trade and investment. In March 2009, the Ministry of Commerce and the State Forestry Administration jointly issued *Guidelines for Chinese Enterprises for Sustainable Forest Business and Use in Foreign Countries*. The guidelines required enterprises to use scientific and reasonable logging methods and operational measures to minimize logging impacts on biodiversity. In February 2013, the Ministry of Commerce and the Ministry of Environmental Protection jointly issued *Environmental Guidelines for International Investment and Cooperation*, which requires enterprises involved to give priority to *in-situ* conservation and to minimize negative impacts on local biodiversity.

(3) Taking biodiversity into consideration in international trade negotiations

Article 145 concerning genetic resources, traditional knowledge and folklore in China-Peru Free Trade Agreement signed in April 2009 clearly provides that "both Parties recognize and re-emphasize the principles and provisions established in the Convention on Biological Diversity adopted on 5 June 1992, and encourage the efforts to establish the mutually supportive relationship between the TRIPS and the CBD."

(4) Undertaking international cooperation in the field of biodiversity

Through bilateral and multilateral channels, the Ministry of Commerce has organized and implemented a number of biodiversity conservation projects, such as China-Norway project on biodiversity and climate change. In November 2012, MOC, UNDP, UNIDO, UNCTAD and UNEP jointly organized a photo exhibition on strengthening South-South cooperation and promoting green development, with more than 120 pictures on display covering energy conservation, environmental protection, responses to climate change, biodiversity conservation and capacity building. The Ministry of Commerce also organized a number of workshops on international research and management of biodiversity, and invited officers and experts from developing countries to introduce transboundary mechanisms, achievements and experiences for biodiversity conservation.

(5) Increasing media coverage and communication of biodiversity

Since 2009, the Ministry of Commerce and its offices overseas and local departments of commerce have collected and reported on news and developments concerning biodiversity conservation, introduced international biodiversity policies and regulations as well as biodiversity aid projects by developed countries. Their work has provided basis for Chinese enterprises and products to go abroad and helped China benefit from the experiences and approaches employed in other countries to improve the public understanding and awareness of biodiversity.

3.9 Customs

The customs, as national authority to manage and supervise imports and exports, have always attached great importance to management of imports and exports of biological resources.

(1) Strengthening supervision of import and export of biological resources

The customs seriously implement the Convention on Biological Convention and the *National Programme for Conservation and Use of Biological Resources* approved and issued by the State Council. According to relevant laws and requirements, the customs exercise the management of the import and export of biological resources by carefully examining documents required for import and export, such as the Certificate of Import /export of Endangered Species, the Certificate of Proof of Species Not in the List of Wild Flora and Fauna Not Allowed for Trade, the Certificate of Approval of Blood Import and Export, the Certificate of Import and Export of Human Genetic Materials and the Certificate for Export of Legally Harvested Products. Meanwhile, the customs also support other relevant departments by assisting with the import of biological resources used for research and conservation purposes.

(2) Participating in development of biosafety regulatory systems

For years, the customs have been devoted to promoting the development of the legal and regulatory system for biodiversity. The customs, together with relevant departments, have adjusted and issued lists of species for import and export in accordance with the *Regulation Concerning Import and Export of Endangered Wild Animals and Plants,* including *List of Wild Animals and Plants Permitted for Import and Export and Their Products* and *List of Dual-use Items and Technology for Import and Export.* The customs provided specific suggestions to the development of regulations such as Implementation Rules for Regulation Concerning Protection of Fossils, Regulation on Management of Genetic Resources and Rules for Environmental Management and Supervision of Invasive Alien Species.

(3) Constantly strengthening training of on-site customs officers

The customs administration has been providing law enforcement capacity building training to on-site customs officers to improve their law-enforcement capacities. The training content mainly included List of Genetic Resources for Import and Export, regulations concerning conservation of genetic resources and professional knowledge needed by customs officers for examination and verification of biological resources.

(4) Investigating relevant cases in import and export

From January 2009 to December 2012, the anti-smuggling department of the customs investigated 406 cases of smuggling of rare animals and plants and their products, and seized 381 of smuggled rare animals and plants and their products, whose value was 5.83 billion yuan RMB. They also investigated 3,573 cases of administrative violations in this regard and seized products worth 130 million yuan RMB. In particular in 2012, the General Customs Administration investigated a number of major cases in smuggling of endangered species, through a nationwide anti-smuggling campaign which called "National Gate Shield". In May 2012, the CITES Secretariat awarded certificates of excellent performance to the General Customs Administration and the State Forestry Administration. In 2013, China took the initiative called "Cobra Action" to crack down smuggling of endangered species involving 22 countries from Asia and Africa. The customs investigated a total of 71 cases of smuggling of endangered species and 85 suspects involved in relevant crimes and activities. The cases handled by the Chinese customs were one third of the total cases handled during the initiative.

3.10 Industry and Commerce Administration

The Industry and Commerce Administration seriously implements relevant regulations such as *Wild Animal Protection Law*, and reinforces market supervision and law enforcement to crack down illegal activities such as illegal purchasing of and business in national key protected animals and plants and their products.

(1) Strengthening leadership and making thorough arrangements

The State Administration for Industry and Commerce (SAIC), while undertaking various market supervision activities, incorporates biodiversity conservation as one of its key work and its various regulations. In holiday market supervisions all the year around, SAIC issued specialized notices many times requiring its local administrations to crack down illegal selling of wild plants and animals and their products. In December 2012, SAIC issued an urgent notice on strengthening protection of wild animals requiring its local administrations to concentrate law enforcement on cracking down those illegal selling and purchasing of national key protected animals and their products. In response, local administrations of industry and commerce rapidly organized one-month campaigns against such illegal activities. During the campaign, Jiangsu Province Administration of Industry and Commerce examined 6,458 markets (times), 78,000 business operators (times). 58 business operators without licenses were banned and 9 illegal cases were investigated with fines worth 18,000 yuan RMB given.

(2) Strengthening routine supervision and law enforcement and maintaining market order

One measure is to manage market access and regulate business qualifications. In accordance with the *Wild Animals Protection Law*, all the groups that do business in purchasing, selling, transporting, domesticating and raising national key protected animals are required to obtain approval documents or licenses from the competent authorities or institutions authorized to issue such documents before they apply for business certificates. Another measure is to regulate business behaviours and to reinforce law enforcement at local levels. Local administrations strictly implement routine patrolling and territorial supervision while enforcing laws related to wild animals and plants and ensure that specific tasks and responsibilities are assigned to persons in relevant positions. Since March 2009, Yunnan Province AIC has inspected 3,068 markets of wild animals and their products, 32,000 restaurants selling wild animals and their products, and 264 companies that reproduce and domesticate wild animals. 120 companies, doing business in wild animals and their products without licenses, have been banned and 3,300 animals of various kinds have been confiscated.

(3) Strengthening investigations into relevant cases and cracking down illegal activities and violations

Local administrations strictly implement relevant laws such as *Wild Animal Protection Law* and crack down illegal activities and crimes against wild animals and plants. Since March 2009, Guangdong Province AIC, while undertaking special inspections for protection of wild flora and fauna, has inspected 35,000 markets of various kinds and 155,000 business operators, investigated 154 cases of violating relevant laws (with value worth 120,000 yuan RMB) with fines of 150,000 yuan RMB given. Guizhou Province AIC examined 2,094 markets of various kinds and 22,000 business operators of various kinds, having seized 857 kg of products of wild animals and confiscated 5,455 animals. Heilongjiang Province AIC investigated more than 100 cases, having confiscated more than 30,000 animals of various kinds, which are worth more than 3 million yuan RMB. Qinghai Province AIC organized a campaign against illegal sales of Huang fish, by having signed more than 3,000 copies of agreements with business operators about accountability for sales of Huang fish and their products, investigated more than 200 illegal cases and given fines of more than 200,000 yuan RMB.

(4) Providing training on law enforcement and strengthening communication and education

First, local administrations increased training of law enforcement personnel through organizing training workshops and disseminating training materials. AICs in Xinjiang, Jilin and

Heilongjiang invited experts many times to provide training to law enforcement personnel and reinforce their capacities to fulfill their duties by law. Second, local administrations make efforts in communicating by various effective ways laws and regulations concerning protection of wild animals and plants, popularizing scientific knowledge of biodiversity and increasing public awareness of ecological conservation, on the occasions of the International Day of Biodiversity, the World Environment Day, National Day for Legal Communication and the International Day for Protection of Consumers' Rights. For example, Liaoning AIC and Forest Department held a joint press conference on actions for protection of wild birds, organized spring actions to love and protect birds and disseminated more than 40,000 copies of education materials.

3.11 Quality Supervision, Inspection and Quarantine

(1) Importance attached to macro-level planning

The quality supervision, inspection and quarantine sector incorporates biodiversity and conservation of biological resources into its sectoral development strategy and plan. In 2011, the General Administration on Quality Supervision, Inspection and Quarantine (GAQSIQ) issued its Twelfth Five-year Plan which addressed biodiversity conservation and inspection and quarantine of imported and exported biological resources from legal, institutional and technical aspects. The plan clearly proposed to establish a system of inspection and quarantine of species and a system of assessment of intercepted species. To implement fully China's Updated NBSAP (2011-2030), GAQSIQ developed an action plan for inspection and quarantine of import and export of biological resources.

(2) Strengthening leadership and organization

To strengthen leadership over conservation of biodiversity and biological resources in this sector, GAQSIQ established a division for supervision and management of biological resources and a division for inspection of tourists and mails. A center for identification and verification of species and biological resources was established in the Chinese Research Academy of Inspection and Quarantine Sciences. In 2012, a leading group on conservation of biodiversity and biological resources was also established to provide guidance on relevant work in this sector.

(3) Providing scientific guidance

To provide guidance for inspection and quarantine of import and export of biological resources, GAQSIQ issued recommendations on strengthening work in this regard. The recommendations required all staff working in the sector to know the situation and huge challenges they face as well as tasks ahead for biodiversity conservation. They should attach strategic importance to strengthening inspection and quarantine of import and export of biological resources including endangered wild flora and fauna. It was also recommended that the whole sector should increase investments, strengthen and upgrade capacities, strengthen coordination and synergies. The sector should give priority to communication and education so as to create a favorable atmosphere. Meanwhile specific responsibilities should be assigned to specific positions and performance evaluation should be strengthened in efforts to upgrade the level and capacity of inspection and quarantine of biological resources.

(4) Undertaking surveys seriously

To know in-depth the import and export of biological resources, since 2008, GAQSIQ has organized surveys in this regard and basically obtained information concerning the varieties and destinations of species imported and exported. It was found that the frequency and varieties of species exported were twice as many as those imported, indicating that the loss of biological resources from China is rather serious.

(5) Undertaking pilot work actively

To get more experience in inspecting biological resources at customs and ports and make preparations for relevant work at broader scale, GAQSIQ, together with relevant departments, undertook some pilot work on inspection of biological resources in sea-ports, land-ports and air-ports in Yantai, Shandong and Xinjiang and got rich results, laying a good foundation for establishing a system of inspection of biological resources in different ports.

(6) Keeping invasive alien species out of border

GAQSIQ and its local administrations strengthened inspection and monitoring of invasive alien species. From 2008 to 2012, the varieties and the frequency of harmful pests constantly increased significantly. The varieties of pests intercepted increased from 2,856 in 2008 to 4,331 in 2012, with annual increase by over 10%. The number of interceptions increased from 229,000 in 2008 to 579,000 in 2012, with annual increase by over 30%. GAQSIQ issued recommendations for strengthening inspection of travellers and mails. By the end of 2012, 31 out of 61 international mail exchange stations have established units for inspection of mails. Since 2007, GAQSIQ has undertaken a lot of research on prevention and control of IASs and developed many standards in this regard. For example, in the field of plant inspection and quarantine, since 2006, 352 sectoral standards, 104 national standards and 2 international standards have been developed and issued, forming China's system of standards for inspection and quarantine of imported and exported plants.

(7) Monitoring GMOs

GAQSIQ implements a system of declaration of imported genetically modified plants, animals and micro-organisms and their products and food. During 2008-2013, GAQSIQ established a technical center for detecting and monitoring imported and exported GMOs and key labs for GMOs. The capacities of more than 30 labs within the sector to detect and monitor GMOs have been fully upgraded and a relatively complete system of techniques and standards has been established. In recent years, GMO detection has been undertaken more than 300 milliont of soybeans, maize, canola and cotton seed and 6 milliont of rice products and other agricultural products. Relevant inspection and quarantine bodies have detected maize and their products imported from USA and identified MIR162, which was not approved by China for import. GMOs were also found from horse feed imported from Ireland and non-GMO soybeans from Taiwan Province. All these imports have been returned to the exporters.

3.12 Forestry

(1) Continuing implementation of key forestry and ecological projects

In the past decade, China has completed reforestation covering an area of 83,000 km^2 through the natural forest resources protection project. Currently, the forest area of 1.049 million km^2 has been effectively protected, with net increase in the forest area by 100,000 km^2, and the forest coverage rate increased by 3.8% and the forest growing stock by 730 million m^3. China has also completed reforestation covering an area of 218,000 km^2 through implementing the project of returning cultivated land to forests, with the forest coverage rate in the project implemented area increased by 3%. Reforestation of additional area of 83,000 km^2 completed through forest belt building in north, northeast and northwest China as well as in the Yangtze River Basin. Among these, forest belt building in north, northeast and northwest China with the forest coverage rate in the project areas increased to 12.4% and ecological conditions in these areas obviously improving. During the decade when the sandstorm control project was implemented, additional area of 89,000 km^2 was reforested and the forest coverage rate in the project area increased by 4.1%, with sound evolutionary changes occurred in the flora of the area. An area of 30,000 km^2 of the rocky regions has been controlled cumulatively with the vegetation coverage rate being 15% higher than that before the project was implemented. During 2006-2010, altogether 205 demonstration projects of wetland conservation and restoration were completed, with 800 km^2 of wetlands restored.

(2) Improving the management of nature reserves

To implement a notice issued by the State Council on improving the management of nature reserves, the State Forestry Administration issued in 2011 a notice on strengthening management of nature reserves within the forestry sector, which contained requirements for master planning, land ownership, management bodies and personnel, funding and institutional arrangements for nature reserves. A number of provinces and municipalities also put in place specific measures to implement this notice. "One law for one area" system is being further implemented. Gansu and Fujian Provinces have approved rules of management for two national-level nature reserves. A legislation is being considered for Hanma national-level nature reserves in Inner Mongolia. Standardized management of nature reserves is being promoted. Ningxia Autonomous Region has issued a notice on further strengthening forestland management for forestry nature reserves, with a view to strengthening land management for nature reserves. Guangxi Autonomous Region is moving forward border determination for 12 local nature reserves and present it for approval by higher authorities. A target has been set that within the next three years master plans will be developed for all local nature reserves. Guizhou Province has undertaken a comprehensive survey on and verification of nature reserves managed by the forestry sector at all levels and developed a directory and an information database of such nature reserves in the province.

(3) Developing and implementing plans for wild flora and fauna protection and development of nature reserves

In 2012, the State Forestry Administration (SFA) issued the *National Twelfth Five-year Plan for Wild Flora and Fauna Protection and Nature Reserves*. The plan proposed that during 2011-2015, China will give priority to protection of more than 60 wild animals, 120 wild plants and establishment of 51 national-level nature reserves. In 2012, SFA and NDRC jointly issued a national plan for rescuing and protecting wild plants with very small populations (2011-2015). Through implementation of this plan and strengthening rescuing and protection of species with very small populations, wild plant protection will be enhanced. In addition, SFA has developed a draft plan for rescuing and protecting wild animals with very small populations.

(4) Exploring nature reserves management mechanisms and ecological compensation mechanisms

Governments of different levels have explored nature reserve management mechanisms. Guangdong Province has classified national-level and province-level nature reserves within the forestry sector as Level A institutions for public benefits. Hainan Province consolidated nature reserve management by putting those province-level nature reserves originally managed by cities and counties directly under management of the province's forest department. In 2011, Hubei Province included the budget and personnel of 5 national-level nature reserves in the province's government budget and personnel quota. In September 2011, Guangxi Autonomous

Region listed 9 national-level nature reserves as institutions as part of its civil service system, providing strong support for long-term development of these nature reserves. Zhejiang Province Forest and Finance Departments jointly issued a notice on renting of collective forests in forest nature reserves at province and above level. In accordance with this notice, core zones and buffer zones in nature reserves at province and above level will be rented, with the price being 50,000 yuan/(km$^2 \cdot$a), further safeguarding the legal rights of the forest owners. Since 2008, pilot work has started in compensating for property and human life losses caused by wild animals in poor areas. Anhui and Gansu Provinces have developed rules for such compensation. In 2012, the central government inspected the pilot work in this regard in those pilot provinces and drafted provisional criteria for the central government to compensate for losses caused by national key protected animals.

(5) Strengthening prevention and control of forest pests

China has established 1,000 central monitoring and warning sites at national level and more than 1,200 sites at provincial level. There are altogether 28,000 sites with those at city and county level counted. This forms a national network of monitoring and early warning of forest pests. China has also established 3,117 institutions and 858 stations of preventing, controlling and quarantine of forest pests. The personnel working in this field is stable and infrastructure for prevention and control has been obviously strengthened, therefore a system of prevention and control is in place with monitoring and early warning, prevention through quarantine, disaster prevention and reduction and technical support at its core.

(6) Strengthening reintroduction of artificially bred populations and restoration of wild animal species

SFA has been strengthening rescuing and breeding of endangered wild animals. As a result, the number of giant pandas raised in captivity has reached 312, the population of artificially bred crested ibises has exceeded 600 and the populations of more than 50 artificially bred wild animals are constantly expanding, such as the Chinese Alligator (*Alligator sinensis*), tigers, golden monkeys and Tibetan Antelopes (*Pantholops hodgsoni*). 8 endangered wild animals have been successfully reintroduced to nature such as Crested Ibis, Asian Wild Horse (*Equus przewalskii*), Pere David's Deer (*Elaphurus davidianus*), Chinese xenosaurs (*Shinisaurus crocodilurus*), Chinese Alligator, Bactrian Red Deer (*Cervus elaphus yarkandensis*) and Yellow-bellied Tragopan (*Tragopan caboti*), with reproduction in nature achieved and new, wild populations established gradually. More than 1,000 rare or endangered or endemic plants from northeast, northwest and southwest China have been conserved *ex-situ*. More than 400 sites for breeding of wild plant germplasm have been established, with protected centers or bases established for protection of germplasms of cycads, Orchids, Magnoliaes and palms.

Research has been undertaken on artificial breeding techniques and seedling for those rare wild plants for which market demands are relatively big, such as Matsutake (*Tricholoma matsutake*), Snow Lotus (*Saussurea involucrata*), Dove Tree (*Davidia involucrata*), Desertliving Cistanche (*Cistanche deserticola*), yews and rare orchids. 280 breeding bases have been established for rare wild plants, stabilizing artificial populations of more than 1,000 wild plants. Reintroduction has been undertaken for endangered endemic orchids in China such as *Doritis pulcherrima* and *Paphiopedilum armeniacum* and some critically endangered species such as *Cycas debaoensis*, *Pachylarnax sinica* and *Cyclobalanopsis sichourensis*.

Case 3.2 Artificial Breeding and Commercialization of Yews

Yews are listed as national level I key protected plant and vulnerable species by IUCN as well as included in Annex II of the CITES. Wild yews are very rare and have important medicinal values. Jiangsu Yew Biotech Company Ltd. began in 1997 to research on seed development and artificial breeding and plantation of yews and made remarkable achievements in technologies for yew cultivation. They have made breakthroughs in rapid reproduction of yews and solved problems of serious shortage of yew seeds. SFA approved in 2008 a high-tech industrial park for yews created by this company as the first demonstration zone for yew-related science and technology development in China. The company's sale income from commercialization of yews in 2010 exceeded 2 billion yuan RMB, with a huge industry chain composed of yew cultivation, bonsai and trees and manufacturing medicines, health and nutrition products from yews.

Zhejiang Haizheng Pharmaceuticals Company Ltd., with the support of Northeast China University of Forestry, established a large base to commercialize plantation and extracting of yews, following the cooperation model with the private sector, university, production base and farmers all involved. By 2010, this company's sale income has reached 1 billion yuan RMB, turning itself into another platform for scientific and technological innovation of the forest sector in China.

(7) Promoting conservation, research, development and use of genetic diversity of cultivated plants, domesticated animals and their wild relatives

Forest departments at all levels have made shifts from use of wild resources to use of artificially bred resources. Farming and cultivation of plants and animals are promoted for use of traditional sectors that had consumed tremendous wild animal and plant resources such as Chinese medicine and musical instrumentation using animal leather. Development of technologies for artificial breeding of wild plants and animals is promoted and guided, and those companies engaging in artificial breeding of wild animals and their products are exempted from corporate income taxation and those artificially domesticating and reproducing wild animals and their products are exempted from administrative fees for animal protection. Meanwhile, tools such as specialized labelling and standards are promoted to reinforce supervision and management and to crack down the smuggling of and illegal business in wild flora and fauna and their products. This has not only effectively reduced pressures on wild resources, but also enhanced development of local unique industries and increased farmers' income.

(8) Undertaking surveys and monitoring of biodiversity

SFA has been gradually surveying and monitoring forest, wetland and desert ecosystems and biodiversity. During 2009-2013, the eighth national survey of forest resources was undertaken, capturing the latest status of forest coverage rate, forest growing stock and diversity of forest types. The fourth national survey on desertification and land degradation completed in 2009 indicates that by the end of 2009, the area of desertification in China is 2.624 million km^2, an annual reduction by 2,491 km^2 compared with the situation five years ago. The current situation is that the overall trend of desertification has been contained while deserts in some regions are still expanding. The average vegetation coverage rate in deserts increased from 17% in 2004 to 17.6% in 2009, with obvious increases in plant species and stability of plant communities in key controlled areas.

(9) Identifying and enclosing desertified land as protected areas and protecting biodiversity in deserts

For those contiguous desertificated land, whose conditions do not allow for any control measures or are not suitable for development or use as required by ecological conservation, they have been enclosed and identified as protected areas, where activities such as land use, grazing, mining and water use will be strictly prohibited. Development activities are strictly controlled as well to protect desert vegetation and promote regeneration of natural desert ecosystems.

(10) Promoting management of forest GMOs and forest genetic resources

In accordance with the SFA regulation on approval of genetic modification of forests and technical specifications for safety assessment of genetically modified forests and their products, SFA undertakes strict risk assessments of the experiment, environmental release and experimental production of genetically modified trees. SFA also monitors the risks of those genetically modified trees allowed for plantation. SFA has been strengthening the management of forest genetic resources, by surveying and cataloguing unique forest genetic resources in China and undertaking pilot work in access to forest genetic resources and associated traditional knowledge and benefit-sharing from their use.

3.13 Intellectual Property Office

The State Intellectual Property Office (SIPO) always attaches great importance to intellectual property right (IPR) protection related to genetic resources, traditional knowledge and folklore.

(1) Actively participating and promoting international consultations concerning a system of protection of genetic resources and traditional knowledge

SIPO has been leading and representing China in the international negotiations in the Intergovernmental Committee on Genetic Resources, Traditional Knowledge and Folklore under the World Intellectual Property Organization (WIPO), as well as international consultations related to genetic resources and traditional knowledge in the WTO TRIPs Council and other relevant multilateral forums. China together with other developing countries jointly proposed that disclosure of the origin of use of genetic resources should be mandatory in patent applications.

SIPO also actively participated in the negotiations of free-trade agreements between China and New Zealand, Peru, Costa Rica and Switzerland. In all these bilateral free-trade agreements that went into effect respectively in 2008, 2009, 2010 and 2013, all of them contain provisions related to protection of genetic resources and traditional knowledge.

SIPO strengthened cooperation and exchanges with its counterparts in other countries in the field of genetic resources and protection of traditional knowledge. In recent years, SIPO organized many training workshops on IPRs for countries from Asia, Africa, Latin America and ASEAN member states. During these workshops introductions were made on the status of conservation of genetic resources, traditional knowledge and folklore as well as developments in relevant international negotiations. Through these workshops, mutual understanding and exchanges were strengthened so as to create good conditions for international cooperation in relevant fields.

(2) Actively participating biodiversity-related legislation, law enforcement, communication and training

SIPO has been actively promoting studies on biodiversity-related IPR legislation and policy development. In 2011, SIPO together with State Administration of traditional Chinese Medicine and other departments, jointly issued recommendations on strengthening IPR protection related to traditional Chinese medicine in an effort to promote establishment of a system of protection of genetic resources and traditional knowledge in the field of traditional Chinese medicine industry. While implementing *China Patent Law* and its implementation rules related to the origin of genetic resources, SIPO required patent applicants to disclose the origin of genetic resources in their patent applications. SIPO also assessed the effectiveness of implementation of relevant laws and policies through relevant research projects.

SIPO organized seminars or training workshops on IPR protection related to genetic resources and traditional knowledge respectively in Hubei, Sichuan and Gansu in 2011, 2012 and 2013. Relevant experts were invited to introduce relevant international and national systems and developments. SIPO strengthened exchanges with the departments of the environment, culture, agriculture, forestry and Chinese medicine, as well as promoted communication, education and relevant research taking into consideration local needs for industry development and the experiences and lessons learned at local levels.

3.14 Tourism

Tourism has become a strategically important industry of national economy, and is also one of human activities with big impacts on biodiversity, so tourism plays an important role in biodiversity conservation.

(1) Incorporating biodiversity conservation into tourism development strategies and plans

In December 2009, *Recommendations on Accelerating Tourism Development* issued by the State Council clearly proposed enhancing environmental protection and energy conservation while developing tourism. In the *Twelfth Five-year Plan for Tourism Development* issued in 2010 there is one chapter particularly on protecting the environment and resources and achieving sustainable development. In recent year in developing trans-region tourism strategies and plans, such as *Plan for Tourism Development in Northeast China*, ecological and biodiversity conservation has been included as important element.

(2) Taking into account CBD requirements while developing sector standards and regulations

Biodiversity is one key indicator in the national standard for classifying tourism spots and assessing their quality. In 2012 the China National Tourism Administration (CNTA) and the Ministry of Environmental Protection (MEP) jointly developed the national standard for establishing demonstration areas for eco-tourism and operation. Meanwhile, guidelines for management of demonstration areas for eco-tourism and detailed rules for evaluating management and operation of such areas (GB/T 26362-2010) have been developed. All this provided a basis for eco-tourism development. In 2013, CNTA and MEP jointly identified 38 national demonstration areas for eco-tourism.

(3) Undertaking thematic activities of Eco-tourism Year to promote the effective implementation of the CBD

CNTA adopted 2009 as "China's Year of Eco-tourism" and selected its theme as "experiencing green tourism and feeling ecological civilization". The Year of Eco-tourism provided eco-tourism areas an opportunity for environmental education and increasing public awareness of ecological conservation. Eco-tourism promotes local economic development through involving local communities and providing local residents job opportunities, making them willing to support and practice eco-tourism.

(4) Integrating conservation into tourism development of geographical ecological regions, in connection with biodiversity priority areas

In connection with 35 biodiversity priority areas identified, in 2012 CNTA developed tourism development plans for Qinling Mountains, Wuling Mountains and Dabie Mountains for 2013-2020, with a view to breaking administrative borders and promoting integrated tourism development using "mountains" and "water systems" as units of geographical ecological regions so that ecological conservation and environmental protection can be integrated into tourism development.

(5) Encouraging innovation and exploring models of coordinated development between tourism and biodiversity conservation

Many famous ecological tourism spots, such as Huangshan Mountain of Anhui Province, Wuyi Mountain of Fujian Province, Jiuzhaigou of Sichuan Province, and Pudacuo of Yunnan Province, have explored and developed a good model of coordination between tourism development and biodiversity conservation, with due consideration to their own circumstances.

3.15 Oceanic Administration

China's oceanic administrations at various levels have incorporated conservation of marine biodiversity into relevant strategies and plans, taken various conservation measures and achieved obvious results.

(1) Improving laws and regulations for conservation of marine biodiversity

China has preliminarily established a legal system for the marine environment consisting primarily of the *Marine Environment Protection Law* and complemented by relevant regulations and local administrative rules for the marine environmental protection. In recent years, China has promulgated *Island Conservation Law* and *Rules for Management of Marine Special*

Protected Areas. Tianjin, Hebei, Zhejiang, Guangdong and Hainan and other provinces have issued their local regulations on the marine environment protection. These laws and regulations have further improved the marine environmental legal system.

(2) Incorporating conservation of marine biodiversity into sectoral strategies and plans

In 2012, the State Council approved the *National 12th Five-year Plan for Marine Development*. This plan proposed that by 2020 land-based pollution will be effectively controlled; the environmental degradation of near-shore marine areas will be fundamentally reversed and the trend of marine biodiversity decline will be basically contained. The plan also identified some important actions such as strengthening conservation of marine biodiversity, enhancing restoration of marine ecosystems and strengthening marine ecological monitoring and management of ecological disasters. The *National Twelfth Five-year Plan for Marine Economy Development,* the *National Plan for Marine Zoning 2011-2020* and the *National Programme for Island Conservation 2011-2020*, all of which have been approved and issued by the State Council, have put marine biodiversity into a very prominent position and identified specific targets and requirements for conservation of marine biodiversity. Local governments in coastal areas also give high importance to marine ecological conservation. Their marine development plans also give high priority to biodiversity conservation and identify protecting and restoring biodiversity as important targets and tasks. They have also implemented a number of projects in preventing and controlling land-based marine pollutants and protecting and restoring marine biodiversity.

(3) Undertaking surveys and monitoring of marine biodiversity

During 2006-2008, China surveyed marine biological resources in coastal and near-shore areas and obtained the baseline information concerning marine biodiversity in China. "*China's Marine Species and Atlas*" published as a result of this survey comprehensively and systematically described the varieties of marine species and their distribution in China. To know dynamic changes in China's marine biodiversity, since 2004, China has established 18 marine ecological monitoring zones in some ecologically vulnerable and sensitive near-shore and coastal areas, with the total area covered reaching 52,000 km^2 and covering typical marine ecosystems such as bays, estuaries, coastal wetlands, coral reefs, mangroves and sea-grass bed. Since 2008, China has been monitoring regularly national-level marine protected areas and marine special PAs every year and has basically known dynamism of biodiversity in marine PAs.

Case 3.3 Coastal Wetland Restoration Project in Wuyuanwan, Xiamen, Fujian

During 2005-2007, the wetland restoration project undertaken in Wuyuanwan, Xiamen, Fujian Province covered opening of coastal dam, dredging of inner bay, coastal conservation and strengthening dams at lower water level and construction of a wetland park. Following ecological recovery and giving ponds back to sea, the hydrological conditions, landscape and environmental quality of Wuyaunwan have significantly improved and biodiversity has gradually recovered. Now Wuyuanwan has become not only a scenic and leisure spot of Xiamen, but also a good educational base for nature and ecology, providing a platform for Xiamen International Ocean Week and the activities of the Earth Day and bird watching. This has strongly enhanced the public awareness of marine biodiversity conservation.

Impacts of coastal wetland restoration in Wuyuanwan, Xiamen
(left is the situation before and right is the situation after restoration)

(4) Strengthening establishment of network of marine protected areas

The State Oceanic Administration (SOA) has issued rules on management of marine special protected areas, established a committee to examine and review marine special PAs and revised technical guidelines for function zoning of marine special PAs and development of master plans. In the past two years, SOA has implemented capacity building projects for 10 PAs with funds from the central government budget appropriated for different marine areas. The cumulative investments exceed 100 million yuan RMB. By now, most marine PAs have established their management bodies, with a certain number of staff members and funds for management and operation in place so that law enforcement capacities in PAs are strengthened. The number of marine PAs in particular national-level ones has increased substantially since 2008. China has established many new national-level marine PAs and marine special PAs. By the end of 2012, China has a total of 240 marine PAs of various types at different levels, with total area covered reaching 87,000 km^2, accounting for nearly 3% of the marine areas under China's jurisdiction.

(5) Undertaking marine ecological conservation and restoration

In 2012, SOA issued provisional rules for management of demonstration areas of marine ecological civilization and a provisional set of indicators for establishment of such demonstration areas. By now, the first group of provinces such as Shandong, Zhejiang, Fujian and Guangdong have applied for establishment of such demonstration areas. SOA is exploring the establishment of marine ecological red line, focusing on important marine biodiversity areas such as important estuaries, coastal wetlands, marine PAs and fishery areas. Shandong Province has established such a red line in Bo Sea, with strict protection provided to over 40% of Bo Sea's marine areas. Since 2010, SOA has supported 180 projects using funds appropriated from the central government budget for different marine areas totalling about 4.43 billion yuan RMB. The projects included coastal restoration, island conservation and restoration, marine ecological restoration, mangrove and tidal flats restoration, covering an area of more than 2,800 km^2.

(6) Responding to climate change impacts on biodiversity

SOA has established a leading group on addressing climate change and developed a plan for adaptation to climate change. SOA has been monitoring regularly climate-change-related phenomenon such as sea water temperature, sea level, sea water erosion and soil salinization. SOA has also strengthened research on how oceans and seas could adapt to climate change, developed methods of calculation of carbon-sequestration and carbon-fixing capacities of coastal wetlands, developed and integrated technologies in this regard.

3.16 Traditional Chinese Medicine Management

Traditional Chinese medicinal resources are core material foundations for Traditional Chinese medicine development, so the Traditional Chinese medicine management sector attaches great importance to sustainable use of Traditional Chinese medical resources.

(1) Taking biodiversity factor into account when developing sector plans and relevant national laws and regulations

The Code of Medicine of China contains the following provisions concerning the traditional Chinese medicinal resources: for rare and expensive medicinal resources, standards for medical use of wild species need to be phased out; for medicinal resources that are in short supply, if possible, medicine will use only parts above the ground, rather than the whole to allow for continued growth of the underground part; farmed or artificially bred species can be used for some traditional Chinese medicine in order to reduce harvesting of wild species. In the *Twelfth Five-year Plan for Traditional Chinese Medicine Development issued* by the State Administration on Traditional Chinese Medicine, goals or targets were set for cultivating wild species and resources, enhancing capacities for research and development and sustainable use. The plan contains tasks such as undertaking consensus on wild species for traditional Chinese medicinal use, accelerating establishment of germplasm banks, strengthening establishment of breeding bases for wild herbal resources and reinforcing macro-adjustment of important and limited wild traditional Chinese medicinal raw materials. The *Programme for Innovation and Development of Traditional Chinese Medicine* (2006-2020) called for protection of rare and endangered species for traditional Chinese medicinal use, studies on alternatives and breeding techniques, establishment of germplasm banks for traditional Chinese herbal medicines and improving key technologies for conservation and sustainable use of traditional Chinese medicinal herbs. The *State Council Recommendations on Supporting and Promoting Traditional Chinese Medicine Development* proposed to strengthen protection, research and development and reasonable use of traditional Chinese medicinal resources, to protect wild flora and fauna for medicinal use, to accelerate establishment of germplasm banks, to establish PAs in areas where wild flora and fauna are mainly distributed, to establish a group of breeding bases and to strengthen protection, breeding and alternative studies of those rare and endangered species to allow medicinal resources to replenish and grow.

(2) Undertaking pilot work in consensus of traditional Chinese medicinal resources

China has initiated pilot work in consensus on traditional Chinese medicinal resources. Such pilot work has begun since 2011 in 698 counties of 25 provinces (autonomous regions, province-level municipalities). The consensus mainly covered: ① the varieties and distribution of traditional

Chinese medicinal resources and reserves of 563 important Chinese herbs; ② traditional knowledge associated with traditional Chinese medicine focusing on knowledge and experiences of medicinal use at local level and by ethnic minorities; ③ establishing 16 breeding bases for traditional Chinese medicinal seedlings and 2 germplasm banks, undertaking studies on artificial breeding of those traditional Chinese medicinal resources with difficulty in reproduction and setting up relevant germplasm banks; ④ establishing a national system of dynamic monitoring and information service for traditional Chinese medicinal resources.

(3) Collecting and conserving traditional Chinese medicinal resources

China has established a technological system for *in-vitro* protection of medicinal plant germplasm resources, with nearly 30,000 accessions of *in-vitro* germplasm of 3,599 species of medicinal plants collected. China has successfully established the first national germplasm bank of medicinal plants and created a system of technologies for *ex-situ* conservation of medicinal plant germplasm, with 5,282 species protected *ex-situ* and the total number of species for medicinal use conserved ranking top in the world.

(4) Undertaking plantation of traditional Chinese medicinal materials to reduce pressures on wild resources

With the joint promotion and support of many ministries, a preliminary progress has been made in developing new technologies for planting traditional Chinese medicinal materials, exploring models of scale plantation and regulating such plantation. Both outputs and harvests have increased so as to reduce pressures on wild resources. For example, areas of plantation of commonly used traditional Chinese medicinal herbs such as *Angelica sinensis, Glycyrrhiza uralensis, Lonicera japonica* and *Rheum officinale* have exceeded 66.7 km^2.

3.17 Poverty Reduction and Development

Biodiversity-rich areas in China are more often poor areas. So China attaches great importance to biodiversity conservation while reducing poverty.

(1) Full consideration given to biodiversity conservation when planning for poverty reduction

In 2011, the Government of China launched *National Programme for Rural Poverty Alleviation and Development* for 2011-2020, which clearly proposed that poverty reduction should be integrated with environmental protection and ecological conservation, and natural resources in poor areas should be fully used for developing environmentally friendly industries and promoting healthy lifestyles and harmony between socio-economic development, the environment and human population. In planning for regional development and poverty reduction in contiguous, extremely poor regions, priority consideration is given to biodiversity and ecological conservation.

(2) Implementing key ecological projects in poor areas, such as returning cultivated land to forests, returning grazing land to grasslands, soil conservation, natural forests protection, building of forest belts and control of desertification and rockiness of land

Ecological compensation system has been gradually established and improved to increase investments into key ecological function zones. Efforts were made in grassland conservation, strengthening establishment and management of protected areas, implementing projects such as returning grazing land to grasslands, including banning, alternating and stopping grazing to restore natural grassland vegetation and ecological functions.

(3) Developing clean energy based on local conditions

Efforts were made to accelerate development of renewable energy in poor areas. Energy projects were implemented in poor areas based on local conditions such as development of small-scale hydropower, solar power, wind power and biomass power, and promoting use of biogas, energy-saving stoves, solid fuels and gas from straw (including central gas supply stations). These projects will bring about changes in water use, improvements in toilets, kitchens and (pig or cow) pens and comprehensive reuse of straws.

(4) Giving more attention to human resources development in poor areas to alleviate conflicts between human beings and natural resources

Core efforts were made in enabling the poor population to get jobs, by providing job training to those middle and high school graduates who cannot continue higher education. Special subsidies such as living expenses and transportation costs were provided to the new labor force from poor rural families who are receiving vocational training. Labor forces in poor rural areas were provided training on practical skills. In recent years as a result of training provided, the labor force from poor areas shifting to other jobs exceeds 1 million annually, lifting more than 4 million people out of poverty and effectively relieving conflicts between human beings and natural resources in poor areas, thus promoting biodiversity conservation in these areas.

(5) Relocating poor people living in extremely hard conditions on a voluntary basis

This helps reduce ecological pressures of those ecologically extremely vulnerable areas. Meanwhile, other migration and relocation projects are guided for implementation in those poor areas that meet migration conditions. Coordination with relocating people to other areas for poverty reduction is strengthened in common efforts to improve the living and production environment of the poor people. Efforts in this regard are well coordinated to address the problems and difficulties of those poor people relocated to ensure that they will stay where they are relocated with job opportunities and livelihood provided. Currently, China is developing a plan for relocating poor people for poverty reduction with a view to reducing population carrying capacity of those regions and improving external conditions for biodiversity conservation.

> **Case 3.4 Poverty Reduction and Biodiversity Conservation in Bijie City, Guizhou Province**
>
> In 1988, Guizhou Province established an experimental zone in Bijie City for poverty reduction, ecological conservation and population control to alleviate conflicts between survival, development, human population and natural resources. For more than two decades, as the only one experimental zone on poverty reduction and ecological conservation in China, faced with the reality of backward socio-economic development, extreme poverty and poor environmental conditions, Bijie has been giving equal importance to development and poverty reduction, ecological restoration and conservation, population control and improvements in human quality, and has successfully found a new path with both poverty reduction and ecological conservation by jumping out of a vicious cycle of "the more people given birth to, the more land used; the poorer the more land used; and the poorer the more people given birth to". From 1988 to 2011, the annual average income of local farmers increased from 182 yuan to 4,300 yuan and the forest coverage rate up from 15% to 41.5%.

3.18 Implementation of Other Related Conventions

3.18.1 The UN Convention to Combat Desertification (UNCCD)

(1) Full consideration given to biodiversity factors when planning desertification prevention and control

The *National Plan for Desertification Prevention and Control* (2011-2020) proposed the principle of "prevention as priority and control in an integrated way". The plan set a target that by 2020, more than a half of controllable desertified land will be controlled, further improving ecological conditions in deserts. Within the plan period, 200,000 km^2 of desertified land will be controlled, with one half to be completed during 2011-2015 and another half during 2016-2020.

(2) Improving policies to support desertification prevention and control

China has developed a series of policy measures to support ecological conservation and industry development in deserts, such as reform in collective forest ownership, compensation for forest ecological benefits, subsidized loans for forestry, subsidies for reforestation, and subsidies for grassland conservation. Preferential policies for desertification prevention and control such as investment, taxation and financing are improved based on local circumstances, significantly mobilizing the enthusiasm of enterprises and individuals of participating in desertification prevention and control and creating the new situation in which all the society participate in and various channels of investment flow to desertification prevention and control.

(3) Promoting implementation of key projects for desertification prevention and control

Since 2007, China has been implementing a series of key ecological projects such as controlling areas of origin of sandstorms affecting Beijing and Tianjin, building forest belts in north, northeast and northwest China, returning cultivated land to forests and grazing land to grasslands, grassland conservation and soil erosion control in small river basins. China has also initiated a number of regional desertification prevention and sand control projects such as those in Talimu Basin and Shiyang River Basin (ecological restoration as well) of Xinjiang and building ecological barriers in Tibet. These projects are intended to control key desertified areas and enhance ecological improvements in degraded or desertified lands across the country. The monitoring results show that during the eleventh five-year plan period, the average area of desertification is reduced by 1,717 km^2 annually. The total reduction within five years in areas of severely, medium and extremely severely desertified land is 36,000 km^2, an indication of decreasing desertification level. Soil erosion in some areas has been effectively controlled. The soil erosion modulus is significantly reduced, with annual reducton of sand into yellow River by more than 300 million tons.

(4) Enhanced support to capacities for desertification prevention and control

One measure is to enhance scientific and technical capacities. SFA has established a research institute on desertification to strengthen scientific and technical support in this regard. Some research results such as "research on evolution of desertification and models of comprehensive control", and "studies on desertification processes in north China and prevention" have won national awards for scientific and technological progress. Some research results and applicable techniques are being promoted in wider areas. Another measure is to develop and improve relevant technical standards. A number of technical standards for desertification prevention and control have been developed and issued, such as *Technical Guidelines for Desertification Prevention and Control, Technical Guidelines for Monitoring Land Degradation and Desertification* and technical standards for controlling areas of origin of sandstorms affecting Beijing, and Tianjin. The third measure is to strengthen the monitoring of land degradation and desertification and emergency responses to sandstorms. The fourth national monitoring of land degradation has been completed and a system of emergency responses to major sandstorms has been established. A monitoring system of sandstorms is also in place, with main support from remote-sensing and on-the-ground monitoring, complemented by informers on the ground.

(5) Strengthened inter-sectoral coordination mechanisms for desertification prevention and control

From the central government to local governments, China has established specialized coordination and leading bodies for desertification prevention and control to strengthen the organization, leadership and coordination of desertification prevention and control. Since 2007, various departments have fulfilled their respective responsibilities, creating a mechanism where all relevant departments work together closely and increase synergies while taking care of their own responsibilities.

(6) Implemented a responsibility system for achieving targets for desertification prevention and control

In accordance with requirements contained in the *Desertification Prevention and Control Law*, during the eleventh five-year plan period, SFA on behalf of the State Council has signed agreements of accountability for achieving the targets for desertification prevention and control with 12 provincial governments of north China and Xinjiang Production Corp. The establishment and implementation of such responsibility system for the first time makes provincial governments accountable for achieving targets for desertification prevention and control, and helps local governments of different levels increase their sense of responsibility for desertification prevention and control, and promote work in this regard across the country. Now agreements of accountability for desertification prevention and control have been signed for the period of 2011-2015.

(7) Encouraging industry development unique to desert areas

To promote industry development unique to desert areas, SFA has developed recommendations for further developing industry in desert areas, which identified the guiding ideology, principles and goals as well as overall layout and priority areas for industry development in deserts. The recommendations require governments of levels to promote industry development in accordance with local circumstances, to guide various entities to use unique resources in deserts to develop unique industries on the basis of effective protection and control, with a view to improving local farmers' income and promoting economic development in desert areas.

Case 3.5 Soil Conservation Project in Anding District, Dingxi City, Gansu

Anding District, Dingxi City, Gansu Province is located at the Loess Plateau, with the total population of 430,000 and covering an area of 3,638 km^2. The area of soil erosion accounted for more than 90% of its total land area. Relevant departments undertook comprehensive control of soil erosion in this region using a small river basin as a unit and following the ecosystem approach.

(1) Building contiguous terraces in cultivated lands on gentle slopes while abandoning land on steep slopes and planting perennial grass for grazing instead. Grass and trees are planted in those bare mountains and slopes. Water is blocked and soil accumulated by building dams in grooves. Water ponds are built behind houses and beside the fields and barns built in support rills. This kind of system has been called by local farmers as "green caps on top of mountain, belts in the middle of mountains and shoes at the foot of mountains".

(2) Soil erosion in more than 90 small river basins covering an area of more than 1,620 km^2 in this district has been effectively controlled. Those fields with three leaks (of water, soil and fertilizers) have now become fields holding water, soil and fertilizers, with the forest and grass coverage rate increased from the original 8% to 43% currently.

(3) In controlling soil erosion attention was given to using biodiversity to maintain the structure and functions of ecosystems. For example, after slope land has been turned into terraces, potatoes are planted in large areas and agricultural commercialization promoted. The project of collecting rainfall water was implemented, the sites for water accumulation were built covering an area of 3.49 million m^2 and more than 50,000 cellars were dug, providing drinking water for more than 200,000 people and more than 300,000 domesticated animals. The district also developed husbandry with its output values increased significantly from 8.78 million yuan RMB in 1982 to 98.64 million yuan RMB. As a result of the measures above, conditions for agricultural production have substantially improved. In the past decade, the food production of this district has maintained at more than 100,000 tons for years and the people's life has improved and poverty in this district reduced.

Impacts of comprehensive soil conservation in Anding District, Dingxi City, Gansu Province

3.18.2 The United Nations Framework Convention on Climate Change (UNFCCC)

In implementing the UNFCCC, China has fully taken into consideration the interrelationship between climate change, ecosystems and biodiversity, and taken a series of measures to mitigate and adapt to climate change. For mitigation the Government of China has committed to a goal that by 2020 the emission of green-house gases (GHGs) per unit of GDP will be reduced by 40% to 50% over the level of 2005, and a binding target that by 2015 CO_2 emission per unit of GDP will decrease by 17% over the level of 2010. To this end, China has undertaken the following main actions:

(1) Adjusting industrial structure and increasing energy efficiency

In 2011, the National Development and Reform Commission revised and issued a guidance list of industrial structure adjustment (2011) to eliminate those projects with high investments, high energy consumption, high pollution and low efficiency, and to encourage those ecological restoration, technological development and infrastructure projects that are environmentally friendly and reuse natural resources comprehensively.

(2) Developing low-carbon energy and improving energy structure

Through policy guidance and investment, efforts are made to enhance coal cleaning, to develop clean energy resources such as coal-bed methane and shale gas, and to support development of renewable energy such as wind, solar, geothermal and biomass power.

(3) Increasing carbon sinks and strengthening ecological conservation

Key ecological projects continue to be implemented, construction of farmland water conservancy facilities and conservation-oriented cultivation strengthened, wetland conservation strengthened, carbons sinks such as forests, grasslands and wetlands increased to further increase their carbon sequestration capacities while conserving biodiversity.

(4) Undertaking pilot and experimental work in low-carbon development

Active efforts are made in pilot work of various forms in developing low-carbon provinces and cities and emission trading to allow for full synergies among climate change adaptation, energy conservation, environmental protection, new energy development and ecological conservation.

In addition to effective control of GHGs emissions, the *National Twelfth Five-year Plan for Economic and Social Development* also required that capacities to adapt to climate change should be strengthened, including adaptation of agriculture, forestry, water resources and other key sectors, and of coastal and ecologically vulnerable areas to climate change. Main actions to this end include:

• Monitoring and assessing responses of ecosystems to climate change, in particular monitoring and assessing sea level change, sea water invasion, soil salinization and coastal erosion in key regions; promoting development of a system of monitoring and calculation of forest sinks; undertaking research on forestry adaptation to climate change and undertaking surveys of status of biodiversity and climate change in key regions including the Yellow River Basin, the Pearl River Basin and the Liaohe River Basin.

• Further improving the system of early warning of extreme weather and climatic events, developing a plan for responses to climate disasters and enhancing capacities to respond to extreme climate events in very bad weather conditions.

3.18.3 The Ramsar Convention on Wetland of International Importance Especially as Habitats of Waterfowls

As one of the countries with richest wetland resources in the world, China has established a specialized wetland conservation management body and improved the system of wetland management. As a result areas of wetlands conserved have increased significantly; many important natural wetlands have been rescued and conserved and wetland ecosystems in many regions effectively restored. Main actions include:

(1) Promoting legislation and planning for wetland conservation

China has issued a series of laws and regulations concerning wetland biodiversity conservation, including *Forest Law, Wild Animal Protection Law, Water Law, Environmental Protection Law, Marine Environment Protection Law* and *Fishery Law*, all of which are important for conservation and use of wetlands. 17 provinces or autonomous regions have issued provincial regulations on wetland conservation. The State Council has issued the *National Plan for Wetland Conservation Projects* (2002-2030) and its *Project Implementation Plan* (2011-2015).

(2) Improving policies for wetland conservation

China has taken various measures to protect wetlands more effectively. The cumulative investment from the central government budget during the eleventh five-year plan period was 1.4 billion yuan RMB, with matching investments from local governments exceeding 1.7 billion yuan RMB. A total of 205 demonstration projects for wetland conservation and restoration have been completed, with nearly 800 km^2 of wetlands restored. So far, China has established 577 wetland nature reserves, 468 wetland parks and 46 wetlands of international importance, effectively protecting about 43.5% of wetlands. The conservation system plays a crucial role in maintaining the health of wetland ecosystems.

(3) Laying a solid foundation for wetland conservation

In 2009 a second national survey on wetland resources was undertaken. Using "3S" technologies and following the criteria set by the Ramsar Convention, all the wetlands with area above 0.08 km^2 have been surveyed. The survey covered wetland types, areas, wild flora and fauna in wetlands, wetland management and threats to wetlands. The survey result was launched in January 2014.

(4) Establishing long-term effective mechanisms for ecological compensation and ecological water supply

In 2009, the central government required that pilot work be initiated for ecological compensation for wetlands. Also in 2009 the National Forestry Conference proposed to establish mechanisms of ecological compensation for wetlands. Meanwhile, it was required that ecological water use for wetlands be incorporated into water use planning to ensure water supply for wetlands and to allow wetlands to play their ecological functions in underground water replenishment, flood regulation and drought control. This would help wetland ecosystems run into sound cycles. Priority was given to water supply to wetlands of national importance which are threatened with serious water shortages.

Though China has achieved considerably in wetland conservation, with obvious improvements in some wetlands in some regions, overall wetlands across the country are still facing serious threats, such as aridity, water shortage, land reclamation, sand and soil sedimentation, water body pollution and overexploitation of natural resources. The trend of ecological function degradation of wetlands has not been yet contained, and wetlands are still the most vulnerable ecosystem easily subject to encroachment and damage.

3.18.4 The Convention on International Trade in Endangered Species of Wild Fauna and Flora (CITES)

China strictly implements the decisions of the CITES and its unique mechanism of Non-detriment Finding (NDF) to ensure that the export of relevant endangered species will not threaten the survival of their wild populations. Meanwhile China cracks down illegal trade in wild flora and fauna. The CITES Office of China together with relevant departments developed a programme of action to implement the decisions of COP 15 of the CITES in which 10 key issues and 6 general issues were identified, with the priorities and direction of implementation identified. These ten key issues include trade in tigers, sharks, ivory, *Saiga tatarica*, rhinoceros, snakes, tropical timber and *Napoleon Wrasse*, import by sea and NDF, and six general issues include labeling of products using crocodiles, sea cucumbers, law enforcement, individual carry-ons and family property, e-trade and coding of sources and uses. In 2011, China established a coordinating group on law enforcement composed of representatives from forestry, agriculture, customs, public security, industry and commerce administration, inspection and quarantine administration to improve the effectiveness of law enforcement and deter crimes. With regard to major timber species and marine species for commercial use, the CITES Office has conveyed to relevant departments the concerns of the CITES and control measures likely to be taken, requesting them to give attention to the conservation of natural resources for commercial use. In routine work of implementing the CITES, as the CITES Office has established a licensing system for import and export, the CITES Office together with the customs identified the scope of control by using the list of commodities from endangered wild species for import and export. Besides the mandatory licensing system, administrative licenses and technical measures are also employed for control such as use of a certificate of proof of species, which has proven effective. However, due to the fact that the scope of the CITES does not overlap with the CBD requirements, and some unique species have not been included in the scope of the CITES, so control measures for some species are yet to be implemented, in particular those new species or insects not identified and amphibians whose population is small and area of distribution is very limited.

The CITES Office also strengthened cooperation with the CITES Secretariat. In September 2010, the CITES Office, the CITES Secretariat and the Secretariat of the Conveation on Migratory Species jointly organized a workshop on conservation and sustainable use of Saiga gazelles in Uyghur, Xinjiang. The workshop invited conservation and management bodies from Kazakhstan, Russia, Uzbekistan and Mongolia. Relevant stakeholders in China such as Chinese Medicine Association were also invited to discuss with these country representatives to increase synergies in implementation of MEAs and involve key stakeholders in conservation and urge them to follow sectoral sustainable development strategies. In April and May 2011, the CITES Office and the CITES Secretariat organized workshops in Guangzhou respectively on snake management in Asia and development of e-licenses for endangered species, discussing about emerging issues for the implementation of the CITES.

China has also undertaken cooperation and exchanges with relevant Parties. China and USA have streamlined management of trade in turtles and developed policies for zero trade in wild species, promoting monitoring and replenishment of wild turtles, and promoting farming to nurture wild populations, based on the current situation of critical endangerment of wild species. These policies have been recognized by COP 16 of the CITES. In September 2012, the Government of China organized a training workshop for African countries, with very good results achieved. China has also strengthened cooperation with African countries. China has further strengthened cooperation with Russia, India, Mongolia, Vietnam, Lao DPR, Indonesia and Thailand in law enforcement and the implementation of the CITES.

Part 4

Progress in Implementation of 2020 Biodiversity Targets and Contributions to Millennium Development Goals

4.1 Indicators for Assessing Progress Towards 2020 Biodiversity Targets

The indicator system for assessing the 2020 biodiversity targets was developed by using the pressure-state-benefit-response framework. In developing these indicators, the principles below were followed: ① Representing various aspects of biodiversity; ② Reflecting truly and timely changes in the status of biodiversity; ③ Easily understood by decision makers, the public and managers, and recognized widely; ④ Providing accurate measurement, while minimizing costs of data collection and using existing data as much as possible; ⑤ Reflecting changes resulting from policy implementation.

The indicator system contains 17 Class A indicators and 42 Class B indicators (see details in Table 4-1).

Table 4-1 Indicators for Assessing China's Progress towards 2020 Biodiversity Targets

Class A Indicators	Class B Indicators and annotations
Biodiversity Status	
1. Macro-structure of ecosystems	(1) This refers to changes in the area and percentage of ecosystems such as forests, grasslands, wetlands and deserts. Remote-sensing data can be used for calculation. Separate analysis undertaken of areas of natural forests. Natural forests refer to forests naturally formed and not pianted by humans
2. Health condition of ecosystems	(2) Living wood growing stock: refers to the totality of stock volume of all living wood grown on land in a certain region. Natural forest growing stock can be also used. It can be calculated using forest consensus data. (3) Annual net primary productivity of forest ecosystems: refers to the remaining part after autotrophic respiration is deducted from the total amount of organic materials generated by photosynthesis of green plants within a unit of time and land area. As a key variable of characterization of vegetation activity, it can play an important role in global carbon balance. It can be calculated using data from remote sensing. (4) **Total output of fresh grass from natural grasslands.** Data from grassland monitoring provided by the Ministry of Agriculture can be used. (5) Marine trophic index: refers to the average nutrition level of marine catches, reflecting the length of marine food chain, and also integrity and resistance to disturbances of marine ecosystems. FAO data can be used.

Class A Indicators	Class B Indicators and annotations
3. Species diversity	(6) Red List Index: refers to changes in level of endangerment and populations of endangered species, and also indicates overall change in endangerment level of certain species. Relevant species can be assessed separately. (7) **Population of endemic fish species in inland waters,** which can reflect changes in inland water fish diversity. Data from the Institute of Aquatic Biology of Chinese Academy of Sciences (CAS) can be used for calculation
4. Genetic resources	(8) Local varieties and breeds, which can reflect the situation of protection of traditional genetic resources
Ecosystem services	
5. Services provided by ecosystems	(9) **Supply of goods:** goods or services provided by ecosystems that human beings can exchange in markets, mainly including foods and husbandry products. Data from remote sensing can be used for calculation" (10) **Regulating function:** regulating water sources, fixing sand and conserving soil. Remote sensing data can be used for calculation. (11) **Supporting function:** ecosystems should provide habitats for wild animals and plants and ensure their reproduction and survival. Remote-sensing data can be used for calculation
6. Changes in community health and well-being that depends directly on goods and services provided by local ecosystems	(12) **Net income per capita of rural households.** (13) **Number of people in poverty.** Data from the National Bureau of Statistics and the monitoring of key ecological projects by the State Forestry Administration (SFA) can be used
Pressures	
7. Environmental pollution	(14) Annual emissions of COD from industrial waste water, SO_2 and soot from effluent gas, and industrial solid wastes etc., which can indicate level of threat to biodiversity from environmental pollution. Data from environmental statistics can be used. (15) **Intensity of pollutant discharging per unit of GDP:** amount of pollutant discharging per unit of GDP. Data from environmental statistics can be used. (16) **Intensity of CO_2 emission per unit of GDP:** amount of CO_2 emission per unit of GDP. Data from relevant statistics can be used. (17) Amount of use of agricultural chemicals: indicating impacts of agricultural activities on biodiversity. High nitrogen input and nitrogen imbalance will pose serious threat to biodiversity. Data from agricultural statistics can be used. (18) **Amount of use of agricultural chemicals per unit of value added from agriculture.** Data from agricultural statistics can be used
8. Climate change impacts on biodiversity	(19) Refers to impacts of climate change on the structure and function of ecosystems, and on distribution and growth of species and genetic resources
9. Level of damage caused by invasive alien species	(20) Number of invasive alien species newly found in every two decades. Data from relevant surveys can be used. (21) Number and frequency of harmful species intercepted by customs and port authorities. Data from customs and port authorities can be used

Class A Indicators	Class B Indicators and annotations
Responses	
10. System of *in-situ* conservation	(22) Number and area covered by nature reserves: coverage rate refers to percentage of nature reserves out of total land area of the country, that can reflect the *in-situ* biodiversity conservation status. Data from environmental statistics can be used. (23) **Number and area of community-based conservation areas.** Data from SFA can be used. (24) **Number and area of scenic spots.** Data from the Ministry of Housing, Urban and Rural Construction can be used. (25) **Number and area of forest parks.** Data from SFA can be used. (26) **Number and area of wetland parks.** Statistics from SFA can be used. (27) **Number and area of national-level germplasm conservation areas.** Statistics from MOA can be used. (28) **Number and area of special marine protected areas.** Data from State Oceanic Administration (SOA) can be used
11. Implementation of policies and programmes	(29) Implementation of national key ecological projects, number of provincial BSAPs launched
12. Conservation and restoration of habitats	(30) **Forest growing stock in key ecological project regions.** (31) **Timber output from key ecological project regions,** indicating adjustments in timber outputs. (32) **Soil erosion in key ecological project regions.** Data from monitoring of key ecological projects provided by SFA can be used for the indicators above
13. Pollution control	(33) **Capacity of flue-gas desulfurization (FGD) units installed and their percentage out of total thermal power facilities.** (34) **Municipal wastewater treatment capacities.** (35) **Solid waste disposal.** Data from environmental statistics can be used for the indicators above
14. Comprehensive reuse of resources	(36) **Amount of use of renewable resources:** number of solar water heaters, solar stoves. (37) **Annual output of projects disposing of agricultural wastes.** (38) **Total capacities of ponds disposing of agricultural wastes.** (39) **Total capacities of biogas ponds disposing of wastewater at village level.** Data from agricultural statistics can be used for the above indicators
15. Public awareness	(40) **Items about China's biodiversity that can be searched through Google or Baidu in different years**
16. Knowledge of conservation and sustainable use of biodiversity	(41) **Number of academic papers on biodiversity conservation.** Data from relevant literature can be used for calculation
17. Biodiversity investment	(42) Investments into projects such as those on natural forest resources protection, conservation of wild animals and plants, establishment of nature reserves and wetland conservation. Data from SFA statistics can be used

Note: Class B indicators with bold black fonts are new indicators used in this report compared with those used in the fourth national report. Relevance of these indicators to the 2020 targets is provided in the table contained in Annex Ⅱ.

4.2 Data Analysis of Indicators for Assessing the 2020 Biodiversity Targets

4.2.1 Biodiversity status

(1) Marco-structure of ecosystems

According to the result from the Project of Remote-sensing Survey and Assessment of National Ecological Changes in the Decade from 2000 to 2010, during the decade from 2000 to 2010, the areas of forests, wetlands and urban ecosystems have increased slightly while those of shrubs, grasslands, deserts and agricultural ecosystems have decreased. The area of forests increased by 29,000 km^2, that of wetlands by 400 km^2, and that of urban ecosystems by 55,000 km^2. The areas of shrubs and grasslands dropped respectively by about 12,000 km^2 and 16,000 km^2, and those of deserts and agricultural ecosystems respectively by 4,400 m^2 and 48,000 km^2 (Figure 4-1). Drastic changes have occurred in the layout of ecosystems in some regions where urbanization, agricultural expansion and the concentrated areas of returning cultivated land to forests took place. Since the 1980's, the area of natural forests in China has constantly increased (Figure 4-2).

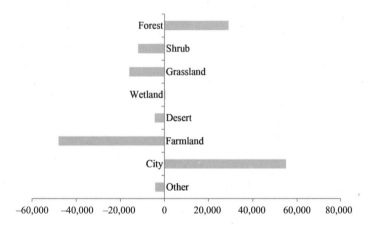

Figure 4-1 Net changes in areas of terrestrial ecosystems in China during 2000-2010 (km^2)
Source: the Project of Remote-sensing Survey and Assessment of National Ecological Changes in the Decade from 2000 to 2010, Courtesy of Ouyang et al.

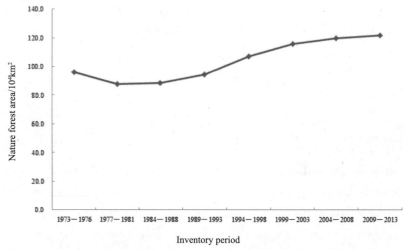

Figure 4-2   Changes in Areas of Natural Forests in China in Different Periods of Time
Source: China's Yearbook of Forestry Statistics.

(2) Ecosystem health conditions

China's forest growing stock has been increasing constantly since the 1980's. If compared with the result of the sixth investigation, the seventh investigation shows that China's total living wood growing stock has a net increase of 1.13 billion m^3, and forest growing stock has a net increase of 1.12 billion m^3, and natural forest growing stock has a net increase of 680 million m^3 (Figure 4-3). During 2000-2010, overall China's annual net primary productivity of forests has been going up (Figure 4-4). Despite the constant increase in the total amount of forest reserves and the recovery of some functions of forest ecosystems, there is still an issue of the inadequate total amount, low quality and imbalanced distribution of forest resources.

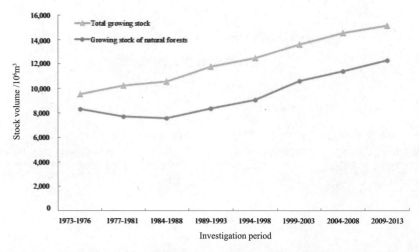

Figure 4-3   Changes in Forest Growing Stock in China in Different Periods of Time
Source: China's Yearbook of Forestry Statistics.

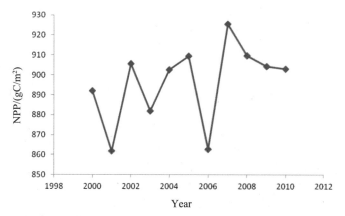

Figure 4-4　China's Annual Net Primary Productivity of Forest Ecosystems
Source: Project of Remote-sensing Survey and Assessment of National Ecological Changes in the Decade from 2000 to 2010, Courtesy of Ouyang et al..

Positive changes have taken place in grassland ecosystem as a result of constant efforts in conservation and restoration of grassland ecosystems. From 2005 to 2012, the total national output of fresh grass from natural grasslands across the country has been going up (Figure 4-5). This indicates that part of the grassland ecosystem is improving. However most parts of the grassland are still being grazed beyond their capacities. Degradation, desertification, salinization and rockiness of grasslands are still very severe and overall ecological conditions of grasslands are serious.

China's marine trophic index (MTI) has been declining due to overfishing from the early 1980's to the middle 1990's, lower than the world's average of the same period. This revealed serious degradation of marine ecosystems. From 1997 to now China's MTI has started to climb up (Figure 4-6), probably due to a positive impact from implementation of summer fishing bans for replenishment of marine fishery resources. However, China's MTI is still at a relatively low level, with low ecological functions, and challenges remain high for marine biodiversity conservation.

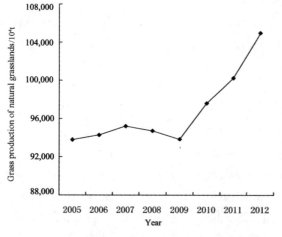

Figure 4-5　Total National Output of Fresh Grass from Natural Grasslands
Source: National Report on Grassland Monitoring.

Figure 4-6　Changes in Marine Trophic Index in Marine Waters of China

(3) Species diversity

Red List Index (RLI) refers to total changes in Red List Categories of certain taxa. When RLI is zero, it means that all species are extinct; and when RLI is one, it means that none of the species are threatened and no need to protect them. From 1998 to 2004, the RLI of China's fresh water fishes was declining. From 1996 to 2008, the RLI of China's mammals was going down. From 1988 to 2012, the RLI of birds went down slightly according to the Equal-steps calculation method, however it went up slightly initially and then turned downward again according to the Extinction-risk calculation method (Figure 4-7). Due to habitat degradation and loss, the threatened status of mammals and fishes has been increasing. Overall, the threatened status of birds as whole is increasing though there are some improvements to some extent in the status of critically endangered species.

China has been monitoring some inland water fishes. From 1997 to 2009 the populations of endemic fishes in the upper reaches of the Yangtze River went down (Liu and Gao, 2012) (Figure 4-8), indicating that biodiversity in the Yangtze River Basin is still declining.

(4) Genetic Resources

It is estimated that loss of China's genetic resources is very serious. However, due to availability of limited data, individual cases can be used to illustrate this. For example, China's main agricultural crops are rice, wheat and maize. In the 1950's, the varieties of rice grown in different parts of China were more than 46,000. However by 2006, this number went down to slightly over 1,000, with most of them being cultivated varieties. The local varieties of maize grown in the 1950's exceeded 10,000 however the varieties being grown now are no longer local varieties. Similarly, the wheat varieties grown in the early 1950's came up to 4,000 or so however in 2000 this number dropped under 400, most of which are cultivated varieties. According to the result of a second national survey on livestock genetic resources, 15 local breeds of livestock cannot be found any more and the populations of more than a half of local breeds are going down (National Committee on Livestock Genetic Resources, 2011).

Part 4　Progress in Implementation of 2020 Biodiversity Targets and Contributions to Millennium Development Goals

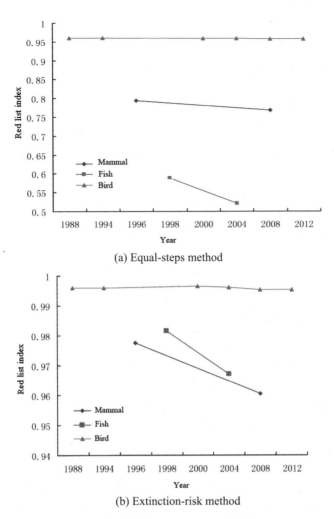

Figure 4-7　Red List Index of China's Vertebrate Species

Source: Birds from Birdlife International, http://www.birdlife.org/, *n*=1208; mammals from IUCN Red List Database: http://www.iucnredlist.org/, *n*=99; fishes from China's Red Data Book on Endangered Animals and China's Species Red List, *n*=81.

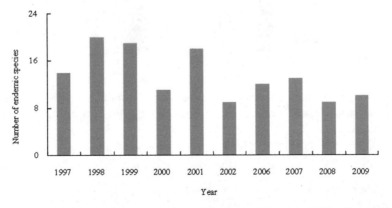

Figure 4-8　Population of Endemic Fishes of the Upper Reaches of the Yangtze River
(Mutong River Section)

Source: Liu and Gao, 2012.

4.2.2 Ecosystem services

(1) Services provided by ecosystems

In accordance with the result of the Project of Remote-sensing Survey and Assessment of Ecological Changes in the Decade from 2000 to 2010, the provision of products from China's terrestrial ecosystems during this decade such as food and husbandry products has been constantly increasing, from 2014.7 trillion kcal to 2,805.2 trillion kcal, an increase by 39.2% within a decade.

In terms of the regulating function of ecosystems, the water regulating functions of China's terrestrial ecosystems increased from 122.6×10^{10} m^3 in 2000 to 123.5×10^{10} m^3, an increase of 0.7% within a decade. The functions of windbreak and sand control increased from 121.2×10^8 t in 2000 to 137.5×10^8 t in 2010, an increase by 13.4% within a decade. Soil holdings increased from 1966.5×10^8 t to 1979.6×10^8 t in 2010, an increase of 0.7% within a decade.

With regard to the supporting functions of ecosystems, the habitat quality index is used to assess changes within a decade in the supporting functions of biodiversity. From 2000 to 2010, the area of ecosystems with low-quality habitats took a bigger percentage, while the area of ecosystems with high or relatively high quality habitats went down constantly (Figure 4-9). This indicates that despite key ecological projects implemented in China, the supporting functions of biodiversity are still going down as the improvement of habitat quality is a gradual and long process.

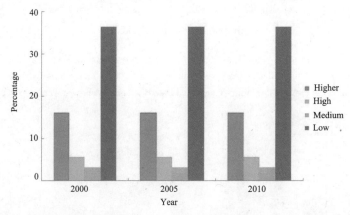

Figure 4-9 Supporting Functions of Biodiversity of China's Terrestrial Ecosystems
Source: Project of Remote-sensing Survey and Assessment of Ecological Changes in the Decade from 2000 to 2010, courtesy of Ouyang et al..

(2) Changes in health and well-being of communities that directly depend on products and services provided by local ecosystems

The net income per capita of rural households in China rose quickly, an increase by 40.8%

in 2011 over that of 2000. This in part benefited from an increase in the provision of ecosystem goods and services (Figure 4-10). For poverty reduction, using the data from the projects on natural forest resources protection and returning cultivated lands to forests as an example, the number of poor people in those sample counties where these projects have been implemented is going down. The number of poor people in those counties where natural forest resources protection projects were implemented has dropped from 3.95 million in 1997 to 1.83 million in 2011, and the number of poor people in those counties where projects on returning farmlands to forests were implemented down from 8.3 million in 1998 to 5.7 million in 2008 (Figure 4-11). While conserving and restoring forest ecosystems, these projects also result in improvement of well-beings of local communities that directly depend on local ecosystem goods and services.

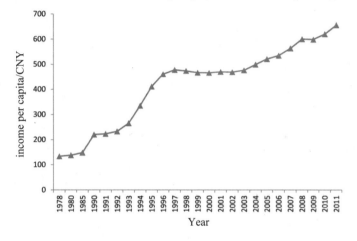

Figure 4-10 Per Capita Annual Net Income of China's Rural Households
Source: China Statistics Yearbook, RMB deflated in 1978

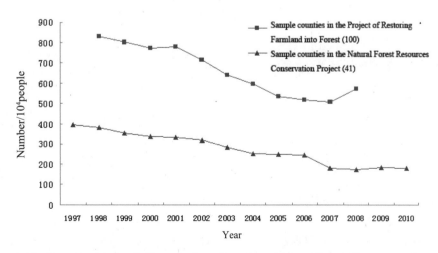

Figure 4-11 Poor Population in Sample Counties where Natural Forest Resources Protection and Returning-farmlands-to-forests Projects were implemented
Source: National Report on Monitoring of Social and Economic Benefits from Key Forestry Projects.

4.2.3 Pressures

(1) Environmental pollution

Since 2006, the emissions of COD in industrial waste water, SO_2 in effluent gas, soot, industrial dust and solid wastes have been going down (Figure 4-12). Despite high-rate economic growth in the past decade, the amount of pollutant discharged per unit of GDP decreased by more than 55% (Figure 4-13). Since 2004, the CO_2 emission amount per unit of GDP dropped by 15.2% (Figure 4-14). From 1991 to 2011, the amount of use of agricultural chemicals more than doubled (Figure 4-15), however after 2003, the amount of use of agricultural chemicals per unit of agricultural added value decreased constantly (Figure 4-16), indicating that the efficiency of use of agricultural chemicals is increasing.

The discharging of waste water is still increasing, and the amount of discharge of pollutants per unit of GDP remains to be at high level. The surface water is lightly contaminated overall. The percentage of river sections with water quality lower than Grade V in the Yangtze River, the Yellow River, the Pearl River, some rivers in Fujian and Zhejiang Provinces and rivers in southwest China is still up to 10.2%. Among 60 lakes (reservoirs) being monitored, 25% of them are in the state of eutrophication. The overall state of the environment of the marine areas under China's jurisdiction is good, however the water pollution of coastal and near-shore marine areas is still very serious. In a word, environmental pollution is posing severe threats to biodiversity.

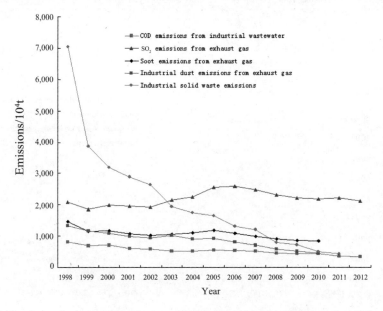

Figure 4-12   Annual amount of emission of COD from industrial wastewater, SO_2, soot and dust from effluent gases and industrial solid wastes

Source: National Environmental Statistics Gazette and China Environmental Statistics Yearbook.

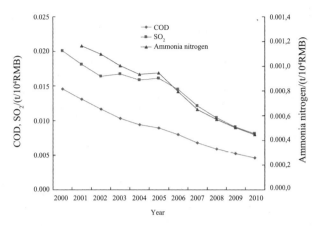

Figure 4-13 Changes in Amount of Emission of Pollutants per unit of GDP
Source: China Environmental Statistics Yearbook and China Statistics Yearbook.

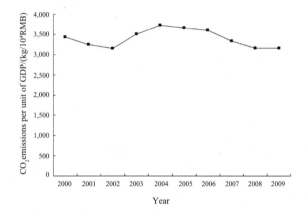

Figure 4-14 Changes in CO_2 Emission per unit of GDP
Source: China Environmental Statistics Yearbook and China Statistics Yearbook.

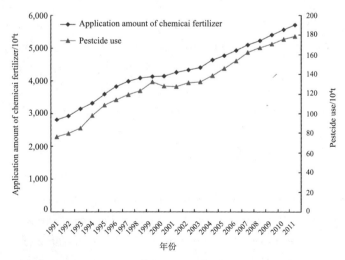

Figure 4-15 Amount of Use of Agricultural Chemicals
Source: China Agricultural Statistics Material.

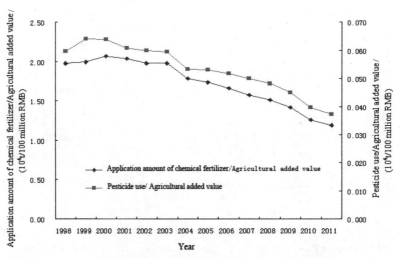

Figure 4-16　Amount of Use of Agricultural Chemicals per unit of agricultural value added(Agricultural value added is adjusted by using CI, with baseline year being 1978)
Source: China Statistics Yearbook 2012.

(2) Climate change

From 1951 to 2009, the average temperature of land surface in China rose by 1.38℃, with the warming rate being 0.23 in the decade. Climate change has had recognizable impacts on China's ecosystems and species. Climate change has aggravated grassland degradation and decreased functions of inland wetlands (Second National Report of Assessment of Climate Change, 2011). In the past several decades, climate warming and drying trend in Hulun Lake region is obvious, with lake area shrinking, grasslands around the lake degraded, land desertified and vegetation coverage decreasing, posing major threats to species living in these habitats (Zhao et al., 2008).

Climate change has also modified the phenology, distribution and migration of species. The spring phenology (initial flowering and leafing periods) of woody plants in northeast and north China and the lower reaches of the Yangtze River has advanced, while the phenology of woody plants in the regions south of Qinling Mountain including the eastern part of southwest China and the middle reaches of the Yangtze River has been postponed (Zheng et al., 2003). Since 1980, the natural phenology of some birds such as *Cuculus canorus* in Qinghai (Qi et al., 2008) and *Cuculus micropterus* in southwest Shandong (Zhang et al., 2011) has advanced, with initial bird singing advanced and final bird singing delayed. Climate change also caused elevation of timberline latitude in some regions. Due to climate warming, the areas of distribution of 120 birds have shifted northward or westward. Climate change has made the original habitats of some species disappear. The warming climate trend in Qinghai Lake region plus impacts from human activities in this region have led to changes in the composition of species in this region, in particular major changes in the composition of bird species. Compared with the 1950's, 26 bird species such as bean goose, gray-headed thrush, Marsh Harrier and quail have disappeared from the lake region (Ma et al., 2006).

Climate change expands the scope of distribution of pests and aggravates their damages. For example climate warming has expanded the scope of distribution of *Solidago canadensis* (Wu et al., 2008) and *Dendrolinus punctatus* (Forest Pest Prevention and Control Station, SFA, 2013).

(3) Invasive Alien Species (IASs)

The distribution of invasive alien species in China is generally divided into three levels, mostly in coastal provinces and gradually less in inland provinces. Most of IASs are found in coastal provinces and Yunnan Province, followed by central China regions and neighboring provinces in eastern and western China (Figure 4-17). Analysis of invasive alien species with records of time of invasion has indicated that the number of new IASs, is gradually going up. Sixty years after the 1950's, 212 new IASs entered China, accounting for 53.5% of the total IASs (Figure 4-18).

With the deepening of opening to the outside and high-rate growth of international trade, plant pests intercepted by customs and port authorities have increased substantially. The varieties of pests intercepted in 2012 were 18.9 times as many as those in 1999, and the frequency was even 230.2 times higher (Figure 4-19). This posed serious threats to agriculture, forestry and ecological security.

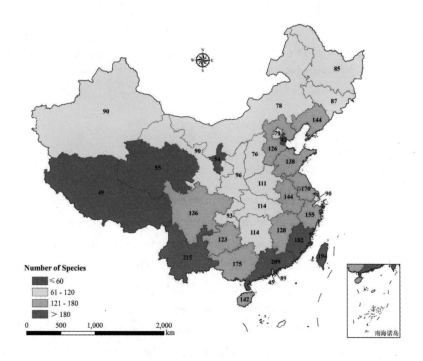

Figure 4-17   Distribution and Number of IASs (Xu and Qiang, 2011)

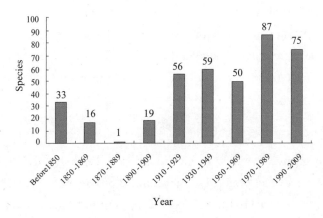

Figure 4-18 Number of Newly-Identified IASs Every Two Decades (Xu and Qiang, 2011)

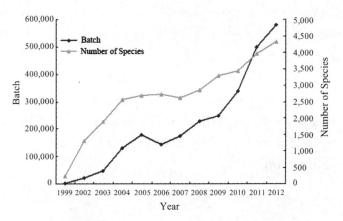

Figure 4-19 Frequency and Varieties of Pests Intercepted by Customs and Port Authorities
Source: Website of General Administration of Quality Supervision, Inspection and Quarantine (www.aqsiq.gov.cn).

4.2.4 Responses

(1) *In-situ* conservation system

In recent years, the number and area of nature reserves in China has been maintained stable (Figure 4-20), with the total areas of nature reserves accounting for 14.8% of the total land area of the country. A network of nature reserves with different types, reasonable layout and relatively sound functions has been basically established. However, their ecological representativeness and management effectiveness need to be improved and more work is needed to improve research and ecological compensation. The number and area of national scenic spots have also increased (Figure 4-21), and shifts are being made from more attention given to protection of scenic landscapes to more integrated protection of scenic landscapes, cultural heritages and biodiversity. However, destruction of scenic spots can be still seen in some regions. The number and area of forest parks have also increased quickly (Figure 4-22), with a system in place for conservation and use of forest scenic resources. However, a lot of rare forest scenic resources have not been effectively protected, and some such resources have been even destroyed in some regions. The number of community-based forestry conservation areas across the country is maintained, however the area is going down constantly (Figure 4-23). The number and area of national-level aquatic germplasm conservation areas and special marine protected areas have been constantly increasing (Figure 4-24 and Figure 4-25).

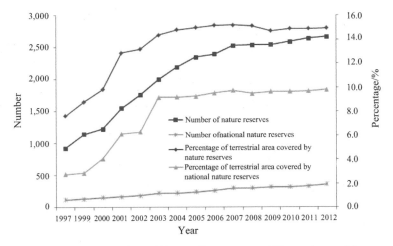

Figure 4-20 Number of Nature Reserves and Percentage of Areas
Source: China Report on the State of the Environment, China Environmental Statistics Yearbook.

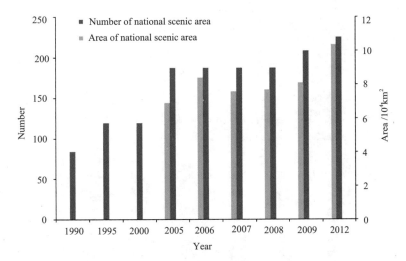

Figure 4-21 Number and Area of National Scenic Areas
Source: China Communique on Development of Scenic Areas.

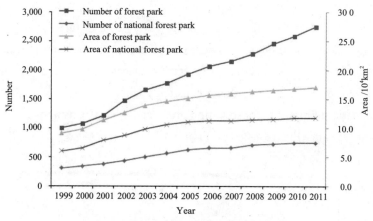

Figure 4-22   Number and Area of Forest Parks
Source: China Forestry Statistics Yearbook.

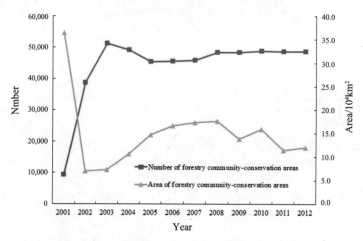

Figure 4-23   Number and Area of Forestry Community-Conservation Areas
Source: China Forestry Statistics Yearbook.

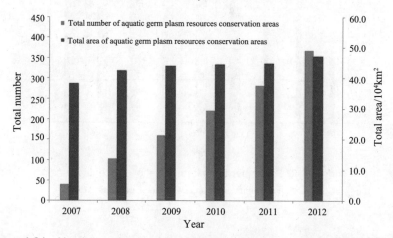

Figure 4-24   Number and area of national aquatic germplasm conservation areas
Source: Ministry of Agriculture.

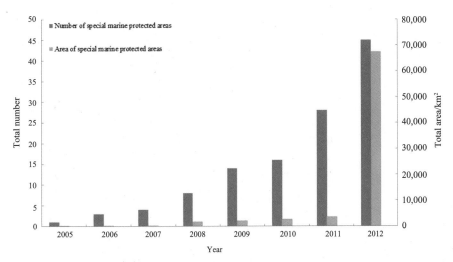

Figure 4-25　Number and area of special marine protected areas
Source: State Oceanic Administration.

(2) Implementation of policies, plans and programmes

China has implemented a series of policies, plans and programmes favorable for biodiversity. China has been implementing a number of major ecological projects such as natural forest resources protection, returning farmlands to forests, returning grazing lands to grasslands, construction of forest belts in north, northeast and northwest China, wetland protection and restoration and soil conservation. The implementation of these projects has enhanced the restoration of degraded ecosystems and habitats for wild species and effectively protected biodiversity. All the provinces (autonomous regions, province-level municipalities) are developing their biodiversity strategies and action plans. So far, 7 provinces have launched their BSAPs.

(3) Habitat protection and restoration

Since 2001, China has made great achievements in key forestry projects, with good results achieved for forest conservation and restoration. Forest reserves in sample counties and enterprises where major ecological projects have been implemented have been going up since 1999 (Figure 4-26 and Figure 4-27). Timber outputs from sample enterprises where natural forest resources projects were implemented have been declining constantly, from 6.243 million m^3 in 1997 to 1.795 million m^3 (Figure 4-28), indicating that adjustments in timber outputs had remarkable achievements. The areas affected by soil erosion in these sample counties are going down overall (Figure 4-29). In sum, these key forestry projects have played a critical role in ecological conservation.

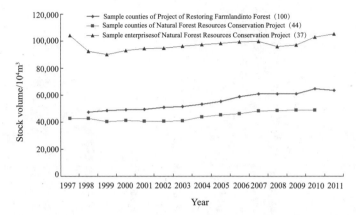

Figure 4-26 Forest reserves of sample counties and enterprises where natural forest resources protection projects and returning-farmlands-to-forest projects were implemented
Data in the figure are sample data.
Source: National Report on Monitoring of Social and Economic Benefits from Key Forestry Projects.

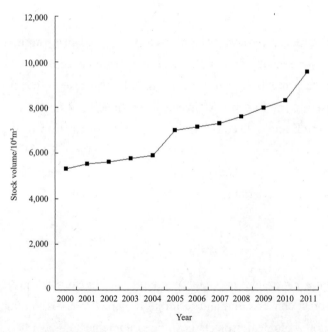

Figure 4-27 Forest reserves of sample counties where projects to control origins of sandstorms affecting Beijing and Tianjin were implemented
Source: National Report on Monitoring of Social and Economic Benefits from Key Forestry Projects.

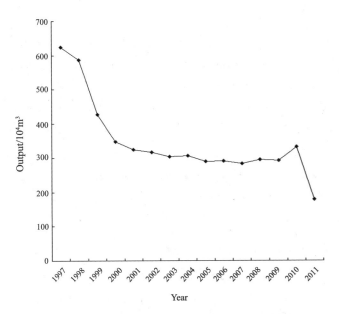

Figure 4-28　Timber outputs of sample enterprises where natural forest resources protection projects were implemented

Source: National Report on Monitoring of Social and Economic Benefits from Key Forestry Projects.

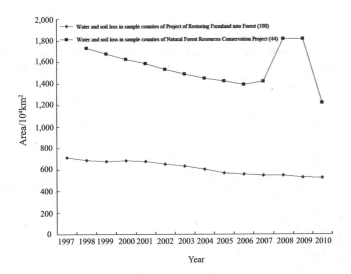

Figure 4-29　Soil erosion in sample counties where natural forest resources protection projects and returning-farmlands-to-forests projects were implemented

Data in the figure are sample data.

Source: National Report on Monitoring of Social and Economic Benefits from Key Forestry Projects.

(4) Pollution control

Remarkable achievements have been accomplished in reducing pollutant emissions. Substantial growth is witnessed in the percentage of flue-gas desulfurization (FDG) units installed out of the total thermal power facilities (Figure 4-30), the rate of municipal waste water treatment (Figure 4-31) and the amount of recycling of industrial solid wastes (Figure 4-32). However the rate of reuse of industrial solid wastes in the past two years dropped slightly.

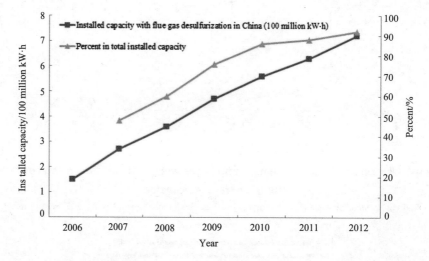

Figure 4-30 Percentage of FDG units installed out of total thermal power facilities
Source: China Report on the State of the Environment and Environmental Statistics Yearbook.

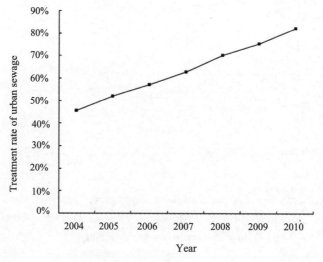

Figure 4-31 Rate of municipal wastewater treatment across China
Source: National Environmental Statistics Gazette.

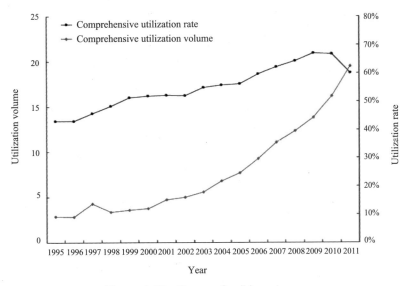

Figure 4-32　Reuse of solid wastes
Source: National Environmental Statistics Gazette.

(5) Comprehensive use of resources

The number of solar water heaters and solar stoves used across the country is increasing year by year (Figure 4-33), with the number of solar water heaters in 2011 reaching 62.32 million, a nearly-10-time increase over that in 1997. The annual amount of gas generated from disposal of agricultural wastes and the total pond capacities for disposing agricultural wastes are also climbing up year by year (Figure 4-34), with the two indicators in 2011 being 60 times and 77 times higher than those in 1997 respectively.

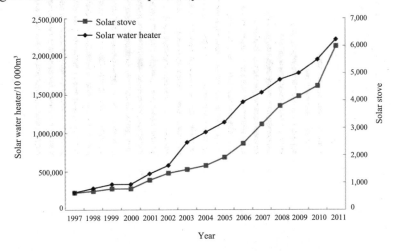

Figure 4-33　Number of solar water heaters and stoves used in China
Source: China Agricultural Statistics Material.

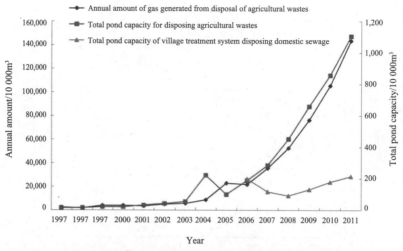

Figure 4-34   Use of agricultural wastes in China
Source: China Agricultural Statistics Material.

(6) Public awareness

By using Google or Baidu advanced search engines to search key words "biodiversity" and "China", the results show that such items are increasing (Figure 4-35 and Figure 4-36), indicating that biodiversity is increasingly attracting public attention.

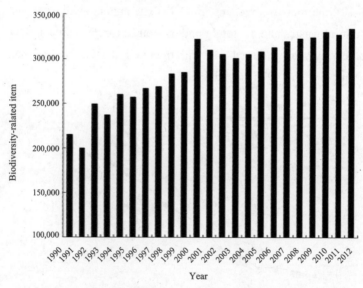

Figure 4-35   China biodiversity items searched through Google in different years

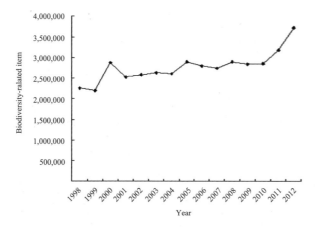

Figure 4-36 China biodiversity items searched through Baidu in different years

(7) Knowledge of conservation and sustainable use of biodiversity

By searching academic papers on biodiversity published annually during 1990-2012, through Chinese VIP database, as well as those on biodiversity in China published every year during the same period through ENSCO & ISI WEB OF SCIENCE, the results show that publications on biodiversity are increasing year by year (Figure 4-37).

(8) Investments into biodiversity conservation

In recent years, China has increased substantially its investments into biodiversity conservation. The investments into projects, such as natural forest resources protection, wild animal and plant conservation, establishment of nature reserves and wetland conservation, have increased from 9.7 billion yuan RMB in 2001 to 21.77 billion yuan RMB in 2011, with the average annual increase of 13.7%, providing financial support to biodiversity conservation (Figure 4-38).

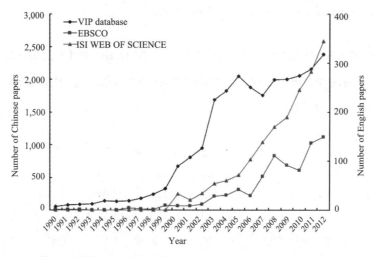

Figure 4-37 Number of academic papers on biodiversity

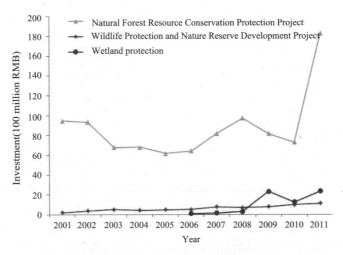

Figure 4-38   Investments into biodiversity conservation
Source: China Forestry Statistics Yearbook.

To sum up, in recent years, the Government of China has reinforced its efforts in biodiversity conservation, through measures such as strengthening conservation policies, expanding networks of conservation, restoring degraded ecosystems, controlling environmental pollution, strengthening science and technology development, promoting public participation and increasing investments. Positive achievements have been made in biodiversity conservation; the ecological damage has been reduced; ecosystem functions in some regions have been restored and the populations of some key protected species increased to some extent. However, the biodiversity decline trend in China has not yet been contained and there are many challenges for biodiversity conservation due to shrinking areas and degrading functions of habitats for wild species, serious pollution of some key river basins and marine areas, growing number and expansion of invasive alien species in some regions, large-scale plantations of single tree species and negative impacts of climate change on biodiversity.

4.3   Overall Assessment of China's Progress in Implementing the Strategic Plan for Biodiversity and Achieving the 2020 Biodiversity Targets

Assessments were made of progress in the implementation of the 2020 Biodiversity Targets by using the biodiversity indicators (see detailed assessments in Annex Ⅱ). For targets 2, 16 and 18, no assessments were made due to lack of relevant national indicators. Various levels of improvements were noted in indicators for targets 1,3,4,5,7,8,10,11,14,15,17,19 and 20, indicating that implementation of these targets is on the right track. In particular, considerable progress has been made in implementing target 3 (incentive measures), target 5 (habitat degradation and loss reduced), target 8 (environmental pollution controlled), target 11 (protected areas and management effectiveness strengthened), target 14 (important ecosystem services restored and ensured), and target 15 (resilience and carbon sequestration of ecosystems reinforced). However, the worsening trends are noted in the majority of indicators for grassland ecosystem conservation in target 5, target 6 (sustainable fishery), target 12 (endangered species protected), and target 13 (genetic resources protected), indicating that more effective strategies and measures are needed to implement these targets, though tremendous work has been done. In conclusion, China is on the right track and making positive progress in implementing the Strategic Plan for Biodiversity and achieving the 2020 Biodiversity Targets, however China still needs to be more determined, take more effective measures and invest more resources to achieve the 2020 targets.

Meanwhile, more indicators need to be developed on biodiversity values, sustainable consumption, ecosystem degradation, impacts of agriculture, forestry and fishery on biodiversity, impacts of climate change on biodiversity, representativeness and management effectiveness of protected areas, and access to genetic resources and benefit-sharing from their use. More attention should be given to conservation of ecosystems as a whole in particular grassland ecosystems and marine ecosystems, conservation of endangered species and genetic resources, and prevention and control of invasive alien species.

4.4 Contributions to Achievement of Millennium Development Goals

In September 2010, the Government of China issued *Progress Report on China's Implementation of the Millennium Development Goals (MDGs)* (2010 edition). The report pointed out that China had made positive progress in achieving MDGs. China has achieved ahead of the time three targets, namely "halving the poor and hungry population", "primary education" and "decreasing the child mortality rate", and one sub-target on "safe drinking water" under the target on environmental sustainability. Other targets are expected to be achieved in time. As said in Part I of this report, biodiversity provides conditions for human survival and material foundations for social and economic sustainable development. The efforts made by the Government of China towards biodiversity conservation contribute significantly to the achievement of MDGs. Certainly China is still facing challenges in fully achieving MDGs, in particular sustainability, including in addressing in imbalance between urban and rural developments and among regions, environmental pressures, and biodiversity loss. To this end, China will continue to implement the "scientific development" approach, deepen various reforms and speed up economic restructuring and transition in growth patterns in efforts to build an ecological civilization and Beautiful China and achieve MDGs.

4.5 China's Experiences in the Implementation of the Convention

Through long-term explorations and practices, China has embarked on a path of biodiversity conservation that suits China's national circumstances, which enriches the global experiences in biodiversity conservation.

(1) Governments taking leading roles, combined with public participation

Biodiversity conservation, which benefits the public, requires investments by governments. Governments of all levels in China have integrated biodiversity conservation into their planning for social and economic development and been increasing investments into biodiversity conservation. Meanwhile, through adjusting relevant policies and measures, they are trying to fully mobilize the civil society participation in biodiversity conservation by increasing their public awareness and their level of participation so that all members of the society care about biodiversity and support biodiversity conservation. For example, a Union of Conservers has been established in Shennongjia Nature Reserve, composed of governments, management bodies, civil society and all other key stakeholders, in common efforts to promote biodiversity conservation in that area.

(2) Inter-departmental interaction and coordination with mechanisms in place for the environment department to coordinate while relevant departments make joint efforts

To further improve the existing mechanisms such as China National Coordinating Group on Implementation of the Convention on Biological Diversity and the Inter-ministerial Joint Conference on Conservation of Biological Resources, China has recently established National Committee on Biodiversity Conservation headed by the Ministry of Environmental Protection and involving 25 departments. These three mechanisms have played an important role in biodiversity conservation in China. Practice has proven that mechanisms with the environment department to coordinate while other relevant departments make joint efforts and implement their respective responsibilities are suitable for China's national circumstances.

(3) Priority given to conservation while combining both conservation and rational use

Priority is given to biodiversity conservation while pursuing social and economic development goals. By following the rule of nature and the principle of "conservation while pursuing development goals, and development in the midst of conservation", important ecosystems, species and genetic resources are effectively protected. With the prerequisite of effective protection, biological resources should be rationally developed and used to increase incomes of local communities and improve their well-beings.

(4) Implementation driven by projects, with breakthroughs made in key regions while promoting implementation at broader areas

While implementing key ecological projects, biodiversity demonstration projects are implemented in important ecosystems such as forests, grasslands and wetlands and ecologically vulnerable regions. Meanwhile, relevant policies and regulations are improved and NBSAP was updated and all of them are being implemented to promote biodiversity conservation across the country.

(5) Implementation enhanced by innovations in both science and technology and management

Professional education and training in the field of biodiversity is strengthened in order to create innovative capacities for conservation and use of biodiversity. A system is in place for development of knowledge and technologies related to biodiversity, with independent intellectual property rights. Meanwhile, relevant policy and regulatory systems, standards and monitoring and early warning systems are improved to promote science-based and rule-based management of biodiversity conservation and to upgrade biodiversity management level.

(6) International cooperation for implementation at both national and global level

China has been fulfilling its international obligations and actively participating in relevant international negotiations and building of multilateral systems in this field. China has been also undertaking international cooperation and exchanges in this field including introducing advanced techniques and concepts of management from other countries and implementing biodiversity projects supported by international donors. Meanwhile, China has been upgrading biodiversity conservation and management, including through allowing governments of all levels to play key roles in biodiversity conservation and integrating biodiversity goals and tasks into the planning processes of various levels of governments and relevant departments.

Part 5
Main Issues and Priority Actions for Biodiversity Conservation in China

Under the leadership of the Government of China and with the support of the whole society, China has achieved tremendously in biodiversity conservation. However, we should be clear that China is still facing a serious situation in biodiversity conservation. Implementing nearly a half of actions proposed in updated NBSAP is still very challenging, and the biodiversity decline trend has not been fundamentally contained, and therefore there is a long way to go for biodiversity conservation in China.

5.1 Main Issues

(1) Relevant regulations and mechanisms need to be further improved

China's existing laws on biodiversity conservation are yet to be improved, with some laws and regulations not adequate to address current social and economic realities. For example, some laws and regulations such as the *Environmental Protection Law*, *Wild Animal Protection Law*, *Wild Plant Protection Regulation* and *Regulation on Nature Reserves* urgently need to be revised or improved. There are no specialized laws or regulations to address issues such as access to genetic resources and benefit-sharing, wetland conservation and prevention and control of invasive alien species. Penalties for violating some laws and regulations are still inadequate, and for this reason some laws and regulations cannot deter relevant crimes or violations.

(2) Awareness of conservation is yet to be further enhanced

The public awareness of biodiversity conservation and of risks of inaction is yet to be further enhanced. Many people have not recognized the importance of biodiversity conservation, therefore their awareness of and participation in conservation is low. Some local governments one-sidedly pursue economic development goals. They promote economic development at the cost of biodiversity where economic development and biodiversity conservation conflict. The private sector is not enthusiastic about getting involved in biodiversity conservation. Decision makers and managers do not have adequate knowledge of biodiversity. Supervisory forces and capacities of civil society are not adequate.

(3) Conflicts between conservation and development and use

With the acceleration of urbanization and industrialization processes in China, biodiversity conservation is facing serious threats. For example, the population in pastoral areas has grown several times and the population density in arid grassland areas of north China has reached 11.2 persons/km^2, 2.2 times as many as 5 persons/km^2, the internationally recognized ecological capacity in arid grassland areas. Demands for wild medicinal herbs are increasing rapidly. As a result the number of some wild species is declining due to overexploitation for a long time. Fishing and harvesting methods are still coarse and predatory, in particular use of electricity for fishing that has caused serious damage to fishery resources and the aquatic environment and ecology. Most of the biodiversity hotspots of China are located in those remote, economically underdeveloped regions, so conflicts between conservation and economic development will be there for long, thus adding further pressures on biodiversity conservation.

(4) Inadequate investments and expenditures

Though the Government of China has made tremendous investments into biodiversity conservation, gaps in funds are still big as China has huge land areas, rich biodiversity and daunting challenges for biodiversity conservation. Capacities for surveying and monitoring biodiversity, establishment and management of nature reserves and restoration of biodiversity are still very weak and funds are seriously inadequate. Due to inadequate law enforcement conditions and lack of adequate infrastructure or equipment in some sites, relevant laws cannot be enforced.

(5) Technological research lacking behind

Due to inadequate investment for long and lack of professional researchers and techniques, researches in some areas such as biodiversity baseline survey and practical technologies and models for conservation are very weak. Some new issues and technologies are yet to be explored, in particular surveying and monitoring of biodiversity, *in-situ* conservation and biodiversity restoration should be given due attention.

5.2 Priority Actions

(1) To improve laws and regulations for biodiversity conservation and reinforce law enforcement

- To revise the *Environmental Protection Law*, *Wild Animal Protection Law*, *Wild Plants Protection Regulation* and *Nature Reserves Regulation*;
- To develop new laws or regulations such as wetland conservation regulation, regulation

on invasive alien species and regulation on access to genetic resources and benefit-sharing;
- To improve rules of ownership of natural resources and rules for their use control, and to establish a system of strictest source protection as well as a system of damage compensation and life-time accountability for ecological damage;
- To establish as early as possible mechanisms of ecological compensation, and integrate biodiversity into ecological compensation policies, in particular compensation mechanisms in biodiversity priority areas, so that policy and financial support will be provided to these areas;
- To strengthen law enforcement capacities and reinforce crackdown on illegal activities destroying biodiversity and examination of the import and export of species.

(2) To promote public participation and enhance public awareness of conservation

- To undertake various forms of biodiversity communication and education activities and to increase public awareness of conservation including through fully playing roles of civil society organizations and the private sector;
- To explore mechanisms and policies for establishing social supervision over biodiversity conservation;
- To develop citizen sciences and promote public participation in biodiversity conservation activities so that the whole society will make joint efforts to promote conservation and sustainable use of biodiversity.

(3) To implement updated NBSAP and National Plan for Major Function Zones

- To implement the *National Plan for Major Function Zones,* establish a system of land development protection, improve spatial layout of land development and propose policies and measures for biodiversity conservation in major function zones;
- To identify red lines for ecological conservation and ensure ecological security of national land;
- To strengthen supervision over biodiversity priority areas and integrate biodiversity into relevant national, sectoral and local planning processes;
- To strengthen environment management of development activities and integrate biodiversity into environmental impact assessments (EIAs) of major projects, regional planning and strategic planning as well as to implement responsibilities for ecological restoration;
- To establish review and supervision mechanisms to promote effective implementation of relevant plans and programmes.

(4) To further improve *in-situ* conservation networks and strengthen conservation efforts

- To improve spatial structure of nature reserves and to establish networks of biodiversity conservation and a system of national parks;
- To strengthen capacities of management of nature reserves, scenic spots, forest parks, wetland parks and aquatic germplasm conservation areas ;
- To continue implementing key ecological projects such as natural forest resources protection, returning cultivated land to forests, returning grazing land to grasslands, construction of forest belts in north, northeast and northwest China and the Yangtze River Basin, control of origins of sandstorms affecting Tianjin and Beijing, control of rockiness in Karst areas, wetland protection and restoration, establishment of protected areas and control of soil erosion;
- To initiate major projects on biodiversity conservation and to establish a mechanism for integrating biodiversity conservation with poverty reduction.

(5) To enhance capacities to meet emerging threats and challenges

- To establish as early as possible a monitoring and early warning system of invasive alien species;
- To take precautionary measures, to undertake risk assessments of alien species intentionally introduced, to implement risk management measures and organize elimination of alien species that have caused major damage;
- To undertake risk assessments of GMOs and fundamental researches on detecting their impacts on the environment;
- To develop technologies for environmental monitoring and detection of GMOs and improve relevant technical standards and specifications.

(6) To strengthen mechanisms and institutional capacities and to upgrade management level

- To strengthen the coordination of China National Committee on Biodiversity Conservation;
- To continue roles of China National Coordinating Group on the Implementation of the Convention on Biological Diversity and the Inter-ministerial Joint Conference on Conservation of Biological Resources;
- To further strengthen capacities of all relevant departments involved in biodiversity conservation in particular capacities of and support to local governments and communities, and to upgrade their management level constantly.

(7) To increase investment

- To broaden channels of investment and increase investments from central and local governments;
- To guide private and international investments and loans to biodiversity conservation so as to create mechanisms of various channels of investment;
- To consolidate scattered funds for biodiversity conservation and increase their efficiency of use;
- To increase investments into biodiversity from government budgets, in particular support to capacity building, fundamental science and research and ecological compensation.

(8) To establish mechanisms for surveying and monitoring of biodiversity and launching of results

- To undertake biodiversity surveys on a regular basis;
- To establish biodiversity monitoring and early warning system to know in time dynamic changes in biodiversity;
- To launch Biodiversity Red Lists in a timely manner;
- To monitor effectively important ecosystems and species.

(9) To strengthen professional education and research

- To strengthen professional education in biodiversity and to further improve research focusing on technical issues such as biodiversity formation mechanisms, causes of loss, models of conservation and restoration, evaluation and ecological compensation;
- To strengthen collection, storage and development of genetic resources;
- To provide scientific and technical support to biodiversity conservation and management.

(10) To strengthen international cooperation and exchanges

- To implement the obligations under the Convention;
- To participate actively in building of multilateral systems;
- To undertake extensively international exchanges and cooperation and to introduce advanced technologies and experiences from other countries to upgrade China's capacities and levels of biodiversity conservation.

Appendix

Appendix I Information Concerning Party and Process of Preparing National Report

A. Party

Party	China
National focal point	
Full name of institution	Department of International Cooperation, Ministry of Environmental Protection
Name and title of contact person	Zhang Jieqing, Director
Mailing address	No. 115 Xizhimennei nanxiaojie, Beijing, China
Telephone	+86-10-66556520
Fax	+86-10-66556513
E-mail	zhang.jieqing@mep.gov.cn
Contact person for national report (if different from above)	
Full name of institution	Division of Biodiversity Conservation, Department of Nature and Ecology Conservation, Ministry of Environmental Protection (Office of Biodiversity Conservation)
Full name and title of contact person	Zhuang Wenguo, Director
Mailing address	No. 115 Xizhimennei nanxiaojie, Beijing, China
Telephone	+86-10-66556309
Fax	+86-10-66556329
E-mail	zhang.wenguo2mep.gov.cn
Submission	
Officer responsible for signature for submission of national report	Zhang Jieqing
Date of submission	March 31, 2014

B. Process of National Report Preparation

1. Developing an outline of the report and establishing an expert group for drafting the report

During January and February 2013, the Ministry of Environmental Protection developed a work programme and an outline for the fifth national report, and invited members of China National Coordinating Group on the Implementation of the Convention on Biological Diversity to recommend relevant experts. All members recommended experts as required and the Ministry of Environmental Protection selected experts after having reviewed their qualifications.

2. First expert group meeting held to initiate the project and identify specific tasks

On 8 March, 2013, the expert group had its first meeting in Beijing to officially initiate the drafting of the report. At this meeting the outline of the report was further improved, and a work plan developed and specific tasks assigned to each expert.

3. Data collection, review and compilation and drafting of sector reports

In April 2013, the experts recommended by relevant departments collected materials and data from their respective departments following requirements specified in the outline. These experts submitted their sector reports during May-August 2013.

4. Expert workshop held to study indicators for assessing the 2020 targets

On 9 June 2013, the Ministry of Environmental Protection held in Beijing a workshop on indicators to assess the 2020 targets. The workshop discussed and improved indicators for assessing the 2020 targets proposed by the expert group. The workshop also requested experts from different sectors to provide data for assessing the 2020 targets.

5. Drafting the first draft of the fifth national report

The expert group prepared the first draft of the fifth national report following the review and integration of sector reports.

6. Three expert workshops held to discuss about the first draft

On 14 November 2013, the Ministry of Environmental Protection held a second meeting of the expert group to discuss about the first draft of the fifth national report. The meeting provided suggestions for revising the first draft. On 12 November and 24 December 2013 respectively, one expert workshop and a third meeting of the expert group on report preparation were held to discuss again about indicators for assessing the 2020 targets and the revised draft. A draft of the fifth national report for consultation was formed after having taken on board suggestions from various experts involved.

7. Consultations with members of China Coordinating Group

In early January 2014, the Ministry of Environmental Protection consulted with the member departments of China National Coordinating Group on the Implementation of the Convention on Biological Diversity. All departments reviewed the national report and provided very good comments and suggestions. The expert group further revised the national report by incorporating suggestions from these departments and prepared the final report and submitted it to the Ministry of Environmental Protection for approval.

8. Approval, translation and submission of the fifth national report

On 7 March 2014, the fifth national report was approved by the Ministry of Environmental Protection. During January-March 2014 the Chinese version of the fifth national report was translated into English, and both versions were submitted to the Secretariat in March 2014.

The following ministries, departments and institutions were involved in the preparation of the fifth national report:

Ministry of Environmental Protection, National Development and Reform Commission, Ministry of Education, Ministry of Science and Technology, Ministry of Finance, Ministry of Land and Resources, Ministry of Housing and Urban-Rural Development, Ministry of Water Resources, Ministry of Agriculture, Ministry of Commerce, General Administration of Customs, State Administration for Industry and Commerce, General Administration of Quality Supervision, Inspection and Quarantine, State Administration of Press, Publication, Radio, Film and Television, State Forestry Administration, State Intellectual Property Office, China National Tourism Administration, Chinese Academy of Sciences, State Oceanic Administration, State Administration of Traditional Chinese Medicine, the State Council Leading Group Office of Poverty Alleviation and Development, National Office of Management of Import and Export of Endangered Species, National Office of Implementation of the Ramsar Convention, Office of National Coordinating Group on Responses to Climate Change, Desertification Prevention and Control Center of SFA, Institute of Geography and Resources, Institute of Botany, Institute of Zoology, Institute of Micro-organisms of CAS, Institute of Crops of Chinese Academy of Agricultural Sciences, Institute of Agricultural Environment and Sustainable Development, Beijing Institute of Animal Husbandry and Veterinary, General Station for Conservation of Agricultural Ecology and Resources of MOA, Chinese Research Institute of Fishery, Chinese Academy of Forestry Sciences, Chinese Institute of Inspection and Quarantine Sciences, Beijing University of Forestry, Central China University of Agriculture, National Marine Environment Monitoring Centre, Third Institute of Oceanography, Chinese Academy of Chinese Medicine, Foreign Economic Cooperation Center of MEP, Nanjing Institute of Environmental Sciences of MEP.

Support from the Global Environment Facility to the preparation of this report is acknowledged with thanks.

Appendix II List of Personnel Involved in the Preparation of China's Fifth National Report on the Implementation of the CBD

The report was finally approved by:

Zhou Shengxian, Minister of the Ministry of Environmental Protection (MEP)
Li Ganjie, Vice-minister of MEP

The drafting work was organized by:

Zhuang Guotai, Director-General, Department of Nature and Ecology Conservation, MEP
Bai Chengshou, Deputy Director-General, Department of Nature and Ecology Conservation, MEP

The draft was compiled by:

Xu Haigen, Professor & Deputy Director-General of Nanjing Institute of Environmental Sciences, MEP
Zhang Wenguo, Director, Division of Biodiversity, Department of Nature and Ecology Conservation, MEP

The report was translated by:

Cai Lijie.

List of Expert Group on Report Preparation

No	Recommended by	Name	Institution	Title
1	Ministry of Environmental Protection	Xu Haigen	Nanjing Institute of Environmental Sciences, MEP	Researcher/ Deputy Director General/Head of expert group
2	Ministry of Environmental Protection	Ouyang Zhiyun	Research Center for Eco-Environmental Sciences, Chinese Academy of Sciences	Researcher/ Secretary
3	Ministry of Environmental Protection	Zhu Liucai	Foreign Economic Cooperation Centre, MEP	Researcher
4	Ministry of Environmental Protection	Wang Zhi	Nanjing Institute of Environmental Sciences, MEP	Director/ Associate researcher
5	Ministry of Environmental Protection	Ding Hui	Nanjing Institute of Environmental Sciences, MEP	Associate researcher
6	Ministry of Environmental Protection	Xu Weihua	Research Center for Eco-Environmental Sciences, Chinese Academy of Sciences	Associate researcher
7	Ministry of Environmental Protection	Wu Jun	Nanjing Institute of Environmental Sciences, MEP	Associate researcher
8	National Development and Reform Commission	Chen Yi	National Center for Strategic Research on Adaptation to Climate Change and International Cooperation	PhD
9	National Development and Reform Commission	Zhu Jianhua	Chinese Academy of Forestry Sciences	Associate researcher

Appendix II   List of Personnel Involved in the Preparation of China's Fifth National Report on the Implementation of the CBD

No	Recommended by	Name	Institution	Title
10	National Development and Reform Commission	Yue Tianxiang	Institute of Geographic Sciences and Natural Resources Research, CAS	Researcher
11	Ministry of Education	Zhang Zhixiang	Beijing University of Forestry	Professor
12	Ministry of Land and Resources	Wang Jun	Land Remediation Center, MLR	Researcher
13	Ministry of Housing, and Urban-Rural Development	Bao Manzhu	College of Gardening, Central China University of Agriculture	Professor/ president
14	Ministry of Water Resources	Wang Jianping	Development Research Center, MWR	Senior Engineer/ Deputy Director
15	Ministry of Agriculture	Yang Qingwen	Institute of Crops sciences, Chinese Academy of Agricultural Sciences	Researcher
16	Ministry of Agriculture	Zhang Guoliang	Institute of Agricultural Environment and Sustainable Development, Chinese Academy of Agricultural Sciences	Researcher
17	Ministry of Agriculture	Shi Rongguang	General Station of Agricultural Ecology and Resource Conservation, MOA	Associate researcher
18	Ministry of Agriculture	Fan Enyuan	Chinese Research Academy of Fishery Sciences	Researcher
19	Ministry of Agriculture	Ma Yuehui	Beijing Institute of Animal Husbandry and Veterinary, Chinese Academy of Agricultural Sciences	Researcher
20	Ministry of Agriculture	Lu Xinshi	Beijing University of Forestry	Professor
21	Ministry of Agriculture	Wang Zhixing	Institute of Biotechnology, Chinese Academy of Agricultural Sciences	Researcher
22	Ministry of Commerce	Li Li	University of International Business and Economics	Assistant Professor/Dean
23	General Administration of Customs	Zhou Yachun	Department of Supervision and Management GAC	Director
24	State Administration of Press, Publication, Radio, Film and Television	Wang Jing	Department of Communication	
25	State Administration of Industry and Commerce	Bai Jinyi	Department of Market Management, SAIC	Deputy Director
26	General Administration of Quality Supervision, Inspection and Quarantine	Li Mingfu	Chinese Academy of Inspection and Quarantine Sciences	Researcher
27	State Forestry Administration	Liu Zengli	State Forestry Administration	Deputy Director

No	Recommended by	Name	Institution	Title
28	State Forestry Administration	Wu Bo	Chinese Academy of Forestry Sciences	Researcher
29	State Forestry Administration	Li Diqiang	Chinese Academy of Forestry Sciences	Researcher
30	State Forestry Administration	Xu Jiliang	Beijing University of Forestry	Associate Professor
31	State Forestry Administration	Zhang Mingxiang	Beijing University of Forestry	Professor
32	State Intellectual Property Office	Zhang Qingkui	Department of Review of Pharmaceutical and Biological Inventions, SIPO	Director
33	China National Tourism Administration	Xi Jianchao	Institute of Geographic Sciences and Natural Resources Research, Chinese Academy of Sciences	Associate researcher
34	State Oceanic Administration	Pan Zengdi	East China Bureau, SOA	Researcher / Deputy Director General
35	State Oceanic Administration	Ma Minghui	National Marine Environment Monitoring Center	Researcher / Division Chief
36	State Oceanic Administration	Chen Bin	The Third Oceanographic Institute of SOA	Researcher / Division Chief
37	Chinese Academy of Sciences	Ma Keping	Institute of Botany, CAS	Researcher
38	Chinese Academy of Sciences	Xie Yan	Institute of Zoology, CAS	Associate researcher
39	Chinese Academy of Sciences	Guo Liangdong	Institute of Micro-organisms, CAS	Researcher
40	State Administration on Traditional Chinese Medicine	Huang Luqi	Chinese Academy of Chinese Medicine	Researcher / Vice President
41	State Council Leading Group Office of Poverty Alleviation and Development	Zhang Liang	Department of Development Guidance	Director
42	National Office of Management of Import and Export of Endangered Species	Lü Xiaoping	National Office of Management of Import and Export of Endangered Species	Director
43		Cui Peng	Nanjing Institute of Environmental Sciences, MEP	Associate researcher
44		Chen Lian	Nanjing Institute of Environmental Sciences, MEP	Associate researcher
45		Cao Mingchang	Nanjing Institute of Environmental Sciences, MEP	Associate researcher
46		Liu Li	Nanjing Institute of Environmental Sciences, MEP	Assistant researcher
47		Liu Xiaoqiang	Nanjing Insfitute of Environmental sciences, MEP	PhD

List of Participants of First Meeting of the Expert Group
(8 March, 2013, Beijing)

Name	Institution	Title
Bai Chengshou	Department of Nature and Ecology Conservation, Ministry of Environmental Protection	Deputy Director General
Zhang Wenguo	Department of Nature and Ecology Conservation, Ministry of Environmental Protection	Director
Lu Xiaoqiang	Department of Nature and Ecology Conservation, Ministry of Environmental Protection	PhD
Xu Haigen	Nanjing Institute of Environmental Sciences, MEP	Researcher/Deputy Director General
Zhu Liucai	Foreign Economic Cooperation Center, MEP	Researcher
Wang Aihua	Foreign Economic Cooperation Center, MEP	Project Officer
Chen Yi	National Center for Strategic Research on Adaptation to Climate Change and International Cooperation	PhD
Yue Tianxiang	Institute of Geographic Sciences and Natural Resources Research, CAS	Researcher
Zhang Zhixiang	Beijing University of Forestry	Professor
Wang Jun	Land Remediation Center, MLR	Researcher
Wang Jianping	Development Research Center, MWR	Senior Engineer/Deputy Director
Yang Qingwen	Institute of Crops sciences, Chinese Academy of Agricultural Sciences	Researcher
Shi Rongguang	General Station of Agricultural Ecology and Resource Conservation, MOA	Associate researcher
Ma Yuehui	Beijing Institute of Animal Husbandry and Veterinary, Chinese Academy of Agricultural Sciences	Researcher
Lu Xinshi	Beijing University of Forestry	Professor
Wang Zhixing	Institute of Biotechnology, Chinese Academy of Agricultural Sciences	Researcher
Li Li	University of International Business and Economics	Assistant Professor/Director
Li Ning	Department of Supervision and Management, General Administration of Customs	Chief Programme Officer
Zhang Guangling	Department of Markets, State Administration of Industry and Commerce	Officer
Li Mingfu	Chinese Academy of Inspection and Quarantine Sciences	Researcher
Liu Zengli	State Forestry Administration	Deputy Director
Wu Bo	Chinese Academy of Forestry Sciences	Researcher
Li Diqiang	Chinese Academy of Forestry Sciences	Researcher
Xu Jiliang	Beijing University of Forestry	Associate Professor
Zhang Mingxiang	Beijing University of Forestry	Professor
Xi Jianchao	Institute of Geographic Sciences and Natural Resources Research, Chinese Academy of Sciences	Associate researcher

Name	Institution	Title
Pan Zengdi	East China Bureau, SOA	Researcher /Deputy Director General
Lan Dongdong	National Marine Environment Monitoring Centre	Assistant researcher
Chen Bin	The Third Oceanographic Institute of SOA	Researcher /Division Chief
Luo Maofang	China Committee for Biodiversity International	Engineer
Xie Yan	Institute of Zoology, CAS	Associate researcher
Guo Liangdong	Institute of Micro-organisms, CAS	Researcher
Chen Meilan	Chinese Academy of Chinese Medicine	Associate researcher
Zhang Liang	Department of Development Guidance, State Council Leading Group Office of Poverty Alleviation and Development	Director
Lu Xiaoping	National Office of Management of Import and Export of Endangered Species	Director
Wu Jun	Nanjing Institute of Environmental Sciences, MEP	Associate researcher

List of Participants of Workshop on Indicators for Assessing 2020 Biodiversity Targets
(9 June, 2013, Beijing)

Name	Institution	Title
Zhang Wenguo	Department of Nature and Ecology Conservation, MEP	Director
Lu Xiaoqiang	Department of Nature and Ecology Conservation, MEP	PhD
Xu Haigen	Nanjing Institute of Environmental Sciences, MEP	Researcher/ Deputy Director General
Liu Jiyuan	Institute of Geology and Natural Resources, Chinese Academy of Sciences	Researcher
Yang Qingwen	Institute of Crops sciences, Chinese Academy of Agricultural Sciences	Researcher
Fan Enyuan	Chinese Research Academy of Fishery Sciences	Researcher
Li Mingfu	Chinese Academy of Inspection and Quarantine Sciences	Researcher
Li Diqiang	Chinese Academy of Forestry Sciences	Researcher
Zhang Qingkui	Department of Review of Pharmaceutical and Biological Inventions, SIPO	Director
Xie Yan	Institute of Zoology, CAS	Associate researcher
Pan Zengdi	East China Bureau, SOA	Researcher/Deputy Director General
Chen Bin	The Third Oceanographic Institute of SOA	Researcher/Division Chief
Xu Weihua	Research Center for Eco-Environmental Sciences, Chinese Academy of Sciences	Associate researcher
Fan Zemeng	Institute of Geological Sciences and Natural Resources, Chinese Academy of Sciences	Researcher
Lan Dongdong	National Marine Environment Monitoring Centre	Assistant researcher
Ding Hui	Nanjing Institute of Environmental Sciences, MEP	Associate researcher
Cao Mingchang	Nanjing Institute of Environmental Sciences, MEP	Associate researcher

Appendix II　List of Personnel Involved in the Preparation of China's Fifth National Report on the Implementation of the CBD

Name	Institution	Title
Chen Lian	Nanjing Institute of Environmental Sciences, MEP	Associate researcher
Wu Jun	Nanjing Institute of Environmental Sciences, MEP	Associate researcher

List of Participants of Second Meeting of the Expert Group
(4 November, 2013, Beijing)

Name	Institution	Title
Zhang Wenguo	Department of Nature and Ecology Conservation, MEP	Director
Lu Xiaoqiang	Department of Nature and Ecology Conservation, MEP	PhD
Zhu Liucai	Foreign Economic Cooperation Center, MEP	Researcher
Ding Hui	Nanjing Institute of Environmental Sciences, MEP	Associate researcher
Chen Yi	National Center for Strategic Research on Adaptation to Climate Change and International Cooperation	PhD
Zhang Zhixiang	Beijing University of Forestry	Professor
Wang Jun	Land Remediation Center, MLR	Researcher
Wang Jianping	Development Research Center, MWR	Senior Engineer/Deputy Director
Yang Qingwen	Institute of Crops Sciences, Chinese Academy of Agricultural Sciences	Researcher
Zhang Guoliang	Institute of Agricultural Environment and Sustainable Development, Chinese Academy of Agricultural Sciences	Researcher
Shi Rongguang	General Station of Agricultural Ecology and Resource Conservation, MOA	Associate researcher
Ma Yuehui	Beijing Institute of Animal Husbandry and Veterinary, Chinese Academy of Agricultural Sciences	Researcher
Lu Xinshi	Beijing University of Forestry	Professor
Wang Zhixing	Institute of Biotechnology, Chinese Academy of Agricultural Sciences	Researcher
Li Li	University of International Business and Economics	Assistant Professor /Dean
Li Ning	Department of Supervision and Management, General Administration of Customs	Chief Programme Officer
Zhang Guangling	Department of Markets, State Administration of Industry and Commerce	Officer
Li Mingfu	Chinese Academy of Inspection and Quarantine Sciences	Researcher
Xu Jiliang	Beijing University of Forestry	Associate Professor
Zhang Mingxiang	Beijing University of Forestry	Professor
Wu Bo	Chinese Academy of Forestry Sciences	Researcher
Li Diqiang	Chinese Academy of Forestry Sciences	Researcher
Xi Jianchao	Institute of Geographic Sciences and Natural Resources Research, Chinese Academy of Sciences	Associate researcher
Ma Keping	Institute of Botany, CAS	Researcher

Name	Institution	Title
Guo Liangdong	Institute of Micro-organisms, CAS	Researcher
Xu Weihua	Research Center for Eco-Environmental Sciences, Chinese Academy of Sciences	Associate researcher
Lan Dongdong	National Marine Environment Monitoring Centre	Assistant researcher
Chen Bin	The Third Oceanographic Institute of SOA	Researcher /Division Chief
Chen Meilan	Chinese Academy of Chinese Medicine	Associate researcher
Zhang Liang	Department of Development Guidance, State Council Leading Group Office of Poverty Alleviation and Development	Director
Lü Xiaoping	National Office of Management of Import and Export of Endangered Species	Director
Chen Lian	Nanjing Institute of Environmental Sciences, MEP	Associate researcher
Cui Peng	Nanjing Institute of Environmental Sciences, MEP	Associate researcher
Liu Li	Nanjing Institute of Environmental Sciences, MEP	Assistant researcher

List of Participants of Expert Workshop on the Preparation of China's Fifth National Report
(12 November 2013, Beijing)

Name	Institution	Title
Zhang Wenguo	Department of Nature and Ecology Conservation, MEP	Director
Lu Xiaoqiang	Department of Nature and Ecology Conservation, MEP	PhD
Xu Haigen	Nanjing Institute of Environmental Sciences, MEP	Researcher /Deputy Director General
Ma Keping	Institute of Botany, CAS	Researcher
Wang Xin	Foreign Economic Cooperation Center, MEP	Director/ researcher
Ouyang Zhiyun	Research Center for Eco-Enviromental Sciences, CAS	Researcher
Cui Guofa	Beijing University of Forestry	Professor
Tang Xiaoping	Institute of Forestry Prospecting Design, SFA	Researcher
Cao Mingchang	Nanjing Institute of Environmental Sciences, MEP	Associate researcher
Chen Lian	Nanjing Institute of Environmental Sciences, MEP	Associate researcher

List of Participants of Third Meeting of the Expert Group
(24 December, 2013, Beijing)

Name	Institution	Title
Zhang Wenguo	Department of Nature and Ecology Conservation, Ministry of Environmental Protection	Director
Lu Xiaoqiang	Department of Nature and Ecology Conservation, Ministry of Environmental Protection	PhD
Xu Haigen	Nanjing Institute of Environmental Sciences, MEP	Researcher/Deputy Director General
Zhu Liucai	Foreign Economic Cooperation Center, MEP	researcher
Wang Xin	Foreign Economic Cooperation Center, MEP	Director/ researcher

Appendix II   List of Personnel Involved in the Preparation of China's Fifth National Report on the Implementation of the CBD

Name	Institution	Title
Wang Jianping	Development Research Center, MWR	Senior Engineer/Deputy Director
Wang Jun	Land Remediation Center, MLR	Researcher
Zhang Zhixiang	Beijing University of Forestry	Professor
Yang Qingwen	Institute of Crops Sciences, Chinese Academy of Agricultural Sciences	Researcher
Zhang Guoliang	Institute of Agricultural Environment and Sustainable Development, Chinese Academy of Agricultural Sciences	Researcher
Fan Enyuan	Chinese Research Academy of Fishery Sciences	Researcher
Lu Xinshi	Beijing University of Forestry	Professor
Wang Zhixing	Institute of Biotechnology, Chinese Academy of Agricultural Sciences	Researcher
Li Li	University of Foreign Trade and Economic Cooperation	Assistant Professor/Dean
Li Ning	Department of Supervision and Management, General Administration of Customs	Chief Programme Officer
Zhang Guangling	Department of Markets, State Administration for Industry and Commerce	Officer
Li Mingfu	Chinese Academy of Inspection and Quarantine Sciences	Researcher
Liu Zengli	State Forestry Administration	Deputy Director
Xi Jianchao	Institute of Geographic Sciences and Natural Resources Research, CAS	Associate researcher
Ma Keping	Institute of Botany, CAS	Researcher
Xu Jiliang	Beijing University of Forestry	Associate Professor
Zhang Mingxiang	Beijing University of Forestry	Professor
Xie Yan	Institute of Zoology, CAS	Associate researcher
Wang Xinge	Institute of Geographic Sciences and Natural Resources Research, CAS	M.S.
Lan Dongdong	National Marine Environment Monitoring Centre	Assistant researcher
Chen Meilan	Chinese Academy of Chinese Medicine	Associate researcher
Lü Xiaoping	National Office of Management of Import and Export of Endangered Species	Director
Ding Hui	Nanjing Institute of Environmental Sciences, MEP	Associate researcher
Cui Peng	Nanjing Institute of Environmental Sciences, MEP	Associate researcher
Liu Li	Nanjing Institute of Environmental Sciences, MEP	Assistant researcher

Annex I Assessment of Progress in Implementing the Updated NBSAP

Strategic Goals
1. By 2015, biodiversity decline in key regions will be effectively contained.
2. By 2020, biodiversity loss will be basically controlled.
3. By 2030, biodiversity will be effectively protected.

Priority Areas	Actions identified	Actions undertaken	Level of progress
1. Improving policy and regulatory systems for conservation and sustainable use of biodiversity	**Action 1** To develop policies to promote conservation and sustainable use of biodiversity	• Incentives measures favourable for biodiversity conservation have been developed and are being implemented (see details in this book section 2.3.10 of Part 2). • However, pricing, taxation, loan, trade, land use and government procurement policies in this regard are yet to be improved	◐
	Action 2 To improve the legal system for conservation and sustainable use of biodiversity	• Legal and regulatory systems for biodiversity conservation have been basically established. (see details in this book section 2.3.1 of Part 2). • In recent years, many studies have been undertaken on legislations related to protected areas, wetland conservation, management of genetic resources, prevention and control of invasive alien species and biosafety management of GMOs, however considerable difficulties exist for development and adoption of these laws or regulations	◐
	Action 3 To establish and improve bodies for conservation and management of biodiversity, and to improve cross-sectoral coordination mechanisms	• China National Coordinating Group on the Implementation of the Convention on Biological Diversity and the Inter-ministerial Joint Conference on Conservation of Biological Resources were established respectively in 1993 and 2003. The Government of China established China National Committee on International Year of Biodiversity (IYB) to coordinate the IYB-related activities in 2010. In 2011, the State Council approved changing this Committee to "National Committee on Biodiversity Conservation". • These three coordination mechanisms headed by the Ministry of Environmental Protection and involving various departments play an important role in promoting biodiversity conservation in China. • The people's governments of most of the provinces (autonomous regions, province-level municipalities) have strengthened departments responsible for the environment, agriculture, forestry and oceanic administration, and established inter-departmental coordination mechanisms	◐

Priority Areas	Actions identified	Actions undertaken	Level of progress
	Action 4 To integrate biodiversity into sectoral and regional planning processes and plans	• Requirements for biodiversity conservation have been taken into account in developing sectoral or cross-sectoral plans by the sectors of development and reform commission, education, science and technology, land and resources, agriculture, commerce, customs, industry and commerce, quality supervision, inspection and quarantine, forestry, tourism, oceanic administration, traditional Chinese medicine and poverty reduction (see details in this book Part 3). • All provinces (autonomous regions, province-level municipalities) are developing their biodiversity strategies and action plans. Among them, 7 provinces have launched their BSAPs. Liaoning Province has issued a biodiversity strategy and action plan for Liaohe River Basin. • However implementation of NBSAP and provincial BSAPs need to be further strengthened	◐
2. Integrating biodiversity into sectoral and regional planning and promoting sustainable use	**Action 5** To ensure that biodiversity will be sustainably used	• Since 2000, the creation of eco-provinces, eco-cities and eco-counties has been organized. Currently, 15 provinces have initiated eco-province building and 13 provinces have issued plans or programmes for building eco-provinces. More than 1,000 counties or cities have started their eco-city or eco-county building. Since 2007, 38 counties or cities have been named as national-level eco-cities or eco-counties, 1,559 towns as national-level eco-towns and 238 villages as national-level eco-villages. • However, pilot work on biodiversity evaluation is yet to be expanded, and the experiences of various sectors in promoting production and consumption patterns favorable for biodiversity are yet to be summarized and disseminated	◐

Priority Areas	Actions identified	Actions undertaken	Level of progress
2. Integrating biodiversity into sectoral and regional planning and promoting sustainable use	**Action 6** To reduce impacts of environmental pollution on biodiversity	• The Government of China has identified significant reduction in the total amount of main pollutants as one of binding targets for social and economic development, in efforts to solve prominent environmental problems. From 2000 to 2010, the overall concentration of main pollutants went down. In the recent decade, the intensity of pollutant emission per unit of GDP has declined by more than 55%. Since 2004, CO_2 emission intensity has dropped by 15.2%. • The plan for water pollution prevention and control in key river basins (2011-2015) has been issued and is being implemented to enhance pollution abatement in rivers and lakes. The percentage of seven major rivers with water quality better than that of Grade III increased from 41% in 2005 to 64% in 2012, and the percentage of those with water quality lower than that of Grade V decreased from 27% to 12.3%. Underground water pollution prevention and control has been also strengthened. A national plan in this regard is being implemented and a plan for preventing and controlling underground water contamination in north China plains has been developed. • The *Twelfth Five-year Plan for Air Pollution Prevention and Control* in key regions has been launched and is being implemented, and a revised *Standard for Air Quality* has been issued. In 2012, the central government provided 1.09 billion yuan RMB to support the renovation of coal-fired boilers in 15 key cities listed in the Plan. As a result coal boilers with capacities of 28,997 t has been renovated. Among them dust removal equipment with capacities of 15,406 t has been renovated, and coal use for equipment with capacities of 13,591 t has been replaced with clean energy. • By the end of 2012, out of 57 facilities for integrated disposal of hazardous wastes identified by National Programme for Building Facilities for Disposal of Hazardous and Medical Wastes, 36 facilities have been built, with capacities of integrated disposal reaching 1.43 million t/a; out of 271 facilities for integrated disposal of medical wastes planned, 231 facilities have been built with capacities of integrated disposal reaching 428,000 t/a. • By the end of 2013, the central government has allocated specialized funds for rural environment improvement totalling 19.5 billion yuan RMB, and 46,000 villages and more than 87 million farmers benefited. In 2012, the central government invested 1.33 billion yuan RMB into building 2.946 million clean toilets in rural areas. By the end of 2012, 3.346 million clean toilets have been built in rural areas across the country, exceeding the original target set for the year. • About 6.7 milliont of chromium was left untreated, most of which were stored for over ten or twenty years and even fifty years. Disposal of chromium started from 2005. By the end of 2012, almost all chromium left from history have been disposed of, with 2.3 million t disposed of in 2012 only, three times higher than the average amount disposed of in the past six years before that. • However, the density of emission of pollutants per unit of GDP is still very high, and the amount of waste water discharged is still increasing	●

Annex I Assessment of Progress in Implementing the Updated NBSAP

Priority Areas	Actions identified	Actions undertaken	Level of progress
3. Undertaking surveys, assessments and monitoring of biodiversity	**Action 7** To undertake baseline surveys of biological resources and ecosystems	• The Seventh forest resources surveys, second national survey on wetlands, second national survey on livestock genetic resources, specialized survey on marine biodiversity, survey and cataloguing of biological resources in southwest China, survey on agricultural biological resources in Yunnan and neighboring areas, and a remote-sensing survey and assessment of ecological changes in the decade from 2000 to 2010 have been completed. • Ongoing surveys include a second national survey on key wild animal resources, a second national survey on key wild plant resources, a fourth survey on giant pandas and a pilot consensus on traditional Chinese medicinal resources. • Based on the above surveys, national biodiversity information system and a network of sharing national specimens and information have been established, however the scope and depth of surveys are yet to be expanded and relevant systems need to be improved	◐
	Action 8 To survey and catalogue genetic resources and associated traditional knowledge	• By the end of 2012, the total agricultural crops collections have reached 423,000 accessions, an increase of 30,000 accessions over those in 2007. To better store collected genetic resources, China has strengthened the construction of storage facilities. On one hand, China has expanded and renovated 1 existing long-term bank, 1 national copy banks, 10 mid-term storage banks and 32 national germplasm nurseries (including 2 plantlets libraries). On the other hand, China has built 7 more national germplasm nurseries. • China has established a network of conservation of livestock genetic resources primarily consisting of conservation farms and complemented by protected areas and gene banks. By 2011, this network has effectively protected more than 100 key resources. China's livestock fiber cell bank has become the world's biggest, with 58,000 cells of 95 local varieties stored in the cell bank. • Surveys have been undertaken of traditional knowledge associated with genetic resources in areas inhabited by minorities, and relevant databases have been established based on these surveys	◐
	Action 9 To undertake monitoring and early warning of biodiversity	• China has developed technical guidelines for monitoring typical ecosystems and important species, however relevant standards have not been issued, so standardization and regulation of monitoring activities are yet to be strengthened. • National biodiversity monitoring network has been designed, and national forest biodiversity monitoring network has been established. Also established a monitoring network of birds and amphibians and demonstration is ongoing, however national long-term biodiversity monitoring network with wide coverage and high level of representativeness is yet to be established. • Studies have been undertaken on the models of biodiversity prediction and early warning, however the technical system of early warning and emergency response mechanisms are yet to be established	◐

Priority Areas	Actions identified	Actions undertaken	Level of progress
3. Undertaking surveys, assessments and monitoring of biodiversity	**Action 10** To promote and coordinate establishment of information systems on biological resources and genetic resources	• For information sharing, three main databases and sharing platforms have been established for genetic resources. • National Platform for Sharing Plant Germplasm Resources (http://icgr.caas.net.cn/pt/) covers agricultural crops, perennial and cloned crops and timbers (including bamboos, rattans and flowers), medicinal plants, tropical crops, important wild plant and grass germplasm resources. • The platform for germplasm resources of domesticated animals (http://www.cdad-is.org.cn/) covers genetic resources of pigs, cows, sheep and other domesticated animals • The platform for microorganisms and fungi (http://www.cdcm.net/index.Action.action) covers information of 162,000 strains of bacteria, accounting for 40%–45% of the total micro-organisms in China	◐
	Action 11 To undertake comprehensive assessments of biodiversity	• In 2012, China completed a national biodiversity assessment, with unit of assessment being at county level. For the first time China has collected data concerning distribution at county level of 34,039 wild vascular plant and 3,865 wild vertebrate species. Through this assessment China has almost known the current status, spatial distribution of and main pressures on biodiversity in land areas. The assessment has identified biodiversity hotspots across the country as well as major gaps of conservation. A report of assessment of China's baseline status of biodiversity has been published. • In 2011, China initiated a Project on Remote-sensing Survey and Assessment of Ecological Changes in China during the Decade from 2000 to 2010. The overall objective of this project is to know fully changes and evolutions in the distribution, layout, quality and services of ecosystems across the country in the last decade. • China has issued technical guidelines for assessing economic values of genetic resources. China is developing technical guidelines for evaluating ecosystem services and functions. China has undertaken tremendous pilot work in the economic evaluation of biodiversity. • In September 2013, China issued China Biodiversity Red List-Higher Plants Volume. Work is on-going on Biodiversity Red List-Vertebrates Volume	◐

Priority Areas	Actions identified	Actions undertaken	Level of progress
4. Strengthening *in-situ* conservation of biodiversity	**Action 12** To improve and implement in a coordinated manner national planning for protected areas	• In 1999, China issued National Programme for Nature Reserves (1996-2010). In 2003, China approved National Programme for Wetland Conservation (2002-2030). China is developing a national plan for development of nature reserves. • China has strengthened establishment of protected areas (PAs) in biodiversity priority areas, improved spatial layouts of PAs and enhanced overall capacities of conservation. • Since 2006, China and Russia have established an intergovernmental working group on transboundary protected areas and biodiversity conservation, which meets on a regular basis every year. So far this working group has had six meetings. Both sides have signed cooperation agreements such as China-Russia strategy for development of networks of transboundary protected areas in Heilongjiang River Basin and China-Russia agreement on establishment of protected areas in Xingkai Lake. with regard to PAs in Sanjiang, Honghe and Bachadao in Heilongjiang Province, China have signed agreements with PAs in Basdak, Daherchel, Xinganski and Bolongski in Russia for cooperation. In 2013, China and Russia signed an agreement on protection of wild tigers. By this agreement both sides will accelerate construction of migratory corridors for tigers and establish transboundary protected areas for tigers. • In 2009, China and Lao DPR established the first transboundary protected areas-Shangyong, Xishuangbanna-South Tananmuha. In early 2012, the two countries decided to establish another transboundary protected area-Menglamanzhuang, China-Fengshali, Lao DPR border areas	◐
	Action 13 To strengthen conservation of biodiversity priority areas	• In the updated NBSAP, China has identified 35 biodiversity priority areas. China has strengthened establishment of protected areas in biodiversity priority areas, improved spatial layouts of PAs and enhanced overall capacities of conservation. • China is studying plans, policies, rules and relevant measures for conservation in priority areas	◐

Priority Areas	Actions identified	Actions undertaken	Level of progress
4. Strengthening *in-situ* conservation of biodiversity	**Action 14** To regulate establishment and management of protected areas and to improve management effectiveness of protected areas	• China has issued a programme for master planning of national-level nature reserves, technical specifications for master planning of nature reserves and for eco-tourism planning in nature reserves, and guidelines for national-level nature reserves management. Based on these guidelines and standards, China has undertaken assessments of management of national-level nature reserves since 2008. By 2012, China has completed assessments of all national-level nature reserves. • China has strengthened standardized construction of national-level nature reserves, improved management facilities and reinforced supervision measures. By the end of 2013, China has established a total of 407 national-level nature reserves, with areas covering about 940,000 km², accounting for 64.3% of the total PAs and 9.8% of the country's total land area. • Departments responsible for nature reserves from the Ministries of Environmental Protection and Agriculture and SFA have organized many training workshops, providing training on PA-related policies, regulations, standardized management, plan development, capacity building project design, supervision of development activities, information system development and status survey. • From 2007 to 2012, the Ministry of Environmental Protection together with relevant departments organized many inspections on law enforcement in nature reserves, with a view to preventing damage from irrational development activities to nature reserves	◐
	Action 15 To strengthen biodiversity conservation outside protected areas	• Key ecological projects continue to be implemented, such as natural forest resources protection, returning cultivated land to forests and grazing land to grasslands, construction of forest belts, conservation of wild fauna, flora and wetland conservation and restoration (see details in this book section 2.3.6 of Part 2). Grassland ecosystem conservation and restoration has been strengthened (see details in this book section 3.7 of Part 3). Conservation of marine biodiversity has also been strengthened and obvious results achieved (see details in this book section 3.14 of Part 3). • In 2012, China initiated a project to rescue wild plants with very small populations. The project lasts five years targeting at 120 wild plants with extremely small populations. The implementation of this project will effectively improve the status of critically endangered, rare plants	◐
	Action 16 To strengthen establishment of livestock genetic resources conservation farms and protected areas	• China has established a network of conservation of livestock genetic resources primarily consisting of conservation farms and complemented by protected areas and gene banks. By August 2012, China has identified 150 national-level conservation farms, protected areas and gene banks. China has rescued a number of livestock species close to extinction, such as Wuzhishan pig, bantam and Jinjiang horse. This network has effectively protected more than 100 key resources. • China has issued a national list of livestock genetic resources for protection, which includes 138 rare and endangered livestock varieties	◐

Priority Areas	Actions identified	Actions undertaken	Level of progress
5. Undertaking *ex-situ* conservation of biodiversity on a scientific basis	**Action 17** To establish *ex-situ* conservation system on a scientific and reasonable basis	• China has established nearly 200 botanical gardens of various kinds at different levels, which collect and store 20,000 species, accounting for two-thirds of China's flora. China has established more than 400 bases for conservation of wild plant germplasms, as well as centers of protection of cycads and orchid germplasms, which have collected and stored over 240 varieties of cycads and over 540 varieties of orchids respectively. • A germplasm bank of wild biological resources in southwest China has been established in Kunming, Yunnan Province. By April 2013, this bank has collected and stored 76,864 accessions of plant seeds of 10,096 species. • According to incomplete statistics, China has established more than 240 zoos including animal demonstration areas, and 250 bases of reproduction and rescuing of wild animals. Various places in China have also established unique farms for conserving local varieties of domesticated animals and national-level key breeding farms, storing 138 varieties of domesticated animals	◐
	Action 18 To establish and improve system of storage of genetic resources	See details in this annex Actions 8, 10 and 17	◐
	Action 19 To strengthen re-introduction of artificially bred species and restoration of wild species	• In March 2012, China issued a national project plan for protection and rescuing wild plants with extremely small population (2011-2015), targeting at 120 plant species with extremely small population in the first phase of the project. Among them there are 36 national Class I protected plants, 26 national Class II protected plants, and 58 provincial protected plants. A similar plan is being developed for protection and rescuing wild animals with extremely small populations. • Research has been undertaken on artificial breeding techniques and seedling for those rare wild plants for which market demands are relatively big, such as Matsutake (*Tricholoma matsutake*), Snow Lotus (*Saussurea involucrata*), Dove Tree (*Davidia involucrata*), Desertliving Cistanche (*Cistanche deserticola*), yews and rare orchidaceae. Studies were also undertaken on techniques of reproduction and conservation of endangered animals, and as a result the artificially bred populations of over 50 wild animals are constantly expanding, such as Giant Panda (*Ailuropoda melanoleuca*), Crested Ibis (*Nipponia nippon*), Chinese Alligator (*Alligator sinensis*), tigers, golden monkeys and Tibetan Antelopes (*Pantholops hodgsoni*). • Preparations and experiments have been made for reintroduction of extremely endangered orchidaceae in China such as *Doritis pulcherrima* and *Paphiopedilum armeniacum*, and some critically endangered species such as *Cycas debaoensis*, *Pachylarnax sinica* and *Cyclobalanopsis sichouensis*. Eight endangered wild animals such as Crested Ibis, Asian Wild Horse (*Equus przewalskii*), Pere David's Deer (*Elaphurus davidianus*), Chinese xenosaurs (*Shinisaurus crocodilurus*), Chinese Alligator, Bactrian Red Deer (*Cervus elaphus yarkandensis*) and Yellow-bellied Tragopan (*Tragopan caboti*) have been reintroduced to nature, and natural reproduction has been realized and new wild populations are being gradually established	◐

Priority Areas	Actions identified	Actions undertaken	Level of progress
6. Promoting access to genetic resources and associated traditional knowledge and benefit-sharing from their use	**Action 20** To strengthen development and use of genetic resources and related innovation research	See details in this book section 2.3.5 of Part 2	◔
	Action 21 To establish rules and mechanisms for access to genetic resources and associated traditional knowledge and benefit-sharing from their use	• To regulate development of genetic resources, the *Patent Law* revised in December 2008 added a provision concerning disclosure of the origin of genetic resources, and clearly provides that patent will not be granted if access to or use of genetic resources violates relevant laws or regulations. • Policies, rules and CHM are under development for access to genetic resources and associated traditional knowledge and benefit-sharing from their use	◔
	Action 22 To establish system of inspection and verification of imported and exported genetic resources	• The General Administration on Quality Supervision, Inspection and Quarantine (GAQSIQ) has established a system of approval of export of biological resources. In 2013, GAQSIQ issued guidance for strengthening inspection and quarantine of imported and exported biological resources, with a view to prevent loss of China's endemic and rare biological resources. • An examination system is being established with two research centers established for identification and verification and six key labs established for pilot inspection. • A lot of research projects have been undertaken and some progress made in developing methods for identifying and verifying genetic resources of animals, plants and micro-organisms as well as human beings. • At the end of 2012, a training workshop was held on the inspection and quarantine of imported and exported biological resources. In early 2013, another training workshop was held for the on-site staff on techniques used for identifying and verifying biological resources. • However, a list is yet to be developed for import and export of genetic resources, and methods, capacities and conditions for rapid identification and verification are still inadequate	◔

Annex I Assessment of Progress in Implementing the Updated NBSAP

Priority Areas	Actions identified	Actions undertaken	Level of progress
7. Strengthening biosafety management of invasive alien species and GMOs	**Action 23** To enhance capacities of early warning, responses and monitoring of invasive alien species	See details in this book section 2.3.8 of Part 2	◕
	Action 24 To establish and improve system and platform for biosafety assessment, inspection and monitoring of GMOs	See details in this book section 2.3.9 of Part 2	◑
8. Increasing capacities to respond to climate change	**Action 25** To develop an action plan to address climate change impacts on biodiversity	• Studies on climate change impacts on biodiversity have been undertaken. Technologies are being developed for monitoring climate change impacts on biodiversity, and action plan is being developed for biodiversity conservation and adaptation to climate change. • There is a need to further investigate positive and negative impacts of climate change on biodiversity and further study relevant adaptation measures	◕
	Action 26 To assess impacts of biofuels on biodiversity	• A study on biofuel plantation impacts on biodiversity has been undertaken, however a system is yet to be established to manage environmental safety of biofuel production	◑

311

Priority Areas	Actions identified	Actions undertaken	Level of progress
9. Strengthening science and technology research and professional education and training	**Action 27** To strengthen scientific research in the field of biodiversity	• See details in this book section 2.3.11 of Part 2, however more investment needed for biodiversity research, infrastructure development and promotion of research results	◐
	Action 28 To strengthen professional education and training in the field of biodiversity	• See details in this book section 3.2 of Part 3. Despite progress in biodiversity professional education, more efforts are needed in this regard, in particular training of taxonomists and leading scientists in this field	◐
10. Establishing mechanisms and partnerships for public participation in biodiversity conservation	**Action 29** To establish mechanisms for public participation	• See details in this book section 2.3.12 of Part 2	◐
	Action 30 To promote establishment of partnerships for biodiversity conservation	• Effective biodiversity partnerships have been established at national and provincial levels (see details in this annex Action 3), however partnerships between relevant international organizations, local communities and NGOs are yet to be strengthened	◐

Notes: ● fully achieved; ◐ significant progress; ◐ considerable progress; ◐ some progress; ● no progress.

Annex II Assessment of China's Progress in Implementing the Strategic Plan for Biodiversity 2011-2020 and the 2020 Biodiversity Targets

Global Targets	National Targets	National Actions	Outcomes Achieved	National Indicators	Overall Assessment and Trends
Strategic Goals: A. Address the underlying causes of biodiversity loss by mainstreaming biodiversity across government and society					
Target 1. By 2020, at the latest, people are aware of the values of biodiversity and the steps they can take to conserve and use it sustainably	• Practical efforts will be made in environmental education and communication, popularizing environmental knowledge and increasing public environmental awareness • By 2030, biodiversity conservation will become voluntary action of the public	• Lecturing on biodiversity-related knowledge provided in primary and middle school classrooms. • Providing biodiversity-related professional education in universities. • Biodiversity communication and education undertaken by using media, such as TV, internet, newspapers and radios, and through organizing training workshops and disseminating training materials and so on	Biodiversity-related knowledge has been incorporated into classroom teaching in primary and middle schools in China. By 2012, 1,908 universities in China have trained 556,000 professionals in the field of biodiversity. To celebrate the International Year of Biodiversity, 40 large-scale communication and education activities were organized at national level and more than 370,000 copies of educational materials were disseminated, with 800 million persons/times influenced by various media. 191 large-scale activities were organized at local levels and more than 350,000 copies of materials disseminated. 25 films with biodiversity themes were developed. About 20,000 institutions, including protected areas, zoos, botanical gardens, environmental education organizations and research institutes as well as media such as TV, newspapers and internet were mobilized to provide a series of communication activities for the public and primary and middle school and university students, whose number came up to 100 million persons/times. Through communication and education, public awareness and participation have increased significantly, and the importance of biodiversity widely recognized	Items concerning biodiversity in China searched through Google or Baidu in different years	Upward trend

Global Targets	National Targets	National Actions	Outcomes Achieved	National Indicators	Overall Assessment and Trends
Target 2. By 2020, at the latest, biodiversity values have been integrated into national and local development, poverty reduction strategies and planning processes and are being incorporated into national accounting, as appropriate, and reporting systems	Resource consumption, environmental damage and ecological benefits will be incorporated into the system of assessing social and economic development, and a system of goals and targets, aswell as related assessment methods and reward/penalty mechanisms that meet requirements for building an ecological civilization, will be established	• Establishing theory and methods for the evaluation of the economic values of biodiversity. • Case studies undertaken in the evaluation of the economic values of biodiversity. • Developing a system of goal and targets, aswell as related assessment methods and reward/penalty mechanisms that meet requirements for building an ecological civilization	China has issued technical guidelines for assessing the economic values of genetic resources, and is developing technical guidelines for assessing ecosystem services and functions. In 1998. China completed the national assessment of the economic values of biodiversity. In 2010. China completed a national assessment of service of forest ecosystems. China has also undertaken assessments of biodiversity values in some typical regions in different periods of time. All these assessments have provided a basis for developing theories and methods for economic evaluation of biodiversity. The Eighteenth National Congress of the Chinese Communist Party held in November 2012 laid out a vision for building an ecological civilization and Beautiful China. The meeting required that national policy of protecting the environment and improving resource use efficiency would continue to be followed, and proposed that priorities would be given to energy conservation, protection and natural recovery of the environment. Future efforts will focus on promoting green, cycling and low carbon development so as to form the industrial structure, production and consumption patterns and spatial layouts favorable to the environment and conservation of natural resources. China is developing a system of and targets, aswell as related assessment methods and reward/penalty mechanisms that meet requirements of an ecological civilization. The biodiversity values will be incorporated into such system of goals and assessment methods	No	

Annex II Assessment of China's Progress in Implementing the Strategic Plan for Biodiversity 2011–2020 and the 2020 Biodiversity Targets

Global Targets	National Targets	National Actions	Outcomes Achieved	National Indicators	Overall Assessment and Trends
Target 3. By 2020, at the latest, incentives, including subsidies, harmful to biodiversity are eliminated, phased out or reformed in order to minimize or avoid negative impacts, and positive incentives for the conservation and sustainable use of biodiversity are developed and applied, consistent and in harmony with the Convention and other relevant international obligations, taking into account national socio economic conditions	Establishment of mechanisms for ecological compensation and increase fiscal transfers to key ecological function zones will be accelerated; and studies will be undertaken on the establishment of national specialized funds for ecological compensation and the system of reserves for sustainable development of resource-consumption enterprises will be promoted	• Eliminated rebates for exports of 553 products with high energy consumption, pollution and resource consumption. • 30 provinces (autonomous regions, province-level municipalities) have required mining operators to deposit funds for environmental and ecological recovery in mining areas, with cumulative funds having reached 61.2 billion yuan RMB. • Subsidies provided to key forestry and ecological conservation projects. • Established funds for compensation for forest ecological benefits. • Preliminarily established ecological compensation mechanisms for national key ecological function zones	(1) The project of returning cultivated land to forests has been implemented since 1999. By 2012, the cumulative investment by the central government has reached 324.7 billion yuan RMB, providing direct benefits to 120 million farmers in 2,279 counties with each household receiving a policy subsidy of 7,000 yuan cumulatively. (2) In 2000, the Government of China initiated the natural forest resources protection project in 17 provinces, with the central government providing subsidies for forest conservation and management, reforestation and social expenditures. By the end of 2010, the cumulative investment for the first phase of the project has reached 118.6 billion yuan RMB and the investment for the second phase will total 244 billion yuan RMB. (3) The project of returning grazing land to grasslands was initiated in 2003 and implemented in 8 provinces, with the central government providing subsidies for setting up fences and supply of forages. The cumulative investment for this project during 2003-2012 has reached 17.57 billion yuan RMB, benefiting more than 4.5 million farmers and herdsmen in 174 counties. (4) China established in 2004 the National Fund for Compensation for Forest Ecological Benefits, with annual payment reaching 3 billion yuan RMB. (5) In 2008, China established funds to be transferred from the central government budget to national key ecological function zones. In 2013, the funds transferred covered 492 counties and 1,367 land zones prohibited for development, with the total funds transferred reaching 42.3 billion yuan RMB. (6) Since 2006, through provision of fiscal subsidies for wetland conservation as proposed in the Eleventh and Twelfth Five-year Plans for Wetland Conservation Project, more than 500 wetlands have been protected and restored, with an area of over 3,000 km² of protected wetlands added annually	Investments in key forestry projects	Investments into natural forest resources protection, wild flora and fauna protection, establishment and management of PAs and wetland conservation increased from 9.7 billion yuan RMB in 2001 to 21.77 billion yuan RMB in 2011, with average annual increase by 13.7%

Global Targets	National Targets	National Actions	Outcomes Achieved	National Indicators	Overall Assessment and Trends
Target 4. By 2020, at the latest, Governments, business and stakeholders at all levels have taken steps to achieve or have implemented plans for sustainable production and consumption and have kept the impacts of use of natural resources well within safe ecological limits	By 2015 considerable progress will be made in building a resource-efficient and environmentally friendly society. Efforts will be made to promote spatial layouts, industrial structure, production and consumption patterns characterized by green, recycling and low-carbon development, conserving natural resources and protecting the environment	• Significant reductions in total amount of main pollutants identified as one of binding targets for social and economic development and pollution reduction projects implemented. • EIA rules strictly implemented	(1) Since 2006, COD discharged from industrial waste water, SO_2 emitted from effluent gas, soot, industrial dust and solid wastes have been going down. In the past decade, the intensity of pollutant emissions per unit of GDP dropped significantly by more than 55%. Since 2004 the intensity of CO_2 emission per unit of GDP dropped by 15.2%. (2) China has been strictly implementing rules for EIA and measures have been taken such as "limited approvals for certain regions and sectors". Since 2008, the central government has refused to approve 332 projects with a total investment of 1.1 trillion yuan RMB, all of which were projects of high pollution, energy consumption, resource use, low-level duplication and exceeding production capacities. These measures have played an important role in adjusting industrial structure and prioritizing economic growth	Reduction in pollutant emission	Reduced overall however wastewater discharging still increasing
				Pollutant emission per unit of GDP	Down by over 55% in the past decade
				CO_2 emission per unit of GDP	Down by 15.2% since 2004
				Sustainable consumption	

Annex II Assessment of China's Progress in Implementing the Strategic Plan for Biodiversity 2011-2020 and the 2020 Biodiversity Targets

Global Targets	National Targets	National Actions	Outcomes Achieved	National Indicators	Overall Assessment and Trends
Strategic Goals: B. Reduce the direct pressures on biodiversity and promote sustainable use					
Target 5. By 2020, the rate of loss of all natural habitats, including forests, is at least halved and where feasible brought close to zero, and degradation and fragmentation is significantly reduced	• By 2015, forest coverage rate will be increased to 21.66% and forest reserves will be increased by 600 million m³ over that in 2010. • By 2020, grassland degradation trend will be basically controlled and grassland ecological environment will be obviously improved • By 2020, the environmental and ecological worsening trends in coastal and near-shore areas will be fundamentally reversed and marine biodiversity decline trend will be basically contained • By 2020, aquatic environment and ecology will be gradually restored and decline of fishery resources and increase in endangered species will be basically contained	• Implemented key forestry and ecological projects. • Great efforts made in comprehensive control of soil erosion. • Undertaken conservation and restoration of grassland ecosystems. • Undertaken conservation and restoration of wetlands. • Undertaken restoration and rebuilding of coastal wetlands	(1) China's forest resources have increased constantly and rapidly since key forestry projects were initiated. China has completed reforestation in areas of 482,000 km², an increase in forest areas by 23% over those a decade ago. The forest coverage rate is 3.8% up over that of a decade ago. The forest reserves are 21.8% higher. All this has enhanced restoration of habitats for wild species and the increase in the number and variety of species. During 2004-2009, areas of land degradation across the country have been reduced by 1,717 km² and areas of intermediately, seriously and extremely seriously degraded land reduced by 35,900 km² in total during these five years, resulting in reductions in annual input of sands into the yellow River by over 300 million t (2) Great efforts were made in soil erosion control in some key regions. During 2009-2012 such projects were implemented in a total of 12,000 small river basins, with areas of soil erosion control reaching 270,000 km². Enclosing mountains for conservation and soil erosion control continues, with cumulative areas enclosed for conservation having reached 720,000 km², and among them ecological conditions in areas of 450,000 km² have begun to recover. (3) Remarkable achievements have been made in conservation and restoration of grassland ecosystems. Compared with areas without such projects implemented, vegetation coverage of grasslands in project areas increased by 11%, grass height increased by 43.1%, and fresh grass output increased by 50.7%. The situation of grassland use is considerably improving, with the rate of overcapacity use of grasslands in 268 counties in 2012 down by 34.5% -36.2% compared with the situation in 2011. However overall most of grasslands are being used beyond their capacities and degradation, desertification and salinization of grasslands have not been effectively controlled. (4) In recent years, areas of wetland conservation added annually have exceeded 3,000 km² and areas of wetlands restored nearly 200 km². The rate of protection of natural wetlands increased by over 1% annually, with more than half of natural wetlands effectively protected	Total growing stock volume	↘ 10.57 billion m³ in 1988, 13.62 billion m³ in 2003, and 14.55 billion m³ currently
				Area of natural forests	↘ 885,000 km² in 1988, 1.158 million km² in 2003, and currently 1.197 million km²
				Area of wetland ecosystems	Increased during 2000—2010
				Area of grassland ecosystems	✗ Decreased during 2000—2010
				Grass output from natural grasslands	Annual increase by 1.6% during 2005—2012
				Area of desertified land	Annual reduction in areas of degraded land by 1,717 km² across the country
				Ecological degradation	⋯

Global Targets	National Targets	National Actions	Outcomes Achieved	National Indicators	Overall Assessment and Trends
Target 6. By 2020 all fish and invertebrate stocks and aquatic plants are managed and harvested sustainably, legally and applying ecosystem based approaches, so that overfishing is avoided, recovery plans and measures are in place for all depleted species, fisheries have no significant adverse impacts on threatened species and vulnerable ecosystems and the impacts of fisheries on stocks, species and ecosystems are within safe ecological limits	• By 2020, aquatic environment and ecology will be gradually restored and decline of fishery resources and increase in endangered species will be basically contained • By 2020, the environmental and ecological worsening trends in coastal and near-shore areas will be fundamentally reversed and marine biodiversity decline trend will be basically contained	• Undertaken conservation and restoration of wetlands. • Undertaken restoration and rebuilding of coastal wetlands. • Increased restocking of aquatic biological resources. • Strengthened construction of marine farms	(1) China has strengthened establishment and management of PAs for aquatic species. More than 200 PAs have been established for conserving various kinds of aquatic wild flora and fauna and types of aquatic ecosystems. (2) China has strengthened establishment of PAs for aquatic germplasm by identifying 368 national-level PAs to protect more than 300 national protected aquatic plants and animals with economic values as well as local rare, endemic aquatic species and their habitats and reproduction sites. (3) China has implemented many projects to protect and restore wetlands. The rate of natural wetland protection has increased by over 1% annually. However studies on fishery impacts on biodiversity are yet to be undertaken	Marine trophic index	◐ Constantly increased since 1997, however still at low level
				Red List Index of fishes (RLI)	⊗ RLI of fresh water fish down during 1998-2004
				Fishery impacts on biodiversity	⋮

Annex II Assessment of China's Progress in Implementing the Strategic Plan for Biodiversity 2011–2020 and the 2020 Biodiversity Targets

Global Targets	National Targets	National Actions	Outcomes Achieved	National Indicators	Overall Assessment and Trends
Target 7. By 2020 areas under agriculture, aquaculture and forestry are managed sustainably, ensuring conservation of biodiversity	• By 2020, national forest holdings will exceed 2.33 million km^2, an increase 223,000 km^2 over that of 2010; and national forest reserves will be increased to 15 billion m^3, an increase of about 1.2 billion m^3 over that of 2010 • By 2020, husbandry production pattern will be changed and grassland sustainability will be effectively enhanced • By 2020, fishing capacities and outputs will be corresponding with carrying capacities of fishery resources	• Provided subsidies to appropriate fertilizer use based on land size. • Undertaken eco-farming and established counties for demonstration in use of new rural energy. • Promoted development of organic farming. • Established eco-provinces, cities and counties. • Implemented natural forest resources protection projects. • Strengthened grassland ecosystem conservation and restoration	(1) A project to subsidize soil testing and formula development for proper fertilizer use has been initiated to address problems of overuse and blind use of fertilizers and low efficiency rate of fertilizer use. Through the project implementation China has basically obtained information concerning soil nutrients and fertilizer needs for main crops in all counties so as to develop a plan for reasonable fertilizer use and popularize techniques for proper fertilizer use, which is important for increasing food production, saving costs and controlling pollution. (2) Eco-farming initiatives and demonstration projects for rural new energy were undertaken to enhance capacities for sustainable agricultural development, with focus on reuse of straw, use of biogas and solar energy in rural areas and establishment of eco-farming bases, with a view to increasing eco-farming efficiency and farmers' income, and improving the rural environment. So far more than 41 million rural households have used biogas and more than 150 million people benefited from this. (3) China has actively promoted development of organic farming. By 2012, China has had 20,000 km^2 of land for eco-farming, ranking top in Asia. (4) 15 provinces (autonomous regions, province-level municipalities) have started their eco-province initiatives and 13 provinces have issued their programmes for eco-province building and more than 1,000 counties (cities, districts) have begun their eco-county development. Since 2005, 38 counties have been awarded as national-level eco-counties, and 1,559 eco-towns and 238 eco-villages have been established. (5) Since the natural forest protection project was implemented, logging has been reduced by 220 million m^3, and forest areas net increased by 100,000 km^2, forest coverage rate up by 3.8% and forest reserves increased by 725 million m^3. (6) Projects such as returning grazing land to grasslands were implemented. Compared with areas without such projects implemented, vegetation coverage of grasslands in project areas increased by 11%, grass height increased by 43.1%, fresh grass output increased by 50.7%. The situation of grassland use has considerably improved	Total growing stock volume Grass output from natural grasslands Impacts of agriculture, forestry and fishery on biodiversity	10.57 billion m^3 in 1988, 13.62 billion m^3 in 2003, and 14.55 billion m^3 currently Annual increase by 1.6% during 2005-2012

Global Targets	National Targets	National Actions	Outcomes Achieved	National Indicators	Overall Assessment and Trends
Target 8. By 2020, pollution, including from excess nutrients, has been brought to levels that are not detrimental to ecosystem function and biodiversity	• By 2015, the total amount of emission of main pollutants will be significantly reduced, with COD and SO_2 reduced by 8%, and ammonia and NO_x reduced by 10% compared with the levels of 2010 • By 2020, energy consumption and CO_2 emission per unit of GDP will decline significantly, with the total amount of main pollutants considerably reduced	• Significant reductions in total amount of main pollutants identified as one of binding targets for social and economic development and pollution reduction projects implemented • Environmental Impact Assessment (EIA) rules strictly implemented. • Undertaken control and comprehensive reuse of wastes	(1) Since 2006, emissions of COD from industrial waste water, SO_2 from effluent gas, soot, industrial dust and solid wastes have been going down. In the past decade, the intensity of pollutant emissions per unit of GDP dropped significantly by more than 55%. Since 2004 the intensity of CO_2 emission per unit of GDP dropped by 15.2%. (2) China has been implementing measures such as "limited approvals for certain regions and sectors". Since 2008, the central government has refused to approve 332 projects with a total investment of 1.1 trillion yuan RMB, all of which were projects of high pollution, energy consumption, resource use, low-level duplication and exceeding production capacities. These measures have played an important role in adjusting industrial structure and prioritizing economic growth. (3) National capacities of flue-gas desulfurization (FGD) units and percentage out of total thermal power capacities, the rate of treatment of municipal wastewater, and the rate of comprehensive reuse of industrial solid wastes have been increasing substantially, however the total amount of pollutant emission is still high and the rate of reuse of industrial solid wastes going down slightly in the past two years	Reduction in pollutant emissions	Overall reduction, however wastewater discharging still increasing
				Pollutant emission per unit of GDP	Down by more than 55% in the past decade
				CO_2 emission per unit of GDP	Down by 15.2% since 2004
				Percentage of FGD unit capacities out of total thermal power capacities	Increased from 48% in 2007 to 92% in 2012, with annual increase of 14% on the average
				Rate of municipal wastewater treatment	Increased from 45.6% in 2004 to 82.3% in 2010, with annual increase of 10.4% on the average
				Rate of comprehensive use of industrial wastes	Annual increase by 2.2% since 2004

Annex II Assessment of China's Progress in Implementing the Strategic Plan for Biodiversity 2011–2020 and the 2020 Biodiversity Targets

Global Targets	National Targets	National Actions	Outcomes Achieved	National Indicators	Overall Assessment and Trends
Target 9. By 2020, invasive alien species and pathways are identified and prioritized, priority species are controlled or eradicated, and measures are in place to manage pathways to prevent their introduction and establishment	• By 2020, forest pest disaster rate will be controlled at 4%	• Preliminarily identified a list of IASs for priority control. • Strengthened capacities for monitoring and early warning. • Undertaken activities to eliminate IASs. • Strengthened studies on prevention and control techniques. • Undertaken communication and education	(1) China has established a cross-sectoral coordinating group on prevention and control of IASs. 18 provinces (autonomous regions or province-level municipalities) have set up offices for IASs management or established joint conference mechanisms. China has developed guidelines for emergency responses to 40 major IASs, issued a second list of IASs and identified IASs for priority control. (2) China has improved its system of inspection and quarantine of imported and exported plants, established a network of monitoring and early warning of forest pests and agricultural IASs. (3) China has undertaken activities eliminating some 20 IASs such as ragweed, which has effectively controlled the expansion of IASs. (4) China has undertaken demonstration projects in surveying IASs and preventing and controlling IASs. (5) China has undertaken communication and education concerning techniques for preventing, controlling and managing IASs by using radios, TV, newspapers and internet. However the trend of increase in the number of IASs has not been effectively contained and the damages caused by IASs are being aggravated	Species of IASs newly found every twenty years	⊗ Number of new IAS increasing, with the total of 212 IASs entering China within 60 years after the 1950's accounting for 53.5% of the total IASs in China

Global Targets	National Targets	National Actions	Outcomes Achieved	National Indicators	Overall Assessment and Trends
Target 10. By 2015, the multiple anthropogenic pressures on coral reefs, and other vulnerable ecosystems impacted by climate change or ocean acidification are minimized, so as to maintain their integrity and functioning	• By 2020, energy consumption and CO_2 emission per unit of GDP will decline significantly • By 2020, a system of nature reserves with reasonable layouts and comprehensive functions will be established, with functions of national-level nature reserves stable, and main targets of protection effectively protected	• Adjusted industrial structure to enhance pollution abatement. • Implemented key forestry projects to protect vulnerable ecosystems. • Strengthened establishment and management of PAs and improved system of *in-situ* conservation	(1) The Government of China identified significant reductions in the total amount of main pollutants as one of binding targets for social and economic development. During 2000-2010, the concentration of main pollutants went down overall. In the past decade the density of pollutant emission per unit of GDP dropped by more than 55%. The density of CO_2 emission per unit of GDP has dropped by 15.2% since 2004. (2) China has implemented key forestry projects resulting in rapid growth in forest resources. The forest area has increased by 23% over that of a decade ago, and the forest coverage rate 3.8% upper over a decade ago. The forest growing sotck also grew by 21.8%. All this has effectively protected vulnerable ecosystems. (3) China has established a system of conservation consisting primarily of nature reserves and complemented by scenic spots, forest parks, community-based conservation areas, protected sites of wild flora, wetland parks, geological parks, special marine protected areas and protected areas for germplasm resources. By the end of 2013, China has established 2,697 nature reserves of various types at different levels, covering areas of 1.463 million km² and accounting for 14.8% of the country's land area	Pollutant emission per unit of GDP	Down by more than 55% in the past decade
				CO_2 emission per unit of GDP	Down by 15.2% since 2004
				Forest growing stock in areas where natural forest protection or returning cultivated land to forests are implemented	Constantly increasing
				Soil erosion in areas where natural forest protection or returning cultivated land to forests are implemented	Down in overall trend
				Biodiversity of coral reefs	
				Climate change impacts on biodiversity	

Annex II Assessment of China's Progress in Implementing the Strategic Plan for Biodiversity 2011-2020 and the 2020 Biodiversity Targets

Strategic Goals: C. Improve the status of biodiversity by safeguarding ecosystems, species and genetic diversity

Global Targets	National Targets	National Actions	Outcomes Achieved	National Indicators	Overall Assessment and Trends
Target 11. By 2020, at least 17 percent of terrestrial and inland water areas, and 10 percent of coastal and marine areas, especially areas of particular importance for biodiversity and ecosystem services, are conserved through effectively and equitably managed, ecologically representative and well connected systems of protected areas and other effective area-based conservation measures, and integrated into the wider landscapes and seascapes	• By 2015, the total area of terrestrial nature reserves will be maintained at 15% or so of the country's land area, protecting 90% of national key protected species and typical ecosystem types • The percentage of the area of marine protected areas out of the marine areas under China's jurisdiction will be increased from 1.1% in 2010 to 3% in 2015 • By 2020, a system of nature reserves with reasonable layouts and comprehensive functions will be established, with functions of national-level nature reserves stable, and main targets of protection effectively protected	• Strengthened *in-situ* conservation mainly through PAs. • Implemented fishing bans in the marine areas, the Yangtze River and the Pearl River to protect aquatic biodiversity	(1) China has established a system of conservation consisting primarily of nature reserves and complemented by scenic spots, forest parks, community-based conservation areas, protected sites of wild flora, wetland parks, geological parks, special marine protected areas and protected areas for germplasm resources. By the end of 2013, China has established 2,697 nature reserves of various types at different levels, covering areas of 1.463 million km² and accounting for 14.8% of the country's land area. Among PAs there are more than 240 marine PAs. By the end of 2012, China has established 2,855 forest parks, with total areas planned covering 174,000 km². China has also established 225 scenic spots covering area of 104,000 km², and more than 50,000 community-based conservation areas, covering an area of over 15,000 km². China has set up 179 national-level protected sites for agricultural wild flora and 468 national-level wetland parks. 368 national-level protected areas for aquatic germplasm resources have been established, covering areas of more than 152,000 km². However the representativeness and management effectiveness of PAs are yet to be improved, and the number and area of marine PAs are still low. (2) Since 1995, marine summer fishing bans have been implemented in the Bo Sea, Yellow Sea, East China Sea and South China Sea north of northern latitude 12°for about three months. Bans were also implemented in the Yangtze River Since 2002, and in the Pearl River since 2011, also for three months every year. The implementation of fishing bans or breaks has strongly helped replenishment of fishery resources and protected aquatic biodiversity	Number of PAs	◔ 606 PAs in 1990, 1,227 PAs in 2000 and 2,697 PAs in 2013, with annual increase of about 8.9% on the average
				Percentage of PA area of the country's total land area	◔ 4% in 1990, 9.9% in 2000 and 14.9% in 2012, with annual increase of 8.1% on the average
				Ecological representativeness of PAs	☹
				Management effectiveness of PAs	☹

Global Targets	National Targets	National Actions	Outcomes Achieved	National Indicators	Overall Assessment and Trends
Target 12. By 2020 the extinction of known threatened species has been prevented and their conservation status, particularly of those most in decline, has been improved and sustained	• By 2015, more than 80% of endangered species whose wild populations are very small and for which *in-situ* conservation capacities are inadequate will be effectively protected • By 2020, functions of national-level nature reserves will be maintained stable, and main targets of protection effectively protected • By 2020, the majority of rare and endangered species and populations will be restored and reproduced, relieving the situation of species endangerment	• Strengthened establishment and management of PAs. • Undertaking *ex-situ* conservation reasonably. • Strengthened research on endangered species. • Promoting international cooperation. • Undertaking public education	(1) For information concerning establishment and management of PAs please see details in this annex Target 11. (2) China has established 200 botanical gardens of various kinds at different levels that have collected and stored 20,000 species, accounting for the two-thirds of China's flora. China has also established more than 240 zoos, and 250 breeding bases for rescuing and reproducing wild animals. These *ex-situ* conservation facilities have played important roles in protecting endangered species. (3) For information concerning scientific research and international cooperation please see details in this annex Target 19. (4) For information concerning public education please see details in this annex Target 1	Red List Index	⊗ RLI of fresh water fish down during 1998-2004; RLI of mammals down during 1996-2008; RLI of birds slightly down during 1988-2012

Annex II Assessment of China's Progress in Implementing the Strategic Plan for Biodiversity 2011-2020 and the 2020 Biodiversity Targets

Global Targets	National Targets	National Actions	Outcomes Achieved	National Indicators	Overall Assessment and Trends
Target 13. By 2020, the genetic diversity of cultivated plants and farmed and domesticated animals and of wild relatives, including other socio-economically as well as culturally valuable species, is maintained, and strategies have been developed and implemented for minimizing genetic erosion and safeguarding their genetic diversity	• By 2020, biodiversity loss will be basically contained, and a system of nature reserves with reasonable layouts and comprehensive functions will be established, with main targets of protection effectively protected • National List of Protection of Livestock Genetic Resources will be revised so as to accord key protection to rare and endangered livestock genetic resources in the list and ensure that protected varieties will not be lost and their economic values will not be decreased	• Developed and implemented plans for protection of genetic resources. • Established *in-situ* conservation sites for genetic resources • Established storage banks for genetic resources(GRs), and undertaken studies on collection, storage and use of GRs	(1) China has developed and issued strategies related to genetic resources, including *National Programme for Conservation and Use of Biological Resources* and *National 12th Five-year Plan for Conservation and Use of Livestock Genetic Resources*. (2) China has established 179 national-level protected sites for agricultural wild plants and 368 national-level protected areas of aquatic germplasms, protecting a group of rare genetic resources. (3) China has established a system of storing crops genetic resources, with 423,000 accessions of crop GR stored. China has also established 150 national livestock seed conservation farms, protected areas and gene banks, effectively protecting more than 100 key livestock resources. China has set up germplasm banks for marine biological resources, such as a big germplasm bank for seaweed and a centre for storage of marine micro-organisms. Though China has done tremendous work in protecting genetic resources, the trend of loss of GRs has not been effectively contained	Number of local varieties	⊗ Trend of loss of GRs not yet effectively contained, according to estimates

Global Targets	National Targets	National Actions	Outcomes Achieved	National Indicators	Overall Assessment and Trends
Strategic Goals: D. Enhance the benefits to all from biodiversity and ecosystem services					
Target 14. By 2020, ecosystems that provide essential services, including services related to water, and contribute to health, livelihoods and well-being, are restored and safeguarded, taking into account the needs of women, indigenous and local communities, and the poor and vulnerable	• By 2020, the stability of ecosystems will be strengthened, and the human environment will be considerably improved • By 2020, grass-herd balance will be achieved in natural grasslands, grassland habitats will be obviously restored and grassland productivity will be significantly enhanced • By 2020, the environmental degradation of the coastal and near-shore marine areas will be reversed, and decline of marine biodiversity will be basically contained	• Strengthened establishment and management of PAs. • Implemented key forestry and ecological projects. • Undertaken conservation and restoration of grassland ecosystems. • Undertaken conservation and restoration of wetlands. • Undertaken restoration and rebuilding of coastal wetlands	(1) For information concerning establishment and management of PAs please see details in this annex Target 11. (2) For information concerning restoration of forest, grassland and wetland ecosystems, please see details in this annex Target 5. (3) As a result of national actions in conserving and restoring biodiversity, well-being of those communities that depend directly on local ecosystem goods and services is also improving. Forest growing stock have constantly been increasing since 1999 in those sample counties and enterprises where projects for protection of natural forest resources, returning cultivated land to forests and controlling areas of origin of sandstorms affecting Beijing and Tianjin have been implemented. Areas of soil erosion in these counties are also going down and net income per capita of rural households is going up rapidly, an increase of 40.8% in 2011 over that of 2000. This has benefited to some extent from increase in the functions of ecosystems to provide more goods and services. The number of poor people living in these areas is going down constantly, with the poor population in areas where natural forest protection implemented decreased to 1.83 million in 2011 from 3.95 million in 1997, and those in areas where returning cultivated land to forests implemented dropped from 8.3 million in 1998 to 5.7 million in 2008	Net income per capita of rural households	Constantly increasing
				Number of poor people in key ecological project areas	Constantly decreasing
				Forest growing stock	Constantly increasing since 1999
				Areas of soil erosion in key ecological project areas	Constantly decreasing

Annex II Assessment of China's Progress in Implementing the Strategic Plan for Biodiversity 2011-2020 and the 2020 Biodiversity Targets

Global Targets	National Targets	National Actions	Outcomes Achieved	National Indicators	Overall Assessment and Trends
Target 15. By 2020, ecosystem resilience and the contribution of biodiversity to carbon stocks has been enhanced, through conservation and restoration, including restoration of at least 15 per cent of degraded ecosystems, thereby contributing to climate change mitigation and adaptation and to combating desertification	• By 2020, forest areas will be increased by 52,000 km² over that in 2010, and forest growing stock net increased by 1.1 billion m³ over that in 2010, and forest carbon sinks by 416 million tons • By 2020, the total areas of control of degraded grasslands will exceed 1.65 million km², with grassland habitats obviously restored and grassland productivity significantly enhanced • By 2020, the aquatic environment and ecology will be gradually restored	• Implemented key forestry and ecological projects. • Undertaken conservation and restoration of grassland ecosystems. • Undertaken conservation and restoration of wetlands. • Undertaken restoration and rebuilding of coastal wetlands	For information concerning restoration of forest, grassland and wetland ecosystems, please see details in this annex Target 5	Forest growing stock	⊘ Constantly increasing since 1999
				Areas of soil erosion in key ecological project areas	⊘ Constantly decreasing

327

Global Targets	National Targets	National Actions	Outcomes Achieved	National Indicators	Overall Assessment and Trends
Target 16. By 2015, the Nagoya Protocol on Access to Genetic Resources and the Fair and Equitable Sharing of Benefits Arising from their Utilization is in force and operational, consistent with national legislation	• By 2020, the system of access to genetic resources and benefit-sharing from their use will be improved	• Promoted development of regulation on access to genetic resources and benefit-sharing. • Support provided to studies on mechanisms for access to genetic resources and benefit-sharing	With the support of relevant research plans China has strengthened information collection concerning access to genetic resources and benefit-sharing as well as studies on ABS mechanisms. China is currently promoting development of a regulation on ABS and ratification of the Nagoya Protocol on ABS	No	

Strategic Goals: E. Enhance implementation through participatory planning, knowledge management and capacity building

Global Targets	National Targets	National Actions	Outcomes Achieved	National Indicators	Overall Assessment and Trends
Target 17. By 2015 each Party has developed, adopted as a policy instrument, and has commenced implementing an effective, participatory and updated national biodiversity strategy and action plan	• Updated NBSAP has been launched	• Implementing updated NBSAP (2011-2030). • Provinces or cities developing their local BSAPs	The updated NBSAP (2011-2030) was approved by the State Council at its 126th regular meeting on 15 September 2010. The updated NBSAP was officially launched by MEP on 17 September 2010. This updated NBSAP reflects wide representativeness and participation, and is a result of joint efforts of all members of China's National Coordinating Group on Implementation of the CBD and members sitting on the Inter-ministerial Joint Conference on Conservation of Biological Resources. It was also an example for cooperation between domestic and international institutions/organizations. All the provinces (autonomous regions, province-level municipalities) are developing their local BSAPs, with 7 provinces having issued their BSAPs so far	Implementation of policies and programmes	◔

Global Targets	National Targets	National Actions	Outcomes Achieved	National Indicators	Overall Assessment and Trends
Target 18. By 2020, the traditional knowledge, innovations and practices of indigenous and local communities relevant for the conservation and sustainable use of biodiversity, and their customary use of biological resources, are respected, subject to national legislation and relevant international obligations, and fully integrated and reflected in the implementation of the Convention with the full and effective participation of indigenous and local communities, at all relevant levels	• By 2020, the documentation of relevant traditional knowledge within China and the intellectual rights protection system will be further improved	• Established relevant projects to compile and document traditional knowledge and to study system for intellectual property rights protection	The Government of China respects traditional knowledge and practice that people of various ethnicities have passed down from generation to generation. Relevant projects have been established to compile and document traditional knowledge and support provided to studies on the system of intellectual property rights protection	No	

Global Targets	National Targets	National Actions	Outcomes Achieved	National Indicators	Overall Assessment and Trends
Target 19. By 2020, knowledge, science base and technologies relating to biodiversity, its values, functioning, status and trends, and the consequences of its loss, are improved, widely shared and transferred, and applied	• By 2020, the percentage of investment in research and development activities will exceed 2.5% of national GDP, with the rate of contributions from science and technology to GDP reaching 60%, and the number of annual patent grants to domestic applicants and the citation of scientific papers by international journals ranking top five in the world. • Environmental education will be undertaken to popularize environmental knowledge and increase public environmental awareness	• Promoted scientific and technical research in the conservation and sustainable use of biodiversity. • Promoted international cooperation in the field of biodiversity conservation. • Great efforts made in communication and educational activities	(1) The Government of China encourages and supports scientific and technical research in the conservation and sustainable use of biodiversity. Projects for conservation and sustainable use of biodiversity have been included in the National Plan for Support to Science and Technology, National Programme for Development of Key Fundamental Research, National High-tech Development Plan, and specialized plans of National Natural Sciences Fund and support to public-benefit sectors. All these research activities have produced a series of valuable and influential research achievements, providing scientific and technical support to biodiversity conservation in China (see details in Section 3.3 of Part 3). (2) China has actively explored multilateral, bilateral and South-South cooperation and accomplished good results. China has been actively participating in negotiations related to the CBD, seriously implementing its obligations under the CBD and actively involved in the construction of the related multilateral systems. China has established with over 50 countries channels of cooperation and exchanges, creating a multiple cooperation system composed primarily of cooperation between Governments. China has actively undertaken South-South cooperation in the field of biodiversity and signed agreements of cooperation with many developing countries in the field of biodiversity and related areas. (3) China has made great efforts in undertaking biodiversity communication and education. As a result of these efforts, the public participation has increased and the public awareness of biodiversity conservation obviously increased, with the importance of biodiversity widely recognized (see details in this annex Target 1)	Academic papers on biodiversity published in different years	Gradually increasing year by year
				Items concerning biodiversity in China searched through the internet in different years	See details in this annex Target 1

Annex II Assessment of China's Progress in Implementing the Strategic Plan for Biodiversity 2011–2020 and the 2020 Biodiversity Targets

Global Targets	National Targets	National Actions	Outcomes Achieved	National Indicators	Overall Assessment and Trends
Target 20. By 2020, at the latest, the mobilization of financial resources for effectively implementing the Strategic Plan for Biodiversity 2011-2020 from all sources, and in accordance with the consolidated and agreed process in the Strategy for Resource Mobilization, should increase substantially from the current levels. This target will be subject to changes contingent to resource needs assessments to be developed and reported by Parties	• Channels of investment will be broadened and investments from local and central governments will be increased and financing from the banking sector, international donors and the civil society will be attracted to biodiversity conservation, with diverse financing mechanisms established	• Significantly increased domestic investments. • Provided as much assistance as China can to some developing countries	China has made huge investments into biodiversity conservation and details can be seen in this annex Target 3. China has provided as much assistance as it can to some developing countries	Investment in key forestry projects	⊙ See details in this annex Target 3

Notes: Overall assessment of national indicators: ⊙ Improving; ⟫ little or no change; ⊗ worsening; ⊙ no adequate data.

331

Annex III Implementation of the Programme of Work on Biodiversity of Arid and Semi-arid Lands

Global targets, sub-targets and activities	National Targets	National Actions	Outcomes Achieved	Global or national indicators used	Overall assessment
See details on the website of the "Convention on Biological Diversity"	By 2020, more than half of controllable desertified land across the country will be controlled, with desert ecology obviously improved, and an area of 200,000 km² of deserts controlled, with one half to be completed during 2011-2015 and another half during 2016-2020. (National Plan for Desertification Prevention and Control 2011-2020)	(1) Improved policies to support desertification prevention and control. China has developed and implemented a series of policies and measures to support ecology conservation and industry development in deserts, including reform in collective forest ownership, compensation for forest ecological benefits, and subsidies for forestry loans, reforestation and grassland ecology conservation. (2) Implemented key projects on desertification prevention and control. China continues implementing a series of key ecological projects, such as controlling areas of origin of sandstorms affecting Beijing and Tianjin, construction of forest belts in north, northwest and northeast China, returning cultivated land to forests and grazing land to grasslands, grassland conservation and comprehensive control in small river basins. China has also initiated a number of projects on regional desertification prevention and sand control such as those projects in Talimu Basin and Shiyanghe River Basin of Xinjiang and building ecological barriers in Tibet. These projects are intended to control key desertified areas and enhance ecological improvements in land degraded or desertified lands across the country. (3) Enhanced support to capacities for desertification prevention and control. One measure is to enhance scientific and technical capacities. SFA has established a research institute on desertification to strengthen scientific and technical support in this regard. Some research results such as "research on evolution of desertification and models of comprehensive control", and "studies on desertification processes in north China and prevention" have won national awards for scientific and technological progress. Some research results and applicable techniques are being promoted in wider areas. Another measure is to develop and improve relevant technical standards. A number of technical standards for desertification prevention and control have been developed and issued, such as Technical Guidelines for Desertification Prevention and Control, Technical Guidelines for Monitoring Land Degradation and Desertification and Technical Standards for Controlling Areas of Origin of Sandstorms Affecting Beijing and Tianjin. The third measure is to strengthen monitoring of land	The monitoring results show that during the eleventh five-year plan period (2006-2010) the average area of desertification is reduced by 1,717 km² annually. The total reduction within five years in areas of severely, medium and extremely severely desertified land is 36,000 km². This is an indication that desertification level is going down. Soil erosion in some areas has been effectively controlled. The soil erosion modulus is significantly reduced, with annual	Percentage and change in areas of desert ecosystems	Improving

degradation and desertification and emergency responses to sandstorms. The fourth national monitoring of land degradation has been completed and a system of emergency responses to major sandstorms has been established. A monitoring system of sandstorms is also in place, with main support from remote-sensing and on-the-ground monitoring, complemented by informers on the ground.

(4) Strengthened inter-sectoral coordination mechanisms for desertification prevention and control. From the central government to local governments, China has established specialized coordination and leading bodies for desertification prevention and control to strengthen the organization, leadership and coordination of desertification prevention and control.

(5) Implemented the responsibility system for achieving targets for desertification prevention and control. In accordance with requirements contained in the Desertification Prevention and Control Law, during the eleventh five-year plan period, SFA on behalf of the State Council has signed agreements of accountability for achieving the targets for desertification prevention and control with 12 provincial governments of north China and Xinjiang Production Corp. The establishment and implementation of such responsibility system for the first time makes provincial governments accountable for achieving targets for desertification prevention and control, and helps local governments of different levels upgrade their sense of responsibility for desertification prevention and control, and promote work in this regard across the country.

(6) Encouraging industry development unique to desert areas. To promote industry development unique to desert areas, SFA has developed recommendations for further developing industry in desert areas, which require governments of levels to promote industry development in accordance with local conditions, with a view to improve local farmers' income and promote economic development in desert areas

erosion of yellow sand cut by more than 300 million tons every year. For example in areas of origin of sandstorms affecting Beijing and Tianjin, according to expert estimate, since the project was initiated a decade ago, the soil erosion modulus in the project area went down by 68.9% on the average, with areas affected by soil erosion down by 39.1%, and the total soil erosion amount down by 29% and the total amount of dust release down by 16.2%

Annex IV Implementation of the Programme of Work on Protected Areas

The reporting framework facilitates capturing the progress in completing assessments on 13 key goals of the programme of work on protected areas (PoWPA) and specific actions taken to implement the results of those assessments, in order to indicate the status of implementation. Progress in completing these assessments is measured from 0~4 (0 – no progress; 1 – planning phase; 2 – initial progress; 3 – substantial progress; 4 – nearly or fully completed). The framework allows Parties to append the results of these assessments, and to optionally describe specific actions taken in three time lines (before 2004; between 2004—2009; and since 2010). If a question is not applicable, the letters N/A should be entered. The PoWPA focal points could upload the information on the CBD website as and when they have undertaken and completed the assessment or following the reporting cycle of national reporting through a user ID and password.

COUNTRY:	The People's Republic of China
Name of person completing survey:	
Email address of person completing survey:	
Date survey completed:	(DATE)
Please briefly describe who was involved in gathering information for this survey	(NAMES AND ORGANIZATIONS)
1) Has a multi-stakeholder advisory committee been formed to implement the PoWPA?	YES
2) Is there an action plan for implementing the PoWPA?	YES
3) If yes, please provide a URL (or attach a PDF) of the strategic action plan:	http://www.zhb.gov.cn/gkml/hbb/bwj/201009/t20100921_194841.htm
4) If yes, which is the lead agency responsible for implementing the action plan?	The Ministry of Environmental Protection
5) If not, are the PoWPA actions included in other biodiversity-related action plans? (please provide a URL or attachment if so)	

1.1 To establish and strengthen national and regional systems of protected areas integrated into a global network as a contribution to globally agreed goals	
1) What progress has been made in assessing the representativeness, comprehensiveness and ecological gaps of your protected area network?	3
2) If available, please indicate the URL (or attach a PDF) of the gap assessment report:	(URL OR ATTACHMENT)
3) Do you have specific targets and indicators for the protected area system?	YES
4) If yes, please provide a URL (or attach a PDF) of the targets and indicators:	http://www.zhb.gov.cn/gkml/hbb/bwj/201009/t20100921_194841.htm

Annex IV   Implementation of the Programme of Work on Protected Areas

5) What actions have been taken to improve the ecological representativeness of the protected area network? Please check all that apply, and provide a brief description:

√	Action	Before 2004	Between 2004-2009	Since 2010
	Created new protected area/s	√	√	√
	Promoted an array of different types of protected areas (e.g., different IUCN Categories, CCAs etc)			
	Expanded and/or reconfigured existing protected area boundaries	√	√	√
	Changed the legal status and/or governance type of protected areas	√	√	√
	Other actions to improve the representativeness and comprehensiveness of the network	√	√	√

To improve the network of protected areas and their ecological representativeness, China had developed national plans for development of protected areas in different periods of time, which identified goals and requirements for spatial layouts, establishment and management of protected areas. Before 2004, China had a total of 1,999 protected areas of various categories and at different levels, with total areas covered by PAs reaching 1.44 million km^2, accounting for about 14.4% of the country's total land area. Among them there were 226 national-level PAs. During 2004-2009, China established 542 new PAs, 93 of which were national-level PAs. From 2010 to the end of June 2013, China has established 128 more PAs, 65 of which are national-level PAs, with total area accounting for 14.9% of the country's total land area.

1.2   To integrate protected areas into broader landscapes- and seascapes and sectors so as to maintain ecological structure and function

1) What progress has been made in assessing protected area landscape and seascape connectivity and sectoral integration?	3
2) If available, please indicate the URL (or attach a PDF) of the assessment of protected area connectivity and sectoral integration:	

3) What actions have been taken to improve protected area connectivity and sectoral integration? Please check all that apply, and provide a brief description:

√	Action	Before 2004	Between 2004-2009	Since 2010
	Changed the legal status and/or governance in key connectivity areas	√	√	√
	Created new protected areas in key connectivity areas	√	√	√
	Improved natural resource management to improve connectivity	√	√	√
	Designated connectivity corridors and/or buffers	√	√	√
	Created market incentives to promote connectivity	√	√	√
	Changed awareness of key stakeholders in key connectivity areas	√	√	√

Improved laws and policies within or around key connectivity areas	√	√	√
Restored degraded areas in key connectivity areas	√	√	√
Changed land use planning, zoning and/or buffers in key connectivity areas	√	√	√
Removed barriers to connectivity and ecological functioning	√	√	√
Integrated protected areas into poverty reduction strategies	√	√	√
Other actions to improve connectivity and integration	√	√	√

To improve the network of protected areas and their ecological representativeness, China had developed national plans for development of protected areas in different periods of time, which identified requirements for spatial layouts of PAs and establishment of ecological corridors. These requirements have been incorporated into relevant sectoral development plans. For example, to implement the national plan for wild flora and fauna protection and protected areas, since 2006 China has invested 2.6 billion yuan RMB into establishment and management of PAs. China has also implemented a project to protect Giant Pandas and their habitats, with a network of Giant Panda protection established in Sichuan, Shaanxi and Gansu Provinces. By 2010, Sichuan Province has invested 200 million yuan RMB which resulted in the increase in the number of Giant Pandas to 41 and the area covered up to 23,000 km^2, protecting more than 50% of the habitats for Giant Panda in the province and more than 60% of wild Giant Pandas in PAs. With the support of Word Wildlife Fund (WWF) and other international organizations, Sichuan Province has implemented a number of projects on ecological corridors and community development to enhance connectivity between PAs and their management effectiveness through implementing relevant management plans and having initiated many poverty reduction programmes. In recent years, China has worked with Myanmar, Vietnam and Lao DPR on the Biodiversity Corridors of the Mekong River sub-region. The cooperation has achieved substantive progress in collaboration in PAs, personnel training, fire prevention in border areas and transboundary protection of Asian elephants. At the end of 2010, the State Council issued a notice on management of protected areas, which required that development activities affecting protected areas would be strictly limited and those activities within PAs would be monitored, inspected and well managed. Currently China is developing another national plan for protected areas, which will propose new requirements for spatial layouts and management of PAs. This plan will be incorporated into broader national plan for social and economic development for implementation.

1.3 To establish and strengthen regional networks, transboundary protected areas (TBPAs) and collaboration between neighbouring protected areas across national boundaries	
1) What progress has been made in identifying conservation priorities and opportunities for establishing transboundary protected areas and regional networks?	3

2) If available, please indicate the URL (or attach a PDF) of the assessment of opportunities for transboundary protected areas and regional networks.	http://politics.people.com.cn/ GB/1026/10568760.html

3) What actions have been taken to strengthen the regional protected area network and foster transboundary PAs? Please check all that apply, and provide a brief description:

√	Action	Before 2004	Between 2004-2009	Since 2010
	Created transboundary protected area/s	√	√	√
	Contributed to the creation of regional-scale conservation corridors	√	√	√
	Participated in the establishment of regional networks	√	√	√
	Created enabling policies to allow for transboundary protected areas	√	√	√
	Established a multi-country coordination mechanism	√	√	√
	Other actions to foster regional networks and transboundary areas	√	√	√

In 1994, China signed with Russia and Mongolia an agreement on joint protected areas, under which the three countries have been undertaking many activities such as joint surveys and monitoring, environmental education and exchange of experiences. Since 2006, China and Russia have established an intergovernmental working group on transboundary protected areas and biodiversity conservation, which meets on a regular basis every year. So far this working group has had six meetings. Both sides have signed cooperation agreements such as China-Russia strategy for development of networks of transboundary protected areas in Heilongjiang River Basin and China-Russia agreement on establishment of protected areas in Xingkai Lake. with regard to PAs in Sanjiang, Honghe and Bachadao in Heilongjiang Province, China have signed agreements with PAs in Basdak, Daherchel, Xinganski and Bolongski in Russia for cooperation. In 2013, China and Russia signed an agreement on protection of wild tigers. By this agreement both sides will accelerate the construction of migratory corridors for tigers and establish protected areas for tigers in border mountain areas. Both sides will also deepen cooperation in joint monitoring and research of wild animals, environmental communication and education, legislation and law enforcement related to PAs, eco-tourism planning and management.

In 2009, China and Lao DPR established the first transboundary protected areas-Shangyong-Xishuangbanna-South Tananmuha, to better protect Asian elephants and other migratory animals. In early 2012 the two countries signed a second agreement on establishing another transboundary protected area-Menglamanzhuang, China-Fengshali, Lao DPR border areas. In December 2012, the two countries signed one more agreement for another transboundary PA. So three transboundary PAs cover Xishuangbanna, China and the three northern provinces of Lao DPR.

In recent years, China has worked with Myanmar, Vietnam and Lao DPR on the Biodiversity

Corridors of the Mekong River sub-region. The cooperation has achieved substantive progress in collaboration in PAs, personnel training, fire prevention in border areas and transboundary protection of Asian elephants.

1.4 To substantially improve site-based protected area planning and management	
1) What progress has been made in developing protected area management plans?	2
2) What percentage of your protected areas has an adequate management plans?	
3) What percentage of the total surface area of protected areas does the management plans cover?	
4) Please provide a URL (or PDF attachment) of a recent example of a participatory, science-based management plan	http://www.doc88.com/p-18967633517.html
5) What actions have been taken to improve protected area management planning? Please check all that apply, and provide a brief description:	

√	Action	Before 2004	Between 2004—2009	Since 2010
	Developed guidelines and tools for developing management plans			
	Provided training and/or technical support in management planning	√	√	√
	Developed management plans for protected areas	√	√	√
	Changed legislation or policy to strengthen management planning	√	√	√
	Improved the scientific basis of existing management plans	√	√	√
	Conducted protected area resource inventories	√	√	√
	Other actions to improve management planning	√	√	√

The "Programme for Master Planning of National-level Nature Reserves" (2002), "Technical Guidelines for Master Planning of Nature Reserves" (2006) and "Technical Guidelines for Eco-Tourism Planning for Nature Reserves" (2006) have provided guidelines, procedures and specific requirements for master planning of nature reserves and eco-tourism planning.

The Guidelines for Management and Standardized Construction of National-level Nature Reserves (Provisional) (2009) proposed requirements for establishment and management of protected areas.

1.5 To prevent and mitigate the negative impacts of key threats to protected areas	
1) What progress has been made in assessing the status of protected area threats, and opportunities for mitigation, prevention and restoration?	3
2) If available, please indicate the URL (or attach a PDF) of the assessment of the status of threats and opportunities for mitigation, prevention and restoration.	
3) What actions have been taken to mitigate or prevent protected area threats, or restore degraded areas? Please check all that apply, and provide a brief description	

Annex IV Implementation of the Programme of Work on Protected Areas

√	Action	Before 2004	Between 2004—2009	Since 2010
	Changed the status and/or governance type of a protected area	√	√	√
	Improved staffing numbers and/or skills to prevent and mitigate threats	√	√	√
	Included measures to address threats in a management plan	√	√	√
	Improved management practices to prevent or mitigate threats	√	√	√
	Increased threat mitigation funding	√	√	√
	Developed a plan to address the impacts of climate change	√	√	√
	Changed market incentives to reduce or prevent threats	√	√	√
	Improved monitoring and detection of threats	√	√	√
	Evaluated the efficacy of threat-related actions	√	√	√
	Improved public awareness and behaviour regarding threats	√	√	√
	Changed laws and policies related to threats	√	√	√
	Restored degraded areas	√	√	√
	Developed and/or implemented strategies to mitigate threats	√	√	√
	Other actions to mitigate and prevent threats	√	√	√

In 2004, the State Environmental Protection Administration issued a notice on strengthening management of nature reserves. The notice required that in undertaking environmental impact assessments of all development projects affecting nature reserves, one special chapter should be included in the assessment report that will predict impacts on the structure, functions and targets of protection and their values, and propose measures for protection (in case projects are approved) as well as how the project implementers can protect, restore and compensate, based on impacts of the project. In 2008, the Ministry of Environmental Protection, together with other relevant ministries, issued another notice in this regard, stressing that these requirements must be met and negative impacts from development projects on PAs must be prevented. In 2011, the Ministry of Environmental Protection issued guidelines for supervision over PAs, which regulates the monitoring and supervision in PAs. MEP together with other relevant departments has organized many inspections on law enforcement in PAs to prevent damage from irrational development activities to nature reserves.

2.1 To promote equity and benefit-sharing	
1) What progress has been made in assessing the equitable sharing of costs and benefits of establishing protected areas?	2
2) If available, please indicate the URL (or attach a PDF) of the assessment of equitable sharing of costs and benefits of establishing protected areas.	(URL OR ATTACHMENT)

3) What actions have been taken to improve equitable benefits sharing? Please check all that apply, and provide a brief description

√	Action	Before 2004	Between 2004-2009	Since 2010
	Developed compensation mechanisms		√	√
	Developed and/or applied policies for access and benefit sharing			
	Developed equitable benefits-sharing mechanisms			
	Diverted PA benefits towards poverty alleviation	√	√	√
	Other actions to strengthen equitable benefit—sharing			

4) What progress has been made in assessing protected area governance?	3
5) What percentage of protected areas has been assigned an IUCN category?	(%)
6) If available, please indicate the URL (or attach a PDF) of the assessment of protected area governance:	http://www.cnki.com.cn/Article/CJF/DTotal-LDGH201006014.htm

7) What actions have you taken to improve and diversify governance types? Please check all that apply, and provide a brief description

√	Action	Before 2004	Between 2004-2009	Since 2010
	Created new protected areas with innovative forms of governance, such as community conserved areas		√	√
	Changed laws or policies to enable new governance types			
	Other actions to diversify governance types			

China has implemented natural forest resources protection projects, established Fund for Compensation for Forest Ecological Benefits and some PAs got support from these projects and funds. China has established funds to be transferred to national key ecological function zones. In 2012, funds were transferred to 466 counties (cities, districts), with the total funds reaching 37.1 billion yuan RMB. The funds played an important role in supporting development of PAs. In 2007, the Ministry of Environmental Protection issued guidance for undertaking pilot work in ecological compensation, which required acceleration of the establishment of ecological compensation mechanisms for PAs. Some regions have established ecological compensation mechanisms on a pilot basis. For example, Jining City has issued rules for ecological compensation for wetland loss in Nansi Lake, Shandong Province.

Besides nature reserves, China has established other types of PAs such as forest parks, scenic spots, wetland parks, geological parks, community-based conservation areas and protected sites for wild agricultural plants (see details in section 2.3.4 of Part 2). These PAs are important components of China's PA networks.

2.2 To enhance and secure involvement of indigenous and local communities and relevant stakeholders

1) What is the status of participation of indigenous and local communities and other key stakeholders in key protected area decisions?		2

2) What actions have been taken to improve indigenous and local community participation? Please check all that apply, and provide a brief description:

√	Action	Before 2004	Between 2004-2009	Since 2010
	Assessed opportunities and needs for local community participation in key protected area decisions			
	Improved laws, policies and/or practices to promote participation	√	√	√
	Developed policies for prior informed consent for resettlement	√	√	√
	Improved mechanisms for participation of indigenous and local communities	√	√	√
	Increased participation of indigenous and local communities in key decisions	√	√	√
	Other actions to promote participation			

China has established public hearing and notice systems and mechanisms for public participation in environmental impact assessments. China has also strengthened capacities for minorities and local communities to allow them to participate effectively in relevant decision/policy making and planning processes.

3.1 To provide an enabling policy, institutional and socio-economic environment for protected areas

1) What progress has been made in assessing the policy environment for creating and managing protected areas?	3
2) If available, please indicate the URL (or attach a PDF) of the assessment of the policy environment:	(URL OR ATTACHMENT)

3) What actions have been taken to improve the protected area policy environment? Please check all that apply, and provide a brief description:

√	Action	Before 2004	Between 2004-2009	Since 2010
	Harmonized sectoral policies or laws to strengthen management effectiveness	√	√	√
	Integrated PA values and ecological services into the national economy	√	√	√
	Improved accountability and/or participation in decision-making	√	√	√
	Developed incentive mechanisms for private protected areas			

	Developed positive market incentives to support protected areas	√	√	√
	Removed perverse incentives that hinder effective management	√	√	√
	Strengthened laws for establishing or managing protected areas	√	√	√
	Cooperated with neighboring countries on transboundary areas	√	√	√
	Developed equitable dispute resolution mechanisms and procedures	√	√	√
	Other actions to improve the policy environment	√	√	√
4) What progress has been made in assessing the contribution of protected areas to the local and national economies?				3
5) What progress has been made in assessing the contribution of protected areas to the Millennium Development Goals?				3
6) If available, please indicate the URL (or attach a PDF) with the assessment of the contribution of protected areas to the local and national economy and to the Millennium Development Goals:				http://www.cnki.com.cn/Article/CJFDTotal-SAHG201206043.htm

7) What actions have been taken to value the contribution of protected areas? Please check all that apply, and provide a brief description:

√	Action	Before 2004	Between 2004-2009	Since 2010
	Implemented a communication campaign to encourage policy makers to recognize the value of protected areas	√	√	√
	Created finance mechanisms linked to protected area values (e.g., payment for ecosystem services)			

In 1997, China issued *National Programme for Development of Nature Reserves* (1996-2010), and goals identified therein have been achieved. In 2003, the State Council approved *National Plan for Wetland Conservation* (2002-2030), and during the eleventh five-year plan period, a total investment of 3.03 billion yuan RMB was made and the tasks planned for the twelfth five-year plan period are being implemented. In 2000 China developed a *National Plan for Wild Flora and Fauna Conservation and Nature Reserves*, which gave priority support to establishment of nature reserves. Currently China is developing national plan for development of nature reserves, which will be submitted to the State Council for approval and incorporated into national plans for social and economic development for different periods of time.

3.2 To build capacity for the planning, establishment and management of protected areas	
1) What progress has been made in assessing protected area capacity needs?	3
2) If available, please indicate the URL (or attach a PDF) of the assessment of capacity needs:	http://www.cnki.com.cn/Article/CJFDTotal-BJLY2011S2012.htm

Annex IV Implementation of the Programme of Work on Protected Areas

3) What actions have been taken to strengthen protected area capacity? Please check all that apply, and provide a brief description:

√	Action	Before 2004	Between 2004-2009	Since 2010
	Created a professional development programme for protected area staff	√	√	√
	Trained protected area staff in key skills	√	√	√
	Increased the number of protected area staff	√	√	√
	Developed a system for valuing and sharing traditional knowledge	√	√	√
	Other actions to improve capacity	√	√	√

Since 1998, the Ministry of Finance has established Specialized Funds for Capacity Building of National-level Nature Reserves. By 2012, the cumulative investment has reached 790 million yuan RMB, which is devoted to strengthening management, conservation, research and educational capacities of PAs. These investments played a very positive role in upgrading management level of PAs. Since 2008, China has also established Specialized Funds for Capacity Building of PAs managed by the forestry sector.

Beijing, Inner Mongolia, Heilongjiang, Zhejiang, Jiangxi, Fujian, Shandong, Hunan, Guangdong, Ningxia and other provinces (autonomous regions or province-level municipalities) have established their own specialized funds for protected areas. From 2000 to 2009 Guangdong Province invested more than 300 million yuan RMB into protected areas. Fujian Province increased standards for ecological compensation as well as investments into infrastructure of nature and forest reserves at provincial and above levels, and strengthened management of protected areas.

The departments of the environment, forestry and agriculture responsible for management of nature reserves have organized many training workshops on nature reserve management, focusing on relevant policies and regulations, standardized management, plan development, capacity building project design, supervision of development activities, establishment of management information systems and survey of status of biological resources.

With the support from the GEF, WWF and other international organizations, China has implemented projects on management of nature reserves, conservation and sustainable use of wetland biodiversity and sustainable forest development. These projects have played a very important role in strengthening capacities for nature reserves.

3.3 To develop, apply and transfer appropriate technologies for protected areas	
1) What progress has been made in assessing the needs for relevant and appropriate technology for protected area management?	3
2) If available, please indicate the URL (or attached a PDF) of the assessment of the technology needs:	(URL OR ATTACHMENT)
3) What actions have been taken to improve the access to and use of relevant and appropriate technology? Please check all that apply, and provide a brief description:	

√	Action	Before 2004	Between 2004-2009	Since 2010
	Developed and/or used appropriate technology for habitat restoration and rehabilitation	√	√	√
	Developed and/or used appropriate technology for resource mapping, biological inventories and rapid assessments	√	√	√
	Developed and/or used appropriate technology for monitoring	√	√	√
	Developed and/or used appropriate technology for conservation and sustainable use	√	√	√
	Encouraged technology transfer and cooperation between protected areas and agencies	√	√	√
	Other actions to improve access to and use of appropriate technologies	√	√	√

The Ministry of Science and Technology has established projects such as "research and demonstration on key techniques for development of protected areas". The study will be undertaken from six aspects: PA system establishment, function zoning, habitat quality and dynamic monitoring of biological resources, protection of endangered species, restoration of affected ecosystems and sustainable use of biological resources and suitable business. These studies will provide technical support to development of PAs in China. The departments of the environment, agriculture and forestry responsible for PAs have been promoting and improving various techniques and innovative approaches for effective management of PAs as well as exchanges of the experiences in this regard, through training activities and meetings. Various PAs have obtained and extensively used techniques for survey, monitoring, conservation and management and upgraded their management level, through collaboration with institutions of higher education, research institutes and NGOs.

3.4 To ensure financial sustainability of protected areas and national and regional systems of protected areas	
1) What progress has been made in assessing protected area finance needs?	3
2) If available, please indicate the URL (or attach a PDF) of the assessment of finance needs:	(URL OR ATTACHMENT)
3) What progress has been made in developing and implementing a sustainable finance plan that incorporates a diversified portfolio of financial mechanisms?	2
4) If available, please indicate the URL (or attach a PDF) of the sustainable finance plan:	(URL OR ATTACHMENT)
5) What actions have been taken to improve the sustainable finance of your protected areas? Please check all that apply, and provide a brief description:	

√	Action	Before 2004	Between 2004-2009	Since 2010
	Developed new protected area funding mechanisms	√	√	√

	Developed protected area business plan or plans	√	√	√
	Developed revenue-sharing mechanism			
	Improved resource allocation procedures	√	√	√
	Provided financial training and support			
	Improved accounting and monitoring	√	√	√
	Improved financial planning capacity			
	Removed legal barriers to sustainable finance			
	Clarified inter-agency fiscal responsibilities			
	Other actions to improve sustainable finance			

Please see details related to 2.1 and 3.2 above.

3.5 To strengthen communication, education and public awareness

1) What progress have you made in conducting a public awareness and communication campaign?	3
2) If available, please indicate the URL (or attach a PDF) of the public awareness and communication plan:	http://www.ynly.gov.cn/news/200810/12064.shtml

3) What actions have you taken to improve public awareness and strengthen education programmes? Please check all that apply, and provide a brief description:

√	Action	Before 2004	Between 2004-2009	Since 2010
	Identify core themes for education, awareness and communication programmes relevant to protected areas	√	√	√
	Conducted an awareness campaign on the value of protected areas to local and national economies and the Millennium Development Goals	√	√	√
	Conducted an awareness campaign on the value of protected areas in climate change adaptation and mitigation	√	√	√
	Established or strengthen communication mechanisms with key target groups, including indigenous and local communities	√	√	√
	Developed protected area curricula with educational institutions	√	√	√
	Produced public outreach materials	√	√	√
	Conducted public outreach programmes	√	√	√
	Other actions to improve communication, education and awareness	√	√	√

China encourages and requires PAs to undertake extensive communication and educational activities to increase public recognition of the importance and benefits of PAs. They put communication boards and slogans in PAs and disseminate PA brochures to tourists and local communities. Various departments and local governments organize various kinds of communication and educational activities to introduce the value and importance of PAs, through activities celebrating biodiversity-related dates such as the International Day of Biodiversity and the Earth Day.

4.1 To develop and adopt minimum standards and best practices for national and regional protected area systems

1) What progress has been made in developing best practices and minimum standards?	3
2) If available, please indicate the URL (or attach a PDF) of examples of protected area best practices and minimum standards:	
3) Is there a system in place for monitoring protected area outcomes achieved through the programme of work on protected areas:	YES

4) What actions have been taken related to best practices and minimum standards? Please check all that apply, and provide a brief description:

√	Action	Before 2004	Between 2004-2009	Since 2010
	Developed standards and best practices for protected area establishment and selection	√	√	√
	Developed standards and best practices for protected area management planning	√	√	√
	Developed standards and best practices for protected area management	√	√	√
	Developed standards and best practices for protected area governance	√	√	√
	Collaborated with other Parties and relevant organizations to test, review and promote best practices and minimum standards	√	√	√
	Other actions related to best practices and minimum standards	√	√	√

In 1999, China developed standards for approving national-level protected areas, in which indicators for planning and management of national-level PAs were proposed.

In 2002, China issued a programme for master planning of national-level PAs to guide the development and implementation of master planning for national-level PAs.

In 2006, China issued technical procedures for master planning of PAs and undertaking eco-tourism in PAs, which provided basic guidelines for PA planning and eco-tourism planning.

In 2009, China issued provisional guidelines for standardized construction and management of national-level PAs, which further regulates the development and management of national-level PAs.

In 2010, China issued guidelines for undertaking scientific research and tours in PAs to regulate such activities in PAs.

4.2 To evaluate and improve the effectiveness of protected areas management

1) What progress has been made in assessing the management effectiveness of protected areas?	3
2) If available, please indicate the URL (or attach a PDF) of the assessment of protected area management effectiveness:	http://www.zhb.gov.cn/gkml/hbb/bgt/201005/w020100524534788478025.pdf
3) In what percentage of the total area of protected areas has management effectiveness been assessed?	62.9%

Annex IV   Implementation of the Programme of Work on Protected Areas

	4) In what percentage of the number of protected areas has management effectiveness been assessed?		13.6%	

5) What actions have been taken to improve management processes within protected areas? Please check all that apply, and provide a brief description:

√	Action	Before 2004	Between 2004-2009	Since 2010
	Improved management systems and processes	√	√	√
	Improved law enforcement	√	√	√
	Improved stakeholder relations	√	√	√
	Improved visitor management	√	√	√
	Improved management of natural and cultural resources	√	√	√
	Other actions to improve management effectiveness	√	√	√
	Have you submitted management effectiveness results to UNEP-WCMC's WDPA	√	√	√

For information concerning the improvement of PA management, please see details in 1.4 above.

Since 2008, the Ministry of Environmental Protection and six other central government departments have jointly organized assessments of management effectiveness of national-level PAs, putting PA management on more standardized track. By 2012, assessments have been completed for more than 300 national-level PAs. MEP and other departments have also inspected law enforcement in PAs to prevent damage from irrational development activities to PAs.

4.3   To assess and monitor protected area status and trends

1) What progress has been made in establishing an effective monitoring system of protected area coverage, status and trends?	3
2) If available, please indicate the URL (or attach a PDF) of a recent monitoring report:	http://www.shidi.org/sf_A4B06758596347D2A155665A2331390C_151_pyh.html

3) What actions have been taken to improve protected area monitoring? Please check all that apply, and provide a brief description:

√	Action	Before 2004	Between 2004-2009	Since 2010
	Assessed the status and trend of key biodiversity	√	√	√
	Monitored the coverage of protected areas	√	√	√
	Developed or improved a biological monitoring programme	√	√	√
	Developed a database for managing protected area data	√	√	√
	Revised management plan based on monitoring and/or research results	√	√	√
	Changed management practices based on the results of monitoring and/or research	√	√	√

Developed geographic information systems (GIS) and/or remote sensing technologies	√	√	√
Other monitoring activities	√	√	√

China encourages and promotes monitoring of PAs. The Ministry of Science and Technology has established key projects such as "monitoring of important biological resources and key techniques for their conservation and demonstration application of these techniques". These projects aim to design monitoring networks and strengthen research on various standards of monitoring. In 2004, the State Oceanic Administration issued "Technical Guidelines for Monitoring of Marine PAs-General Principles", which provided the content, technical requirements and methods for monitoring MPAs. The Ministry of Environmental Protection is developing technical guidelines for monitoring of species. In 2011, Henan Province Forestry Department issued a provisional programme for research and monitoring in PAs, which further regulates relevant work in PAs in the province.

Since 2009, the Ministry of Environmental Protection initiated the first baseline survey of PAs in China, with a view to identifying all types of PAs at all levels and their status and management. In 2011, MEP established a remote-sensing monitoring system of PAs by using the environment satellite, which can monitor in time through satellite remote-sensing and undertake on-site inspection based on information provided by the satellite remote-sensing. An integrated monitoring of PAs from the sky and on the ground has been established. In 2012, MEP and CAS initiated a Project on Remote-sensing and Assessment of Ecological Changes in the decade 2000-2010. In this project, there is a sub-project on changes in national-level PAs, which will study the environmental issues and threats/pressures/drivers faced by more than 300 national-level PAs, and comprehensively assess the effectiveness of PAs in China.

Most of PAs in China have certain capacities for monitoring, with some of them undertaking long-term monitoring of biodiversity. PAs such as those in Changbai Mountain, Dongling Mountain, Shennongjia, Gutian Mountain, Dinghu Mountain and Xishuangbanna have established big sample sites for monitoring.

4.4 To ensure that scientific knowledge contributes to the establishment and effectiveness of protected areas and protected area systems			
1) What progress has been made in developing an appropriate science and research programme to support protected area establishment and management?	3		
2) If available, please indicate the URL (or attach a PDF) of a recent research report:			
3) What actions have been taken to improve protected area research and monitoring? Please check all that apply, and provide a brief description:			

√	Action	Before 2004	Between 2004-2009	Since 2010
	Identified key research needs	√	√	√
	Assessed the status and trends of key biodiversity	√	√	√

Developed or improved a biological monitoring programme	√	√	√
Conducted protected area research on key socio-economic issues	√	√	√
Promoted dissemination of protected area research	√	√	√
Revised management plan based on monitoring and/or research results	√	√	√
Changed management practices based on the results of monitoring and/or research	√	√	√
Other research and monitoring activities	√	√	√

After more than 50 years of surveys of China's biota and more than 100 years of information collection, the Chinese Academy of Sciences has published *Flora of China*, *Fauna of China*, *China Spore Plants Annals* and large volumes of local plant and animal annals. All these publications provide scientific basis for establishment of PAs and improvement of their management.

To further develop science related to PAs, Beijing University of Forestry has established an institute of PAs; Nanjing Institute of Environmental Sciences under MEP has established a research centre on PAs, and the Forestry Department of Guangdong Province and South China University of Agriculture have jointly established Guangdong Provincial Research Centre on PAs. All these research centres and institutes have made considerable progress in sciences related to PAs.

With the support of the GEF, China has implemented projects on PA management, conservation and sustainable use of wetland biodiversity, and sustainable forest management. These projects help introduce advanced concepts and methods of nature conservation from other countries.

The departments of the environment, agriculture and forestry responsible for PAs have organized many training workshops and seminars to promote theory, technical and innovative approaches for improving PA management, and to help upgrade PA management level.

Annex V Implementation of the Capacity-building Strategy for the Global Taxonomy Initiative and the Global Strategy for Plant Conservation

Relevant COP decisions, programmes of work and suggested activities	National Implementation and Contributions	Assessment of Progress
The Capacity-building Strategy for the Global Taxonomy Initiative(GTI)		
Action 1: By the end of 2013, at the latest, review taxonomic needs and capacities at national, subregional and regional levels and set priorities to implement the Convention and the Strategic Plan for Biodiversity 2011-2020	A preliminary assessment was made of capacities for Chinese and Asian plant taxonomy. A report on progress in plant conservation in Asia-the implementation of the Global Strategy for Plant Conservation in Asia was published in 2011, however needs assessments are yet to be undertaken	Partially completed
Action 2: By the end of 2013, organize regional and subregional workshops aimed at informing Parties and their CBD/GTI national focal points, representatives of ministries of science, education and conservation, and other relevant sectors about the importance of taxonomy and the need for cooperation in this field to implement the Convention and the Strategic Plan for Biodiversity 2011-2020	China has actively undertaken training activities related to taxonomy and introduced the importance of taxonomy to biodiversity conservation. For example, the Institute of Botany of the Chinese Academy of Sciences organized a training workshop on botanical taxonomy in September 2011. This workshop introduced history of taxonomy, basic theory and methods of taxonomy, focusing on systems of classification of vascular plants as well as their taxonomic traits of various vascular plants and techniques for identifying and classifying common populations and those difficult to identify. Through training, trainees have mastered classical methods of plant taxonomy as well as the importance of plant taxonomic studies by using new techniques and methods. A training workshop was organized by Shanghai Chenshan Botanical Garden (Plant Research Centre in Chenshan, Shanghai, under CAS) in October 2012 for those personnel working on the site. The training content included history and literature of taxonomy, plant taxonomic research methodologies, seed plant taxonomy, moss plant classification and application, fern plant taxonomy, plant specimen museum establishment and associated management techniques, molecular phylogeny analysis and population genetics. Nearly 130 participants from more than 40 institutions attended the workshop. China Fungi Research Society organized three workshops on fungi taxonomy, diversity and systemic evolution respectively in 2010, 2011 and 2013. The workshops invited taxonomic experts from home and abroad to introduce fundamental theory, methods and application of new technologies concerning fungi taxonomy and population genetics. Collection of specimens in the wild and taxonomic identification were organized for participants. The workshops were attended by more than 150 participants from over 50 institutions	Fully completed

Annex V Implementation of the Capacity-building Strategy for the Global Taxonomy Initiative and the Global Strategy for Plant Conservation

Relevant COP decisions, programmes of work and suggested activities	National Implementation and Contributions	Assessment of Progress
Action 3: By 2014, organize additional technical workshops and academic training to improve taxonomic skills and the quality of taxonomic knowledge and information, as well as the contribution of taxonomy for the implementation of the Convention	Being planned	
Action 4: By 2015, produce and continue to share taxonomic tools (e.g., field guides, online tools such as virtual herbaria, genetic and DNA sequence-based identification tools such as barcoding) and risk-analysis tools in the context of invasive alien species and biosafety, taking into account the identified needs of users; and facilitate the use of those tools to identify and analyse: (1) threatened species; (2) invasive alien species; (3) species and traits that are useful to agriculture and aquaculture; (4) species subject to illegal trafficking; and (5) socio-economically important species, including microbial diversity	China has launched many tools for taxonomy. *Flora of China* can be searched on line. Its English version database was launched at the same time as hard copies were published. Flora of Pan-Himalaya Region under development will be available on line before hard copies are printed, and will contain more information. Since 2008, the Ministry of Science and Technology has been strongly promoting the establishment of national platform for sharing specimens. Currently the number of specimens shared on-line has reached 8 million pieces, with 1 million pages of documents, and 4 million pictures taken in the wild and 15,060 holotypes compiled. Technological system using digital survey in the wild and information management as well as relevant information systems have been established. In September 2012, CD of China Species List 2012 was developed by Species 2000 China Node and launched by the Science Press Ltd. In 2010, the Ministry of Environmental Protection issued *Provisional Technical Guidelines for Surveying Plant Species and Resources*. And MEP is developing technical guidelines for monitoring species. All this has provided a good basis for undertaking the survey and monitoring of species in the wild. The agriculture and quality supervision departments have developed tools for risk assessment of pests, and they have been undertaking risk assessments of invasive alien species	Mostly completed
Action 5: By 2015, review and enhance human capacity and infrastructure to identify and to assist monitoring of biodiversity, particularly on invasive alien species, understudied taxa, threatened and socio-economically important species among others. The review might be undertaken with regional networks and coordinated with national and international activities	China has assessed the existing capacities and facilities for biodiversity monitoring and proposed a programme for establishing national biodiversity monitoring network. China has also established a national network of monitoring of forest biodiversity. China is developing technical guidelines for species monitoring, covering many populations including those populations not well studied	Mostly completed

Relevant COP decisions, programmes of work and suggested activities	National Implementation and Contributions	Assessment of Progress
Action 6: To the extent possible, support existing efforts to establish capacity for national and thematic biodiversity information facilities, build and maintain the information systems and infrastructure needed to collate, curate and track the use of biological specimens, in particular type specimens, and provide free and open access to the relevant biodiversity information for the public by 2016	*Flora of China* can be searched on line. Its English version database was launched at the same time as hard copies were published. Flora of Pan-Himalaya Region under development will be available on line before hard copies are printed. The establishment of national platform for sharing specimens and other relevant information systems will make it possible to get specimens, documents and pictures from *Flora of China*. The information therein can be accessible to the public free of charge	Completed
Action 7: By 2017, establish the human resources and infrastructure sufficient to maintain the existing collections and build further collections of biological specimens and living genetic resources. This action may strengthen and facilitate: (1) *ex-situ* conservation of microorganisms; (2) engagement of academics; (3) internships, exchanges and cooperation of experts; (4) job opportunities for becoming specialized and continuing to work in taxonomy; (5) allocation of public-funds for establishment and maintenance of collections infrastructure; (6) business-case for investment in human resources and infrastructure; (7) access to information; (8) coordinated global systems of biological collections	As a result of efforts of many years, China has stored more than 30 million specimens. To better store collected crop genetic resources, China has expanded and renovated 1 national long-term banks, 1 national copy bank, 10 national mid-term banks ad 32 national germplasm nurseries. China has built 7 new national germplasm nurseries. The total number of agricultural crops stored in these facilities has reached 423,000 accessions, mainly being local varieties and wild relatives. Many research institutions and universities in China have established plant specimen museums, with relevant professionals and facilities provided for these museums. They have also collected, stored and identified plant specimens. Animal specimens in China should be fully collected and stored in specimen museums in the future	Mostly completed
Action 8: By 2019, improve the quality and increase the quantity of records on biodiversity in historic, current and future collections and make them available through taxonomic and genetic databases to enhance resolution and increase confidence of biodiversity prediction models under different scenarios	Since 2008, the Ministry of Science and Technology has been strongly promoting the establishment of national platform for sharing specimens. Currently, the number of specimens shared on-line has reached 8 million copies, with 1 million pages of documents, and 4 million pictures taken in the wild and 15,060 holotypes compiled. The system allows for high-speed browsing and statistical analysis. Technological system using digital survey in the wild and information management as well as relevant information systems have been established. The establishment of this platform will provide basic data for biodiversity scenario modelling in different circumstances	Mostly completed

Annex V Implementation of the Capacity-building Strategy for the Global Taxonomy Initiative and the Global Strategy for Plant Conservation

Relevant COP decisions, programmes of work and suggested activities	National Implementation and Contributions	Assessment of Progress
Action 9: Facilitation of all-taxa inventories in targeted national, regional and subregional priority areas such as biodiversity hot spots, key biodiversity areas, protected areas, community-conserved areas, sustainable biodiversity management zones, and socio-ecological production landscapes considered under the *Satoyama* Initiative and other programmes in which biodiversity inventories are a priority for decision-making	In September 2012, CD of *China Species List 2012* was developed by Species 2000 China Node and launched by the Science Press. This list provides a good baseline data for taxonomic research as well as core taxonomic programmes for management and application of biodiversity information. Based on this list, China has completed national biodiversity assessments. For the first time China has collected information from county level concerning distribution of 34,039 wild vascular plants in 2,376 counties. China has established biodiversity information system. Using this system you can easily identify national hotspots for plant biodiversity and develop a list of vascular plants in hotspots and biodiversity priority areas	Mostly completed
Action 10: Between 2018 and 2020, using, *inter alia*, the Aichi Biodiversity Target indicators relevant to taxonomy, evaluate the progress in the GTI Capacity-building Strategy at the national, subregional, regional and global levels with a view to sustaining them beyond 2020	Being planned	
The Global Strategy for Plant Conservation (2011-2020)		
Target 1: An online flora of all known plants	*Flora of China* can be searched on line. Its English version database was launched at the same time as hard copies were published. Flora of Pan-Himalaya Region under development will be available on line before hard copies are printed. The establishment of national platform for sharing specimens and other relevant information systems will make it possible to get specimens, documents and pictures from plant annals	Fully completed
Target 2: An assessment of the conservation status of all known plant species, as far as possible, to guide conservation action	*China Biodiversity Red List-Higher Plant Volume*, which is going to be published soon, assesses 34,450 species (including sub-species) in accordance with IUCN standards, by following four steps including basic list development, searching species information, panel pre-review and expert review and examination. The result shows that there are 52 species that are extinct, extinct in the wild or regional extinct, and 3,767 species that are critically endangered, endangered and vulnerable. The publication of this list will provide important baselines for plant conservation in China	Fully completed
Target 3: Information, research and associated outputs, and methods necessary to implement the Strategy developed and shared	China has promoted the achievement of this target through compiling plant annals, establishment of network of PAs and the national platform for sharing specimens	Mostly completed

Relevant COP decisions, programmes of work and suggested activities	National Implementation and Contributions	Assessment of Progress
Target 4: At least 15 percent of each ecological region or vegetation type secured through effective management and/or restoration	China has established a network of PAs including 2,697 nature reserves, with areas covered accounting for 14.8% of the country's land area and protecting effectively 90% of the terrestrial ecosystems, 65% of higher plant communities, 25% of primitive natural forests, more than 50% of natural wetlands and 30% of typical desert areas. Meanwhile, restoration of various ecosystems has been enhanced through implementing key ecological projects such as natural forest resources protection, returning cultivated land to forests, wetland conservation and desertification control	Fully completed
Target 5: At least 75 percent of the most important areas for plant diversity of each ecological region protected with effective management in place for conserving plants and their genetic diversity	China has established 407 national-level nature reserves protecting the most important regions including forest, grassland, desert and wetland and other ecosystems	Mostly completed
Target 6: At least 75 percent of production lands in each sector managed sustainably, consistent with the conservation of plant diversity	China has a tradition of intensive and rotational cultivation and intercropping, which is very favorable to biodiversity conservation. Since 2000, China has organized creation of eco-provinces, eco-cities and eco-counties. The goal of this action is consistent with that of GSPC. So far, 15 provinces (cities) have started eco-province initiatives, and 13 provinces have issued programmes for building eco-provinces. More than 1,000 counties (cities, districts) have started eco-county initiatives. Since 2007, 38 counties (cities, districts) have been awarded national-level eco-counties, and 1,559 towns as national-level eco-towns and 238 villages as national-level eco-villages	Partially completed
Target 7: At least 75 percent of known threatened plant species conserved *in situ*	85% of national key protected plants have been protected according to relevant statistics. In 2012, SFA initiated a project to protect and rescue wild plants with extremely small populations, including a five-year campaign to protect and rescue 120 plant species with extremely small populations. The implementation of this campaign will effectively improve conditions for critically endangered, rare wild plants	Mostly completed
Target 8: At least 75 percent of threatened plant species in *ex situ* collections, preferably in the country of origin, and at least 20 per cent available for recovery and restoration programmes	Southwest China Germplasm Bank of Wild Resources was established in Kunming, Yunnan, which is the first germplasm bank established for wild flora and fauna and micro-organisms. It started operation from 29 October 2008. By April 2013, the bank has collected and stored 76,864 copies of plant seeds of 10,096 species. Meanwhile, a network of botanical gardens with those in Beijing (under CAS), Wuhan, Kunming and South China as core botanical gardens has well implemented *ex-situ* conservation of plants	Mostly completed
Target 9: 70 percent of the genetic diversity of crops including their wild relatives and other socio-economically valuable plant species conserved, while respecting, preserving and maintaining associated indigenous and local knowledge	National Crops Germplasm Bank was established in 1986 for storing seeds of germplasm of crops and wild relatives. So far the number of germplasms stored has exceeded 423,000 accessions. The national bank plays a very important in storing and using genetic resources of crops in China	Mostly completed

Annex V Implementation of the Capacity-building Strategy for the Global Taxonomy Initiative and the Global Strategy for Plant Conservation

Relevant COP decisions, programmes of work and suggested activities	National Implementation and Contributions	Assessment of Progress
Target 10 : Effective management plans in place to prevent new biological invasions and to manage important areas for plant diversity that are invaded	China has established a relatively sound system of inspection and quarantine to prevent invasion of new alien species from international trade. In the national networks of PAs, the management of endangered, rare plants and plant communities distributed within PAs and relevant habitats within PAs as well as alien plants has been integrated into the routine management plan of various PAs	Partially completed
Target 11 : No species of wild flora endangered by international trade.	China is effectively implementing the CITES and regulating international trade in endangered wild plants so no wild plant species are being endangered due to regular international trade	Mostly completed
Target 12 : All wild harvested plant-based products sourced sustainably	Currently, the percentage of products extracted from wild plants for social and economic uses in China is small. Major herbal materials such as Ginseng, Honeysuckle and Tianma have been artificially cultivated so that sustainable use of biological resources is achieved while protecting wild resources	
Target 13 : Indigenous and local knowledge innovations and practices associated with plant resources maintained or increased, as appropriate, to support customary use, sustainable livelihoods, local food security and health care	China actively collects, stores and uses local traditional knowledge. The Institute of Biotechnology and Germplasm Resources of Yunnan Province Academy of Agricultural Sciences has developed advanced and practical methods for surveying, collecting, storing and assessing traditional knowledge, and encouraged farmers holding traditional knowledge to publish articles on traditional knowledge. And the institute has stored and documented more than 300 pieces of traditional knowledge related to agricultural biodiversity. The institute has created a training model where farmers give lectures and resources surveyors and analysts participate. A total of over 600 farmers and technicians participated in the training, with their capacities and level to protect and use agricultural biological resources and associated traditional knowledge upgraded and the protection of warts grain wild rice and development of unique industries in ethnic-minority-residing areas in Yunnan promoted	Partially completed
Target 14 : The importance of plant diversity and the need for its conservation incorporated into communication, education and public awareness programmes	Plant biodiversity knowledge has become an important element of national science popularization programmes, and is disseminated to the public through TV, radio, newspapers and other media, on the National Day for Science Popularization and in botanical gardens. For education, besides biology in middle schools, science teaching is common in many primary schools across the country. The percentage of knowledge concerning plant biodiversity and conservation is gradually increasing in science teaching. A number of biodiversity information systems established by CAS such as information-sharing platforms for natural specimens museums and specimens for teaching contain a large of information concerning plants. These systems provide the public effective ways to gain knowledge concerning plant biodiversity	Mostly completed

Relevant COP decisions, programmes of work and suggested activities	National Implementation and Contributions	Assessment of Progress
Target 15: The number of trained people working with appropriate facilities sufficient according to national needs, to achieve the targets of this Strategy	MEP, together with CAS and PAs has organized training workshops on techniques for surveying species in the wild. In particular since 2009, the Institute of Botany of CAS has organized consecutively five training workshops on application of digitalization and land marking techniques in wild surveys and plant photographic techniques. The training workshops covered GPS positioning, information collection through digital photography, and internet information management. Through these training activities, equipment such as digital camera, GPS loggers have been gradually used in the surveys and monitoring of plants. Some protected areas have begun to try using internet information systems to manage data from surveys and monitoring and for science popularization. In sum, the number of trained and adequately equipped staff for plant conservation in PA management is gradually increasing, however the number and capacities of such personnel need to be further increased	Partially completed
Target 16: Institutions, networks and partnerships for plant conservation established or strengthened at national, regional and international levels to achieve the targets of this Strategy	China has established the National Coordinating Group for the Implementation of the CBD, the Inter-ministerial Joint Conference on Conservation of Biological Resources and China National Committee on Biodiversity Conservation. This indicates full, strategic importance China attaches to biodiversity conservation. Meanwhile, the Ministry of Agriculture and SFA are responsible for protection of wild plants and forests. Most of the provinces (autonomous regions, province-level cities) have established inter-sectoral coordinating mechanisms to coordinate biodiversity conservation and management within their jurisdictions	Mostly completed

References

[1] An Jiandong, Chen Wenfeng. 2011. Assessment of economic values of insect pollinators for fruits and vegetables in China. Insects Journal, 54(4): 443-450.

[2] Committee on Second National Assessment of Climate Change. 2011. Report of Second National Assessment of Climate Change. Beijing: Sciences Press.

[3] Fan Jiangwen, Zhong Huaping, Yuan Xujiang. 2002. Grassland reclamation in the past five decades in China and its ecological impacts. Grasslands of china, 24(5): 69-72.

[4] Fishery Bureau of MOA. 2011. China Fishery Statistics Yearbook. Beijing: China Agriculture Press.

[5] Forest Pests Prevention and Control Station under SFA. 2013. Studies on Climate Change Impacts on Forest Pest Disasters and Responses. Beijing: China Forestry Press.

[6] Gao Zhiqiang, Zhou Qixing. 2011. Pollution from open-air mining of rare earth and impacts on ecology and natural resources Chinese. Journal of Ecology, 30(12):2915-2922.

[7] Garibaldi LA, Aizen MA, Klein AM, et al. 2011. Global growth and stability of agricultural yield decrease with pollinator dependence. PNAS, 108(14): 5909-5914.

[8] Grassland Monitoring and Management Center under MOA. 2012. Statistical Analysis Report on Illegal Cases related to Grasslands in China in 2011. can be downloaded from the following website: http://www.grassland.gov.cn/Grassland-new/Item/3550.aspx.

[9] Jing Zhaopeng, Ma Youxin. 2012. Dynamic assessment of values of ecosystem services of Xishuangbanna, Yunnan Province. Journal of Central South China University of Forestry Science and Technology, 32(9):87-93.

[10] Liu HZ and Gao Xin. 2012. Monitoring Fish Biodiversity in the Yangtze River, China. *In The Biodiversity Observation Network in the Asia-Pacific Region: Toward Further Development of Monitoring* (Shin-ichi Nakano et al. eds.), Ecological Research Monographs, DOI 10.1007/978-4-431-54032-8_12, Springer, Japan.

[11] Liu Ruiyu. 2011. Progress in research on marine biodiversity in China. Biodiversity science, 19(6):614-626.

[12] Lu Lijun, Wang Jiaxue. 2009. A summary analysis of research on water environmental pollution of Dianchi Lake. Water Sciences and Engineering Technologies, 5:65-68.

[13] Ma Ruijun, Jiang Zhigang. 2006. Impacts of environmental degradation of Qinghai Lake Basin on wild terrestrial vertebrates. Acta Ecologica Sinica, 26(9): 3061-3066.

[14] Millennium Ecosystem Assessment. 2005. Ecosystem and Human Well-being: Biodiversity Synthesis. World Resources Institute, Washington, DC.

[15] Ministry of Agriculture. 1991-2012. China Yearbooks of Rural Statistics (1991-2012). Beijing: China Agriculture Press.

[16] Ministry of Agriculture. 1997-2012. China Agriculture Statistics Materials (1997-2012). Beijing: China Agriculture Press.

[17] Ministry of Agriculture. 2005-2012. National Reports on Grassland Monitoring (2005-2012).

[18] Ministry of Environmental Protection, Chinese Academy of Sciences. 2013. China Biodiversity Red List-Volume on Higher Plants.

[19] Ministry of Environmental Protection. 1985-2012. Reports on the State of the Environment (1985-2012).

[20] Ministry of Environmental Protection. 1997-2012. China Environmental Statistics Yearbooks (1997-2012). Beijing: China Environmental Sciences Press.

[21] Ministry of Environmental Protection. 2011. China's Updated National Biodiversity Strategy and Action Plan. Beijing: China Environmental Sciences Press.

[22] Ministry of Housing, Urban and Rural Development. 2012. Report on the State of Scenic Spots in China (2012).

[23] Ministry of Housing, Urban and Rural Development. 2006-2012. China Yearbooks of Statistics on Urban Construction (2006-2012). Beijing: China Planning Press.

[24] National Committee on Livestock Genetic Resources. 2011. China's Livestock Genetic Resources-Volume on Pigs. Beijing: China Agriculture Press.

[25] National Committee on Livestock Genetic Resources. 2011. China's Livestock Genetic Resources-Volume on Cattles. Beijing: China Agriculture Press.

[26] National Committee on Livestock Genetic Resources. 2011. China's Livestock Genetic Resources-Volume on Sheep. Beijing: China Agriculture Press.

[27] National Committee on Livestock Genetic Resources. 2011. China's Livestock Genetic Resources-Volume on Domesticated Animals. Beijing: China Agriculture Press.

[28] National Committee on Livestock Genetic Resources. 2011. China's Livestock Genetic Resources-Volume on Horses, Donkeys and Camels. Beijing: China Agriculture Press.

[29] National Committee on Livestock Genetic Resources. 2011. China's Livestock Genetic Resources-Volume on Endemic Livestock. Beijing: China Agriculture Press.

[30] National Committee on Livestock Genetic Resources. 2011. China's Livestock Genetic Resources-Volume on Honeybees. Beijing: China Agriculture Press.

[31] National Statistics Bureau. 2000-2012. China Yearbooks of Statistics (2000-2012). Beijing: China Statistics Press.

[32] Ouyang Zhiyun, Zhao Tongqian, Zhao Jingzhu, et al. 2004. Ecological functions of ecosystems in Hainan Island and their ecological economic values. Chinese Journal of Applied Ecology, 15(8):1395-1402.

[33] Qi Ruiying, Qi Yongting, Guo Weidong, et al. Initial and final singing dates of Cuckoos in Eastern Qinghai and responses to climate change. Advances in Climate Change Research, 4(4):225-229.

[34] Research Center on Economic Development under SFA, Department of Development Planning and Fund Management of SFA. 2003-2012. Reports on Monitoring of Socio-economic Benefits from National Key Forestry Projects (2003-2012). Beijing: China Forestry Press.

[35] State Forestry Administration. 1995-2012. China Forestry Statistics Yearbooks (1995-2012). Beijing, China Forestry Press.

[36] State Oceanic Administration. 2000-2012. Report on the State of the Marine Environment of China (2000-2012).

[37] Wang Song, Xie Yan. 2004. China Species Red List (Volume I). Beijing: Higher Education Press.

[38] Wu Chunxia, Liu Ling. 2008. Analysis of global climate change background for invasion of *Solidago canadensis*. Agricultural Environment and Development, 25(5):95-97.

[39] Wu Jun, Xu Haigen, Chen Lian. 2011. A summary analysis on research on climate change impacts on species. Journal of Ecology and Rural Environment, 27 (4): 1-6.

[40] Xie Gaodi, Zhang Yili, Lu Chunxia, et al. 2001. Values of ecosystem services of natural grasslands in China. Journal of Natural Resources, 16(1): 47-53.

[41] Xu Cunze. 2006. A brief analysis of impacts of flood control projects on fishes and responses. Journal of Yunnan University of Agriculture, (12): 31-32.

[42] Xu Haigen, Cao Mingchang, Wu Jun, et al. 2013. Report on Baseline Assessment of China's Biodiversity. Beijing: Sciences Press.

[43] Xu Haigen, Qiang Sheng. 2011. Invasive Alien Species in China. Beijing: Sciences Press.

[44] Xu Haigen, Wang Jianmin, Qiang Sheng, et al. 2004. Research on Hot Issues Related to the CBD: IAS, Biosafety and Genetic Resources. Beijing: Sciences Press.

[45] Xu Haigen, Wu Jun, Chen Jiejun. 2011. Environmental Risk Assessments of IAS and Studies on IAS Control. Beijing: Sciences Press.

[46] Zhang Cuiying, Li Ruiying, Zhao Chendao. 2011. Initial and final singing dates of cuckoos in southwestern Shandong and responses to climate change. Meteorology Sciences, 39(1): 114-117.

[47] Zhao Huiying, Wu Lijie, Hao Wenjun. 2008. Climate change impacts on ecological evolutions in wetlands of Hulunbei'er Lake and adjacent areas. Acta Ecologica Sinica, 28(3): 1064-1071.

[48] Zhao Tongqian, Ouyang Zhiyun, Zheng Hua, et al. 2004. Assessment of services and values of forest ecosystems of China. Journal of Natural Resources, 19(4):480-491.

[49] Zheng Jingyun, Guo Quansheng, Zhao Huixia. 2003. Studies on responses of China's plant phenology to climate change. China Rural Meteorology, 24(1):28-32.

(The original of this book of which the chinese version is the authentic text.)